# POLYFUNCTIONAL CYTOKINES: IL-6 AND LIF

The Ciba Foundation is an international scientific and educational charity. It was established in 1947 by the Swiss chemical and pharmaceutical company of CIBA Limited—now CIBA-GEIGY Limited. The Foundation operates independently in London under English trust law.

The Ciba Foundation exists to promote international cooperation in biological, medical and chemical research. It organizes about eight international multidisciplinary symposia each year on topics that seem ready for discussion by a small group of research workers. The papers and discussions are published in the Ciba Foundation symposium series. The Foundation also holds many shorter meetings (not published), organized by the Foundation itself or by outside scientific organizations. The staff always welcome suggestions for future meetings.

The Foundation's house at 41 Portland Place, London W1N 4BN, provides facilities for meetings of all kinds. Its Media Resource Service supplies information to journalists on all scientific and technological topics. The library, open five days a week to any graduate in science or medicine, also provides information on scientific meetings throughout the world and answers general enquiries on biomedical and chemical subjects. Scientists from any part of the world may stay in the house during working visits to London.

101678

B/QH 604

Ciba Foundation Symposium 167

BOCK, G.

# POLYFUNCTIONAL CYTOKINES: IL-6 AND LIF

*A Wiley–Interscience Publication*

1992

## JOHN WILEY & SONS

Chichester · New York · Brisbane · Toronto · Singapore

Published in 1992 by John Wiley & Sons Ltd
Baffins Lane, Chichester
West Sussex PO19 1UD, England

*Other Wiley Editorial Offices*

John Wiley & Sons, Inc., 605 Third Avenue,
New York, NY 10158-0012, USA

Jacaranda Wiley Ltd, G.P.O. Box 859, Brisbane,
Queensland 4001, Australia

John Wiley & Sons (Canada) Ltd, 22 Worcester Road,
Rexdale, Ontario M9W 1L1, Canada

John Wiley & Sons (SEA) Pte Ltd, 37 Jalan Pemimpin #05-04,
Block B, Union Industrial Building, Singapore 2057

Suggested series entry for library catalogues:
Ciba Foundation Symposia

Ciba Foundation Symposium 167
x + 279 pages, 43 figures, 28 tables

*Library of Congress Cataloging-in-Publication Data*
Polyfunctional cytokines : IL-6 and LIF.
    p.   cm. — (Ciba Foundation symposium ; 167)
  Editors, Gregory R. Bock (organizer) and Kate Widdows.
  "A Wiley–Interscience publication."
  Includes bibliographical references and index.
  ISBN 0-471-93439-9
  1. Interleukin-6 — Congresses.   2. Leukemia inhibitory factor —
Congresses.   I. Bock, Gregory.   II. Widdows, Kate.   III. Series.
  [DNLM: 1. Growth Inhibitors — pharmacology — congresses.
2. Interleukin-6 — pharmacology — congresses.   3. Leukemia —
physiopathology — congresses.   W3 C161F v.167]
QR185.8.I56P65   1992
616.07′9 — dc20
DNLM/DLC
for Library of Congress                                                92-5732
                                                                             CIP

*British Library Cataloguing in Publication Data*
A catalogue record for this book is
available from the British Library

ISBN 0 471 93439 9

Phototypeset by Dobbie Typesetting Limited, Tavistock, Devon.
Printed and bound in Great Britain by Biddles Ltd, Guildford.

# Contents

# Participants

**L. A. Aarden**  Central Laboratory of The Netherlands Red Cross Bloodtransfusion Service and Laboratory of Experimental & Clinical Immunology, University of Amsterdam, Plesmanlaan 125, NL-1066 CX Amsterdam, The Netherlands

**S. Akira**  Division of Immunology, Institute for Molecular & Cellular Biology, Osaka University, 1-3 Yamada-oka, Suita, Osaka 565, Japan

**J. T. Arnold**  Department of Physiology, University of Melbourne, Parkville, Melbourne, Victoria 3052, Australia

**H. Baumann**  Department of Molecular & Cellular Biology, Roswell Park Cancer Institute, Elm and Carlton Streets, Buffalo, NY 14263, USA

**J. F. Bazan**  Department of Biochemistry & Biophysics, School of Medicine, University of California, San Francisco, CA 94143-0448, USA

**G. Ciliberto**  IRBM-Istituo di Ricerche di Biologia Molecolare, Via Pontina km 30,600, I-00040 Pomezia (Roma), Italy

**T. M. Dexter**  Department of Haematology, Kay Kendall Laboratory, Paterson Institute for Cancer Research, Christie Hospital & Holt Radium Institute, Wilmslow Road, Manchester M20 9BX, UK

**G. Fey***  IMM 14, Department of Immunology, Research Institute of Scripps Clinic, 10666 North Torrey Pines Road, La Jolla, CA 92037, USA

**J. Gauldie**  Molecular Virology & Immunology Program, Department of Pathology 2N16, McMaster University, 1200 Main Street West, Hamilton, Ontario, L8N 3Z5, Canada

---

*Present address:* Institute of Microbiology, Biochemistry & Genetics, University of Erlangen, Staudstrasse 5, D-852 Erlangen, Germany

**D. P. Gearing**   Immunex Research & Development Corporation, University Building, Suite 600, 51 University Street, Seattle, WA 98101, USA

**N. M. Gough**   Cancer Research Unit, Walter and Eliza Hall Institute of Medical Research, Post Office, Royal Melbourne Hospital, Parkville, Melbourne, Victoria 3050, Australia

**J. K. Heath**   Department of Biochemistry, University of Oxford, South Parks Road, Oxford OX1 3QU, UK

**P. C. Heinrich**   Institute of Biochemistry, Neuklinikum der Rheinisch-Westfälischen Technischen Hochschule (RWTH), Aachen, Pauwelsstrasse 30, D-5100 Aachen, Germany

**T. Hirano**   Division of Molecular Oncology, Biomedical Research Center, Osaka University Medical School, 2-2 Yamada-oka, Suita, Osaka 565, Japan

**T. Kishimoto**   Osaka University Medical School, Department of Medicine III, 1-1-50 Fukushima, Fukushima-ku, Osaka 553, Japan

**F. Lee**   Department of Molecular Biology, DNAX Research Institute of Molecular & Cellular Biology, 901 California Avenue, Palo Alto, CA 94304-1104, USA

**J. Lotem**   Department of Molecular Genetics & Virology, The Weizmann Institute of Science, PO Box 26, Rehovot 76100, Israel

**T. J. Martin**   St Vincent's Institute of Medical Research and University of Melbourne Department of Medicine, St Vincent's Hospital, 41 Victoria Parade, Melbourne 3065, Australia

**D. Metcalf**   Cancer Research Unit, The Walter and Eliza Hall Institute of Medical Research, Post Office, Royal Melbourne Hospital, Parkville, Melbourne, Victoria 3050, Australia

**N. A. Nicola**   Molecular Regulatory Laboratory, Cancer Research Unit, The Walter and Eliza Hall Institute of Medical Research, Post Office, Royal Melbourne Hospital, Parkville, Melbourne, Victoria 3050, Australia

**P. H. Patterson**   Biology Division 216-76, California Institute of Technology, Pasadena, CA 91125, USA

**R. Romero*** Department of Obstetrics & Gynecology, Yale University School of Medicine, 339 Farnam Memorial Building, 333 Cedar Street, PO Box 3333, New Haven, CT 06510-8063, USA

**P. B. Sehgal** Departments of Microbiology & Immunology and Medicine, New York Medical College, Basic Sciences Building, Valhalla, NY 10595, USA

**C. L. Stewart** Department of Cell & Developmental Biology, Roche Institute of Molecular Biology, Nutley, NJ 07110, USA

**N. T. Williams** Department of Physiology, University of Melbourne, Parkville, Melbourne, Victoria 3052, Australia

*Present address:* Wayne State University, Department of Obstetrics & Gynecology, 4707 St Antoine Boulevard, Hutzel Hospital, Detroit, MI 48201, USA

# Introduction

Donald Metcalf

*The Walter and Eliza Hall Institute of Medical Research, Post Office, Royal Melbourne Hospital, Parkville, Melbourne, Victoria 3050, Australia*

The two biological regulators interleukin 6 (IL-6) and leukaemia inhibitory factor (LIF) are both appropriate subjects for discussion in a Ciba Foundation symposium because they are regulators with complex and intriguing biological actions and are candidates for therapeutic use in human disease. The reason that it is particularly appropriate to link them in a joint symposium is not simply because they share a number of different functional activities but because both present common conceptual difficulties in understanding why it should be advantageous for the body to use polyfunctional regulator molecules with such an astonishing array of diverse activities.

Both IL-6 and LIF have a quite unusual history of discovery and rediscovery by workers in quite separate biological disciplines. This is the origin of the multiplicity of the names attached to each (Table 1).

IL-6 was first encountered as an agent able to stimulate the proliferation *in vitro* of murine myeloma cells and was later shown to be identical to B cell-stimulating factor 2, characterized and cloned as a factor able to influence the maturation and function of normal B lymphocytes. In a quite separate series of studies, an antiviral agent (β2-interferon) and a 26 kDa protein were shown by sequence analysis to be identical to IL-6. Workers analysing the nature of factors capable of stimulating hepatocytes to produce acute-phase proteins identified as a major factor hepatocyte-stimulating factor 1; when purified and sequenced, this molecule proved to be IL-6. Similarly, workers investigating factors which induced differentiation in the murine myeloid leukaemia line M1 characterized an active protein, macrophage–granulocyte inducer 2, which again on sequencing proved to be IL-6.

The history of LIF has some striking parallels. Workers analysing factors which induced differentiation in a separate subline of murine M1 leukaemic cells characterized a differentiation factor (DIF) as the active agent. A potentially similar molecule was purified and cloned on the basis of its ability to induce differentiation in and inhibit the proliferation of the M1 leukaemic cell line. This factor was termed leukaemia inhibitory factor (LIF); sequencing of purified DIF showed it to be identical to LIF. The molecular properties of LIF suggested that it was similar to a differentiation-inhibiting agent which prevented

1

**TABLE 1    Alternative names for interleukin 6 and leukaemia inhibitory factor**

| IL-6 | LIF |
|------|-----|
| Myeloma/plasmacytoma growth factor | Differentiation-inducing factor (DIF) |
| B cell-stimulating factor 2 (BSF-2) | Differentiation-stimulating factor (D-factor) |
| β2-interferon (IFN-β2) | Differentiation inhibitory activity (DIA) |
| 26 kDa protein | Differentiation-retarding factor (DRF) |
| Hepatocyte-stimulating factor I (HSF-I) | Human interleukin for DA cells (HILDA) |
| Cytotoxic T cell differentiation factor | Hepatocyte-stimulating factor III (HSF-III) |
| Macrophage–granulocyte inducer 2 (MGI-2) | Cholinergic neuronal differentiation factor (CNDF) |
| | Melanoma-derived lipoprotein lipase inhibitor (MLPLI) |

differentiation commitment in normal embryonic stem cells (differentiation inhibitory activity, DIA), and the identity of LIF and DIA was subsequently confirmed. A human-derived factor capable of stimulating the proliferation of the murine haemopoietic cell line DA1 (HILDA) was shown on sequencing to be LIF, as was an agent purified on the basis of its ability to induce the switching of autonomic nerve signalling from adrenergic to cholinergic mode (CNDF). Workers studying regulation of the production by hepatocytes of acute-phase proteins had a second misfortune in that, on sequencing, hepatocyte-stimulating factor III (HSF-III) proved also to be LIF. Finally, the cachexia-inducing factor released by a human melanoma was identified as an inhibitor of lipoprotein lipase and, on sequencing, this protein proved to be LIF.

These remarkable series of multiple rediscoveries may be a presage of things to come. The search for the biological regulators of cells of all types is now at its height. For the eight major lineages of haemopoietic cells, 16 distinct regulators have already been identified, and in each case the pattern emerging is that multiple regulators exist for each cell type. If this basic pattern holds true for other tissues, one might postulate the existence of several hundred such regulators—most yet to be discovered. Against this are the examples of IL-6 and LIF, where apparently distinct regulators proved to be identical. Does this suggest we are approaching an asymptote in discovery, with relatively few regulators remaining to be discovered? Time will tell.

Until recently, most workers accepted the notion that exclusive regulators probably existed for each distinct cell type. At least for haematologists, the early example of erythropoietin as an agent acting exclusively on erythroid cells at a particular stage of differentiation seemed to be the prototype for what might be expected of the regulators yet to be discovered. This view of course ignored the complexity of earlier work on classical hormones, which showed clear promiscuity in the range of responding target cells. Equally, it ignored other prototype regulators such as nerve growth factor, fibroblast growth factor and epithelial growth factors, where the broad range of actions uncovered soon made the names for these factors somewhat inappropriate. The subsequent generations of haemopoietic growth factors—the colony-stimulating factors and the interleukins—seemed initially to fit the erythropoietin prototype well enough. However, with time, their actions have been recognized to be quite wide-ranging within the haemopoietic lineages, and suspicions have grown that they might also possess significant effects on other types of cells.

The advent of IL-6 and LIF has changed the situation radically. Now one must face the unpalatable fact that single, molecularly defined agents can have actions at the same concentration on a bewildering variety of target cells. Concepts of the private worlds of organs and cell types are now in disarray. Are our notions of tissue specificity and organization in need of a major reappraisal? One can envisage few situations in which it might be necessary or desirable to alter simultaneously the functioning of cell populations as radically different as haemopoietic cells, neurons, liver cells or bone cells. Does this simply reveal that there are body responses controlled by agents like IL-6 or LIF whose subtlety and complexity have yet to be recognized?

A major issue in this symposium will therefore be the attempt to answer this question. Why would the body choose to use the same molecule to regulate the functions of such a bizarre array of cell types? What makes IL-6 or LIF such unusually useful regulators?

A partial escape from the dilemma is to propose a concept of local regulator production and action—a design system in which it is never intended that the regulator concerned should leave its site of production and local responding target cells. This at least avoids the problem of unwanted side actions of such molecules, but, to be effective, it would require some selectivity in induction signals for regulator production and, ideally, a locally based origin of such inductive signals.

These latter considerations might appear to suggest that neither IL-6 nor LIF will ever be clinically useful, other than when locally delivered. This view is over-pessimistic and, as is the case for all therapeutically administered agents, careful pharmacological testing is required to establish whether particular effects dominate when the agents are delivered systemically and whether such effects are, on balance, beneficial in certain abnormal states, despite any significant associated side effects.

The decision to link IL-6 and LIF in one symposium was not based only on the problems of polyfunctionality. The significance of why IL-6 and LIF share so many functional actions needs to be examined. From studies so far, IL-6 has not been implicated in embryonic development as LIF has. This exception apart, we need to discuss why it is that both emerged as major molecules regulating M1 leukaemic cell differentiation, hepatocyte function and megakaryocyte and platelet formation. There is a distinct possibility that these two regulators may be designed to function in tandem in some situations. In support of the idea of some biological link between IL-6 and LIF are the production of both molecules by certain cells, the existence of common inducing signals and, indeed, the ability of one to induce the production of the other.

Is it possible to assign a 'major' function to IL-6 or LIF—a function that is not merely part of some redundant regulatory network, but a unique function irreplaceably lost in the absence of the molecule? The answer to this question may depend on the age of the animal, because LIF appears to have functions in the early embryo quite different from those in adult life.

An intriguing aspect of the biology of both IL-6 and LIF is their role in differentiation commitment. As far as myeloid leukaemic stem cells are concerned, it might be argued that the behaviour of M1 leukaemic cells is quite idiosyncratic and that neither IL-6 nor LIF has a comparable action on stem cells from other types of myeloid leukaemias. In the case of LIF, a further intriguing question is raised. Why is it that a factor purified on the basis of its ability to induce differentiation in leukaemic cells actually prevents differentiation commitment in normal embryonic stem cells (DIA)? Is this a difference between developmental stages of cells, or between normal and neoplastic cells, or does the phenomenon indicate that the induction or suppression of differentiation are alternative results of a very simple biological switching event, whose actual outcome is determined by the affected cell itself?

IL-6 and LIF therefore pose problems of interest to a broad range of biological scientists, from those interested in embryonic development to those involved in attempts to correct disease states. The biological problems posed are very unlikely to be unique and can be expected to recur. If the functions of these two agents can be resolved, an important advance will have been made in our concepts of cell and organ biology that will be of value in understanding the principles involved when further agents of this type are discovered.

# The molecular biology of interleukin 6 and its receptor

Tadamitsu Kishimoto*, Masahiko Hibi†, Masaaki Murakami†, Masashi Narazaki†, Mikiyoshi Saito† and Tetsuya Taga†

*Department of Medicine III, Osaka University Medical School, 1-1-50 Fukushima, Fukushima-ku, Osaka 553 and †Institute for Molecular and Cellular Biology, Osaka University, 1-3 Yamada-oka, Suita, Osaka 565, Japan

*Abstract.* Functional pleiotropy and redundancy are characteristic features of cytokines. Interleukin 6 (IL-6) is a typical example: IL-6 induces cellular differentiation or expression of tissue-specific genes; it is involved in processes such as antibody production in B cells, acute-phase protein synthesis in hepatocytes, megakaryocyte maturation, cytotoxic T cell differentiation, and neural differentiation of PC12 (pheochromocytoma) cells. It promotes growth of myeloma/plasmacytoma cells, T cells, keratinocytes and renal mesangial cells, and it inhibits growth of myeloid leukaemic cell lines and certain carcinoma cell lines. The IL-6 receptor consists of two polypeptide chains, a ligand-binding chain (IL-6R) and a non-ligand-binding, signal-transducing chain (gp130). Interaction of IL-6 with IL-6R triggers the association of gp130 and IL-6R, and the signal can be transduced through gp130. Association of gp130 with IL-6R is involved in the formation of high affinity binding sites. This two-chain model has been shown to be applicable to receptor systems for several other cytokines, such as granulocyte-macrophage colony-stimulating factor (GM-CSF), IL-3, IL-5 and nerve growth factor (NGF). The pleiotropy and redundancy of cytokines may be explained on the basis of this unique receptor system.

*1992 Polyfunctional cytokines: IL-6 and LIF. Wiley, Chichester (Ciba Foundation Symposium 167) p 5–23*

Interleukin 6 (IL-6) was originally identified as a T cell-derived lymphokine which induced final maturation of activated B cells into antibody-producing cells and was called B cell differentiation factor (BCDF) or B cell stimulatory factor 2 (BSF-2). Its cDNA was isolated in 1986 (Hirano et al 1986). At about the same time, the cDNAs for the molecules called β2-interferon and 26 kDa protein were cloned, and a partial amino acid sequence of hybridoma growth factor was reported (Zilberstein et al 1986, Haegeman et al 1986, Van Damme et al 1987). The results revealed that these molecules were identical, suggesting that IL-6 has multiple biological functions. Indeed, subsequent studies with recombinant IL-6 and anti-IL-6 antibodies showed that IL-6 has a wide variety of biological functions on various tissues and

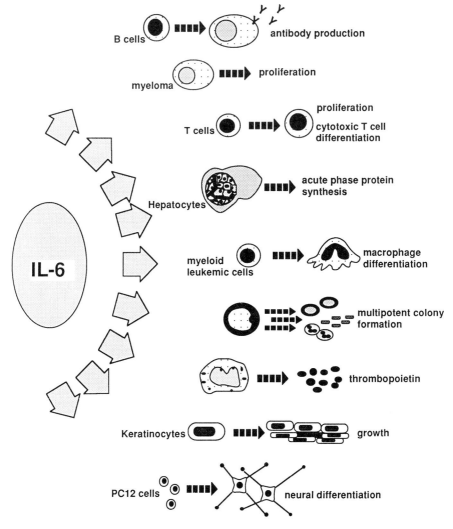

FIG. 1.   Schematic illustration of the polyfunctional nature of IL-6 action.

cells, as summarized schematically in Fig. 1 (reviewed by Kishimoto 1989, Hirano & Kishimoto 1990).

## The structure of IL-6

Both human and mouse IL-6 are glycoproteins with a molecular mass ranging from 21 to 28 kDa, depending on the degree of glycosylation. The amino acid sequence deduced from the cDNA clone indicated that human IL-6 consists of

212 amino acids, including a hydrophobic signal sequence of 28 amino acids. Biological activities of IL-6 are not affected by deletion of amino acids 1–28 at the N-terminus, whereas removal of four amino acids at the C-terminus resulted in a complete loss of biological activity (Brakenhoff et al 1989, Krüttgen et al 1990). The amino acid sequence of IL-6 showed significant identity to that of granulocyte colony-stimulating factor (G-CSF); the positions of the four cysteine residues in IL-6 match those in G-CSF (Hirano et al 1986, Nagata et al 1986a). This suggests there is similarity in the tertiary structures of these two molecules, and may indicate some functional similarity. As will be discussed later, the IL-6 receptor signal-transducing protein, gp130, and the G-CSF receptor have a high level of sequence identity (Hibi et al 1990, Fukunaga et al 1990). The genomic DNAs encoding IL-6 and G-CSF also have similar structures; both are composed of five exons and four introns, and the sizes of the exons are strikingly similar, suggesting that the two genes evolved from a common ancestor gene (Nagata et al 1986b, Yasukawa et al 1987).

**The pleiotropic functions of IL-6**

A study with recombinant IL-6 confirmed our previous study showing that IL-6 functions as a B cell differentiation factor. *In vivo* administration of IL-6 augments production of anti-sheep red blood cell antibodies more than 30-fold in sheep red blood cell-primed mice (Takatsuki et al 1988). Antibodies against IL-6 inhibited *in vitro* immunoglobulin induction in pokeweed mitogen-stimulated human peripheral mononuclear cells without having any effect on the growth of mitogen-stimulated B cells (Muraguchi et al 1988). These results showed the essential role of IL-6 in immunoglobulin induction in B cells.

   IL-6 acts on liver cells as a hepatocyte-stimulating factor and induces various acute-phase proteins such as C-reactive protein, $\beta_2$-fibrinogen, haptoglobin and haemopexin (Andus et al 1987, Gauldie et al 1987). *In vivo* administration of IL-6 into rats induces a typical acute-phase reaction similar to that caused by the injection of terpen oil (Geiger et al 1988). A nuclear factor responsible for controlling IL-6 gene expression (NF-IL6) is also involved in the regulation of various acute-phase genes in hepatocytes (Akira et al 1992, this volume).

   IL-6 has various actions on haemopoietic cells. (a) It activates haemopoietic progenitor cells at the $G_0$ stage, causing them to enter the $G_1$ phase, in which they are responsive to other growth factors such as IL-3. Thus, IL-6 and IL-3 show synergistic effects on blast colony formation *in vitro* (Ikebuchi et al 1987). (b) The bone marrow of transgenic mice carrying the IL-6 gene was found to contain a large number of mature megakaryocytes, suggesting that IL-6 acts on megakaryocytes and functions like thrombopoietin (Suematsu et al 1989). Studies *in vitro* and *in vivo* confirmed that IL-6 induces maturation of megakaryocytes and an increase in platelets (Ishibashi et al 1989a,b). (c) IL-6 blocks the proliferation of a myeloid leukaemic cell line, M1, and induces differentiation of the cells into macrophages (Shabo et al 1988, Miyaura et al 1988). LIF (leukaemia inhibitory factor) has exactly the same effects on M1 cells (Gearing et al 1987). It is noteworthy that the sequence of the LIF receptor is highly similar to that of the IL-6 receptor signal transducer, gp130 (Hibi et al 1990, Gearing et al 1991).

IL-6 is a potent growth factor for human myeloma and mouse plasmacytoma cells, and it has been suggested that abnormal production of IL-6 is responsible for the generation of myeloma/plasmacytoma (Hirano et al 1992, this volume). As summarized in Fig. 1, IL-6 also acts on nerve cells (Satoh et al 1988, Naitoh et al 1988), epidermal keratinocytes (Grossman et al 1989, Yoshizaki et al 1990) and T cells (Lotz et al 1988, Houssiau et al 1988, Takai et al 1988), inducing their growth or differentiation.

## The IL-6 receptor

As expected from the pleiotropic functions of IL-6, the IL-6 receptor protein (IL-6R) is expressed in various tissues and cells including lymphoid cells, hepatocytes and epidermal keratinocytes, although the number of receptors per cell is extremely small (reviewed by Taga & Kishimoto 1990). Resting B cells do not express the IL-6 receptor, but $\mu^+/\delta^-$ activated B cells do. In contrast, resting T cells express the receptor (Taga et al 1987, Hirata et al 1989). The expression of the receptor coincides with the functioning of IL-6; IL-6 acts on activated B cells as a differentiation factor and on resting T cells as an activation factor. Responding cells express both high and low affinity binding sites, as observed with other cytokine receptors.

cDNA encoding human IL-6R was cloned and the deduced amino acid sequence showed that IL-6R is composed of 468 amino acid residues, including a signal peptide of 19 amino acid residues and a single transmembrane domain (Yamasaki et al 1988). There are six potential $N$-glycosylation sites and the mature receptor protein has a relative molecular mass ($M_r$) of 80 000, indicating that IL-6R is a highly glycosylated membrane protein. The region between amino acid residues 20 and 110 fulfils the criteria for the constant 2 ($C_2$) set of the immunoglobulin superfamily. However, deletion of this domain does not affect the IL-6-binding activity of IL-6R (H. Yawata, K. Yasukawa, S. Natsuka, M. Murakami, K. Yamasaki, M. Hibi, T. Taga & T. Kishimoto, unpublished work).

The rest of IL-6R did not show structural similarity to any other proteins at the time of its cloning. However, in 1989 and 1990 more cytokine receptors were molecularly cloned and the results revealed the existence of a 'cytokine receptor family'; features of this family include conservation of four cysteine residues in the N-terminal portion and an extracellular Trp-Ser-X-Trp-Ser motif just before the transmembrane region (Bazan 1989, 1990, Cosman et al 1990). Thus, the extracellular portion of IL-6R consists of two domains, one belonging to the immunoglobulin superfamily and the other to the cytokine receptor family (Fig. 2).

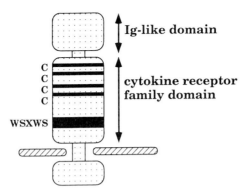

FIG. 2.   Model of the structure of IL-6R, the IL-6-binding component of the IL-6 receptor, as deduced from the presumed amino acid sequence of the cloned human IL-6R and comparison with the members of the immunoglobulin superfamily and the cytokine receptor family. The Ig-like domain represents the region between amino acids 20 and 110. Features shared with other members of the cytokine receptor family include the four conserved cysteine residues (C) and the Trp-Ser-X-Trp-Ser (WSXWS) motif.

## The IL-6 receptor-associated signal transducer

The cytoplasmic portion of IL-6R consists of only 82 amino acids and contains no sequence that might mediate signal transduction, such as a tyrosine kinase domain. Interestingly, a mutated IL-6R lacking the cytoplasmic region was shown to bind IL-6 and to transduce signals, suggesting the existence of an associated molecule for signal transduction. When cell lysates from a surface-labelled human myeloma cell line, U-266, were precipitated with an anti-IL-6R antibody, only the 80 kDa IL-6R chain could be detected in the precipitate. However, when the lysates were prepared after IL-6 stimulation of the cells, another polypeptide chain, molecular mass 130 kDa, was co-precipitated, indicating that binding of IL-6 to IL-6R triggered its association with a second polypeptide chain, termed gp130 (Taga et al 1989). This interaction between IL-6R and gp130 was shown to occur through their extracellular domains, and the human IL-6R could also associate with mouse gp130. The IL-6 receptor system is schematically summarized in Fig. 3. Recombinant soluble IL-6R (which is prepared by introducing a termination codon at the C-terminal end of the extracellular region) was able to bind to gp130 in the presence of IL-6. Thus, the soluble receptor is not an antagonist of IL-6—rather, it can augment the activity of IL-6 by forming a complex with IL-6 that is capable of associating with gp130 (Taga et al 1989, Yasukawa et al 1990).

Hibi et al (1990) cloned a cDNA encoding human gp130 by screening a placental λgt 11 library with an anti-human gp130 monoclonal antibody. The deduced amino acid sequence revealed that it is composed of 918 amino acids with a single transmembrane domain. The extracellular region contains six units

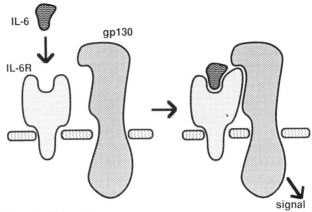

FIG. 3.   Schematic illustration of the IL-6 receptor system. IL-6 binds to IL-6R, the ligand-binding component of the receptor system. This allows association of IL-6R with gp130, the signal-transducing component of the receptor system.

of a fibronectin type III module and a region of about 200 amino acids which has features typical of the cytokine receptor family (Fig. 4). The overall structure of gp130 is highly similar to that of the G-CSF receptor (Fukunaga et al 1990). However, ligands, including IL-6, G-CSF, IL-3 and IL-2, were found not to bind to gp130. The association of IL-6R with gp130 may stabilize the binding of IL-6 to IL-6R, resulting in the formation of high affinity IL-6-binding sites. This mechanism may be involved in other cytokine receptor systems in which cloned cDNAs generate receptors with only low affinity binding sites although both high and low affinity binding sites are detected on the responding cells—for example, the IL-3, GM-CSF, IL-5 and nerve growth factor (NGF) receptors. Indeed, the associated molecules for these receptors have been identified (Hayashida et al 1990, Devos et al 1991, Takaki et al 1991, Hempstead et al 1991).

This mechanism of transduction could explain the functional pleiotropy and redundancy of cytokines: utilization of the same signal transducer by several different receptors would explain the functional redundancy of cytokines, and the presence of structural variants of the signal transducer would explain functional pleiotropy. Recent studies have shown or implied that IL-3, IL-5 and GM-CSF receptors use the same signal-transducing molecule (Lopez et al 1989, Hayashida et al 1990, Takaki et al 1991).

We have cloned a cDNA for the mouse homologue of gp130 (Saito et al 1992). It is highly similar to human gp130 and contains a 116 amino-acid stretch including the transmembrane and cytoplasmic domains that is identical to the equivalent sequence in human gp130. The transcripts of mouse gp130 were ubiquitously expressed, being found in all tissues tested, some of which did not express IL-6R. This finding suggests that gp130 might act as a signal transducer

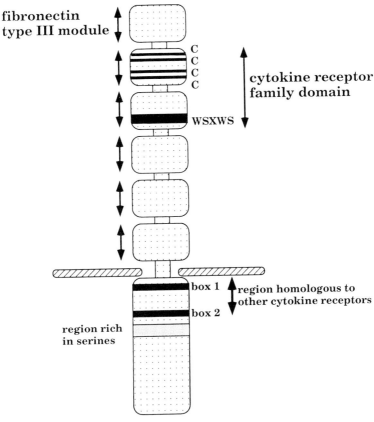

FIG. 4.    Model of the structure of gp130, the signal-transducing component of the IL-6 receptor system. The extracellular region contains six fibronectin type III modules and a 200 amino acid region that has features typical of members of the cytokine receptor family (see Fig. 2). The intracellular portion contains two short segments—box 1 and box 2—also showing similarity to other cytokine receptor family members.

for certain other cytokines. *In vivo* administration of IL-6 in the mouse caused up-regulation of gp130 mRNA in several tissues. mRNAs for both gp130 and IL-6R were up-regulated by IL-6 in the liver; this may make the cells more sensitive to IL-6 (Saito et al 1992).

## Signal transduction through gp130

So that the cytoplasmic region of the IL-6 signal transducer (gp130) critical for mediating the IL-6-induced growth signal could be examined, truncations or amino acid substitutions were introduced into gp130. The mutant cDNAs were transfected into a mouse IL-3-dependent cell line (BAF-B03) and transfectants

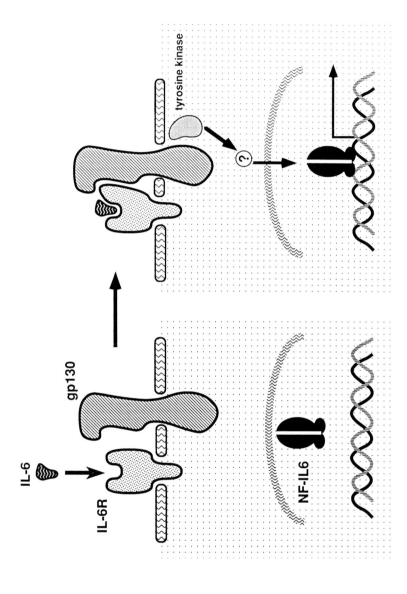

FIG. 5. A model of IL-6 action in hepatocytes. Binding of IL-6 to IL-6R triggers association of IL-6R and gp130 and consequent activation of a tyrosine kinase activity. Through unknown intermediate steps, this leads to binding of a transcription factor termed NF-IL6 to IL-6-responsive elements, and initiation of transcription of various acute-phase protein genes (see Akira et al 1992, this volume, for further details).

were stimulated with the soluble IL-6 receptor and IL-6. In the 277 amino acid cytoplasmic region of gp130, a sequence of 61 amino acids proximal to the transmembrane domain was shown to be sufficient for generation of the growth signal. Within this region were two short segments showing significant sequence identity to other members of the cytokine receptor family (Murakami et al 1991). One of these (box 1) includes a proline-X-proline sequence, which is found in all members of the family (see Fig. 4). When the two proline residues in this sequence were substituted with serines, gp130 completely lost its signal-transducing activity. Substitution of the prolines by serines may have changed the tertiary structure of the cytoplasmic region of gp130. It is noteworthy that the hinge region of immunoglobulins is rich in proline residues, which are thought to affect the flexibility of the molecules and their ability to bind complement. Another conserved region (box 2; residues 50–61, numbering the first cytoplasmic residue as 1) was also shown to be essential for signal transduction; gp130 lost its activity when a termination codon was introduced at residue 54 (Murakami et al 1991).

When transfectant cells were stimulated with soluble IL-6 receptor and IL-6, we observed a tyrosine-specific phosphorylation of gp130 immunoprecipitated from cell lysates with anti-human gp130 antibody. Tyrosine phosphorylation of gp130 was also observed in an *in vitro* kinase assay with the anti-gp130-precipitated complex. These results indicate the association of a cytoplasmic tyrosine kinase with gp130 after stimulation. At present, we do not know anything about down-stream molecules involved in the signal transduction; some of these may be targets of the activated tyrosine kinase. Hepatocytes may provide a useful model for the study of signal mediation through gp130, because IL-6 induces post-translational modification of NF-IL6, which then binds to the IL-6-responsive elements of the various acute-phase protein genes (Fig. 5) (Akira et al 1992, this volume).

*Acknowledgements*

This study was supported by a grant-in-aid for specially promoted research from the Ministry of Education, Science and Culture, Japan.

**References**

Akira S, Isshiki H, Nakajima T, Kinoshita S, Nishio Y, Natsuka S, Kishimoto T 1992 Regulation of expression of the IL-6 gene: structure and function of the transcription factor NF-IL6. Wiley, Chichester (Ciba Found Symp 167) p 47–67
Andus T, Geiger T, Hirano T et al 1987 Recombinant human B cell stimulatory factor 2 (BSF-2/IFNβ2) regulates β-fibrinogen and albumin mRNA levels in Fao-9 cells. FEBS (Fed Eur Biochem Soc) Lett 221:18–22
Bazan JF 1989 A novel family of growth factor receptors: a common binding domain in the growth hormone, prolactin, erythropoietin and IL-6 receptors, and the p75 IL-2 receptor β-chain. Biochem Biophys Res Commun 164:788–795
Bazan JF 1990 Structural design and molecular evolution of a cytokine receptor superfamily. Proc Natl Acad Sci USA 87:6934–6938

Brakenhoff JP, Hart M, Aarden LA 1989 Analysis of human IL-6 mutants expressed in *Escherichia coli*: biologic activities are not affected by deletion of amino acids 1–28. J Immunol 143:1175–1182

Cosman D, Lyman SD, Idzerda RL et al 1990 A new cytokine receptor superfamily. Trends Biochem Sci 15:265–270

Devos R, Plaetinck G, Van der Heyden J et al 1991 Molecular basis of a high affinity murine interleukin-5 receptor. EMBO (Eur Mol Biol Organ) J 10:2133–2137

Fukunaga R, Ishizaka-Ikeda E, Seto Y, Nagata S 1990 Expression cloning of a receptor for murine granulocyte colony-stimulating factor. Cell 61:341–350

Gauldie J, Richards C, Harnish D, Lansdorp P, Baumann H 1987 Interferon β2/B-cell stimulatory factor type 2 shares identity with monocyte-derived hepatocyte-stimulating factor and regulates the major acute phase protein response in liver cells. Proc Natl Acad Sci USA 84:7251–7256

Gearing D, Gough NM, King JA et al 1987 Molecular cloning and expression of cDNA encoding a murine myeloid leukaemia inhibitory factor (LIF). EMBO (Eur Mol Biol Organ) J 6:3995–4002

Gearing D, Thut CJ, VandenBos T et al 1991 Leukemia inhibitory factor receptor is structurally related to the IL-6 signal transducer, gp130. EMBO (Eur Mol Bio Organ) J 10:2839–2848

Geiger T, Andus T, Klapproth J, Hirano T, Kishimoto T, Heinrich PC 1988 Induction of rat acute-phase proteins by interleukin-6 *in vivo*. Eur J Immunol 18:717–721

Grossman RM, Krueger J, Yourish D et al 1989 Interleukin-6 is expressed in high levels in psoriatic skin and stimulates proliferation of cultured human keratinocytes. Proc Natl Acad Sci USA 86:6367–6371

Haegeman G, Content J, Volckaert G, Derynck R, Tavernier J, Fiers W 1986 Structural analysis of the sequence encoding for an inducible 26-kDa protein in human fibroblasts. Eur J Biochem 159:625–632

Hayashida K, Kitamura T, Gorman DM, Arai K-I, Yokota T, Miyajima A 1990 Molecular cloning of a second subunit of the receptor for human granulocyte-macrophage colony-stimulating factor (GM-CSF): reconstitution of a high-affinity GM-CSF receptor. Proc Natl Acad Sci USA 87:9655–9659

Hempstead BL, Martin-Zanca D, Kaplan DR, Parada LF, Chao MV 1991 High-affinity NGF binding requires coexpression of the *trk* proto-oncogene and the low-affinity NGF receptor. Nature (Lond) 350:678–683

Hibi M, Murakami M, Saito M, Hirano T, Taga T, Kishimoto T 1990 Molecular cloning and expression of an IL-6 signal transducer, gp130. Cell 63:1149–1157

Hirano T, Kishimoto T 1990 Interleukin-6. In: Sporn MB, Roberts AB (eds) Handbook of experimental pharmacology, vol 95/I: Peptide growth factors and their receptors. Springer-Verlag, Berlin, p 633–665

Hirano T, Yasukawa K, Harada H et al 1986 Complementary DNA for a novel human interleukin (BSF-2) that induces B lymphocytes to produce immunoglobulin. Nature (Lond) 324:73–76

Hirano T, Suematsu S, Matsusaka T, Matsuda T, Kishimoto T 1992 The role of interleukin 6 in plasmacytomagenesis. In: Polyfunctional cytokines: IL-6 and LIF. Wiley, Chichester (Ciba Found Symp 167) p 188–200

Hirata Y, Taga T, Hibi M, Nakano N, Hirano T, Kishimoto T 1989 Characterization of IL-6 receptor expression by monoclonal and polyclonal antibodies. J Immunol 143:2900–2906

Houssiau FA, Coulie PG, Olive D, Van Snick J 1988 Synergistic activation of human T cells by interleukin 1 and interleukin 6. Eur J Immunol 18:653–656

Ikebuchi K, Wong GG, Clark SC, Ihle JN, Hirai Y, Ogawa M 1987 Interleukin-6 enhancement of interleukin-3-dependent proliferation of multipotential hemopoietic progenitors. Proc Natl Acad Sci USA 84:9035–9039

Ishibashi T, Kimura H, Shikawa Y et al 1989a Interleukin 6 is a potent thrombopoietic factor *in vivo* in mice. Blood 74:1241–1244

Ishibashi T, Kimura H, Uchida T, Kariyone S, Friese P, Burstein SA 1989b Human interleukin 6 is a direct promoter of maturation of megakaryocytes *in vitro*. Proc Natl Acad Sci USA 86:5953–5957

Kishimoto T 1989 The biology of interleukin-6. Blood 74:1–10

Krüttgen A, Rose-John S, Möller C et al 1990 Structure–function analysis of human interleukin-6: evidence for the involvement of the carboxy-terminus in function. FEBS (Fed Eur Biochem Soc) Lett 273:95–98

Lopez AF, Englinton JM, Gillis D, Park LS, Clark S, Vadas MA 1989 Reciprocal inhibition of binding between interleukin 3 and granulocyte-macrophage colony-stimulating factor to human eosinophils. Proc Natl Acad Sci USA 86:7022–7026

Lotz M, Jirik F, Kabouridis R, Tsoukas C, Hirano T, Kishimoto T 1988 BSF-2/IL-6 is a co-stimulant for human thymocytes and T lymphocytes. J Exp Med 167:1253–1258

Miyaura C, Onozaki K, Akiyama Y et al 1988 Recombinant human interleukin 6 (B-cell stimulatory factor 2) is a potent inducer of differentiation of mouse myeloid leukemia cells (M1). FEBS (Fed Eur Biochem Soc) Lett 234:17–21

Muraguchi A, Hirano T, Tang B et al 1988 The essential role of B cell stimulatory factor 2 (BSF-2/IL-6) for the terminal differentiation of B cells. J Exp Med 167:332–344

Murakami M, Narazaki M, Hibi M et al 1991 Critical cytoplasmic region of the interleukin 6 signal transducer gp130 is conserved in the cytokine receptor family. Proc Natl Acad Sci USA 88:11349–11353

Nagata S, Tsuchiya M, Asano S et al 1986a Molecular cloning and expression of cDNA for human granulocyte colony-stimulating factor. Nature (Lond) 319:415–418

Nagata S, Tsuchiya M, Asano S et al 1986b The chromosomal gene structure and two mRNAs for human granulocyte colony-stimulating factor. EMBO (Eur Mol Biol Organ) J 5:575–581

Naitoh Y, Fukata J, Tominaga T et al 1988 Interleukin-6 stimulates the secretion of adrenocorticotropic hormone in conscious, freely-moving rats. Biochem Biophys Res Commun 155:1459–1463

Saito M, Yoshida K, Hibi M, Taga T, Kishimoto T 1992 Molecular cloning of a murine IL-6 receptor-associated signal transducer, gp130, and its regulated expression *in vivo*. J Immunol 148(12)

Satoh T, Nakamura S, Taga T et al 1988 Induction of neural differentiation in PC12 cells by B cell stimulatory factor 2/interleukin 6. Mol Cell Biol 8:3546–3549

Shabo Y, Lotem J, Rubinstein M et al 1988 The myeloid blood cell differentiation-inducing protein MGI-2A is interleukin 6. Blood 72:2070–2073

Suematsu S, Matsuda T, Aozasa K et al 1989 IgG1 plasmacytosis in interleukin-6 transgenic mice. Proc Natl Acad Sci USA 86:7547–7551

Taga T, Kishimoto T 1990 IL-6 Receptor. In: Cochrane CC, Gimbrone MA Jr (eds) Cellular and molecular mechanisms of inflammation. Academic Press, New York, vol 1:219–243

Taga T, Kawanishi Y, Hardy RR, Hirano T, Kishimoto T 1987 Receptors for B cell stimulatory factor 2 (BSF-2): quantitation, specificity, distribution, and regulation of their expression. J Exp Med 166:967–981

Taga T, Hibi M, Hirata Y et al 1989 Interleukin-6 triggers the association of its receptor with a possible signal transducer, gp130. Cell 58:573–581

Takai Y, Wong GG, Clark SC, Burakoff SJ, Herrmann SH 1988 B cell stimulatory factor-2 is involved in the differentiation of cytotoxic T lymphocytes. J Immunol 140:508–512

Takaki S, Mita S, Kitamura T et al 1991 Identification of the second subunit of the murine interleukin-5 receptor: interleukin-3 receptor-like protein, AIC2B is a component of the high-affinity interleukin-5 receptor. EMBO (Eur Mol Biol Organ) J 10:2833–2838

Takatsuki F, Okano A, Suzuki C et al 1988 Human recombinant interleukin 6/B cell stimulatory factor 2 (IL-6/BSF-2) augments murine antigen-specific antibody responses *in vitro* and *in vivo*. J Immunol 141:3072–3077

Van Damme J, Opdenakker G, Simpson RJ et al 1987 Identification of the human 26-kDa protein, interferon β2 (IFNβ2), as a B cell hybridoma/plasmacytoma growth factor induced by interleukin 1 and tumor necrosis factor. J Exp Med 165:914–919

Yamasaki K, Taga T, Hirata Y et al 1988 Cloning and expression of the human interleukin-6 (BSF-2/IFNβ2) receptor. Science (Wash DC) 241:825–828

Yasukawa K, Hirano T, Watanabe Y, Muratani K, Matsuda T, Kishimoto T 1987 Structure and expression of human B cell stimulatory factor 2 (BSF-2/IL-6) gene. EMBO (Eur Mol Biol Organ) J 6:2939–2945

Yasukawa K, Saito T, Fukunaga T et al 1990 Purification and characterization of soluble human IL-6 receptor expressed in CHO cells. J Biochem 108:673–676

Yoshizaki K, Nishimoto N, Matsumoto K et al 1990 Interleukin-6 and expression of its receptor on epidermal keratinocytes. Cytokines 2:381–387

Zilberstein A, Ruggieri R, Korn JH, Revel M 1986 Structure and expression of cDNA and genes for human interferon-beta-2, a distinct species inducible by growth-stimulatory cytokines. EMBO (Eur Mol Biol Organ) J 5:2529–2537

## DISCUSSION

*Hirano:* Do you think that the region of the cytoplasmic domain of gp130 which you have shown to be essential and sufficient for signal transduction in IL-3-dependent BAF-B03 cells is also essential or sufficient in other cells, such as hepatocytes?

*Kishimoto:* These two portions are also critical in the hepatocyte system for the induction of acute-phase proteins. The tyrosine kinase activation also occurs in hepatocytes. The question to be answered now is how in myeloma or IL-3-dependent cells the IL-6 can provide a growth signal whereas in hepatocytes it induces acute-phase proteins without growth.

*Fey:* There are two groups of acute-phase protein genes. One of the two groups is induced by IL-1 and by combinations of IL-1 and IL-6 through the transcription factor NF-IL6/IL-6DBP/LAP. The other group of genes is activated mainly by IL-6 and LIF, but not by IL-1, in an unknown way. The effects on growth and differentiation may be mediated by different pathways within one cell—there is not necessarily just one sole pathway through which IL-6 exerts all its effects.

*Kishimoto:* I did not wish to suggest that there is only one pathway; this is one model of acute-phase protein gene activation involving IL-6 and NF-IL6.

*Sehgal:* In your figures you depicted a single monomer of IL-6 interacting with the receptor. Is the stoichiometry really well established? Is it clear that one IL-6 polypeptide interacts with the IL-6 receptor, which then interacts with one gp130 monomer?

*Kishimoto:* We haven't studied the stoichiometry in detail, but cross-linking experiments suggest that dimers may bind to the receptor.

*Gearing:* Could you clarify what you mean by 'dimer'—do you mean a homodimer of the IL-6 ligand, or a heterodimer of the ligand and the soluble receptor?

*Kishimoto:* I was referring to cross-linking experiments using recombinant IL-6 which show the presence of 100 kDa and 120 kDa complexes (Taga et al 1987); it is possible that IL-6 dimers bind to the (80 kDa) IL-6 receptor.

*Gearing:* Fukunaga et al (1990, 1991) have suggested that G-CSF interacts with a homodimeric receptor complex.

*Kishimoto:* The G-CSF receptor is a single-chain receptor, so formation of a homodimer of the receptor molecule may be involved in its activation mechanism. The association of the IL-6 receptor with gp130 may be comparable to the dimerization of two G-CSF monomers.

*Sehgal:* One of the things that Professor Metcalf mentioned in his introduction was the idea of local action as opposed to systemic effects, and Professor Kishimoto talked about pleiotropic effects. Something we should think about is the way in which the cytokine is actually carried in the blood, and if this determines whether or exactly how it interacts with the receptor, that is, how it is presented to particular cell types. In cell culture experiments with recombinant cytokines you see all these different effects; the key question is, how does the cytokine really work in the body?

I now have difficulty accepting that the concentration of IL-6 in normal serum is of the order of pg/ml, though six months ago I would have accepted this. Most assays on the market will give you levels in the low pg/ml range, but other assays indicate that IL-6 levels in normal plasma are of the order of nanograms per ml, 1000-fold higher than the values that I have seen published and that we ourselves have published. Our assays on plasma and serum fractions taken off gel filtration columns suggest that the traditional hybridoma growth factor assays are extremely poor at measuring IL-6 in human body fluids and they suggest that all the hybridoma growth factor activity detected is in the monomeric form. Certain ELISAs that have been calibrated using recombinant IL-6 preferentially pick up the low molecular weight monomeric form of IL-6 in human serum. Other ELISAs suggest that most of the IL-6 in human serum or plasma is in very high molecular weight complexes. If you are lucky and happen to find the right monoclonal antibody, you find out that IL-6 exists in human blood in 100 kDa and 500 kDa forms—the bulk of the IL-6 is in high molecular mass structures and you can prove that it is IL-6 with associated proteins by direct amino acid sequencing.

The manner in which IL-6 circulates and the manner in which different assay techniques capture different fractions of the circulating IL-6 are major issues. Some of the conceptual issues that are being discussed might be resolvable once we have understood the structural aspects of circulating IL-6. I'm struggling to reconcile the notions of multiple biological effects, a single uniform description of the receptor, and our experimental observation that different activity profiles can be derived from one gel filtration run.

*Lotem:* With which bioassay do you detect activity of the 500 kDa form?

*Sehgal:* The high molecular mass form appears to have activity in the hepatocyte stimulation assay with Hep3B cells. The fractions from gel filtration columns have no activity in the B9 hybridoma growth assay. Thus, we have to ask whether IL-6, as it circulates in the human bloodstream, has any effect on B cells.

*Gough:* Human tissues have been reported to show high levels of mRNA for IL-6 (Tovey et al 1988), but in our experience mouse tissues don't. One possible explanation for that is that the mice come from nice, clean boxes whereas humans are not such clean-living beasts and there might well be inductive stimuli.

*Heinrich:* Dr Schgal, have you checked whether the high molecular mass forms are complexes of IL-6 with the receptor? A soluble IL-6 receptor has been identified in normal human urine (Novick et al 1989).

*Sehgal:* By direct sequencing, the 100 kDa IL-6 'complexes' from serum, as purified further by immunoaffinity chromatography, were found to contain a fragment of complement C3, a fragment of C4, a fragment of C-reactive protein and serum albumin. My objective at this point is to raise questions. At the moment, I have difficulty in accepting that IL-6 exists as a free-floating molecule in the bloodstream.

*Ciliberto:* Another possible explanation for a discrepancy between the results from the hybridoma assay and the ELISA might be the presence of a natural inhibitor, as in the case of IL-1. Does anyone have evidence for a natural inhibitor of IL-6 or LIF?

*Aarden:* We have looked in the B9 assay but have never found any evidence for such an inhibitor.

*Sehgal:* We have done extensive mixing experiments with the fractions from gel filtration and exogenous IL-6, but have found no inhibitor.

*Gauldie:* If we treat a rat with a high dose of endotoxin we measure a large amount of IL-6 in the circulation. When we fractionate the serum we find only one form of IL-6, using both the hepatocyte and the B9 assay, which is about the size of the native single monomer—we don't find activity in any higher molecular weight entities or lower molecular weight entities. I assume that IL-6 travels in the circulation as a monomer, but whether or not it has to dimerize before it interacts with the receptor, I don't know.

*Kishimoto:* We can detect basal levels of IL-6 and the soluble IL-6 receptor in serum. The amount of soluble IL-6 receptor is 1000-fold greater than the amount of IL-6, so IL-6 and the soluble receptor might form a complex and stimulate other types of cells which do not even express IL-6 receptors.

*Fey:* Under what conditions did you see this?

*Kishimoto:* In normal healthy humans we find that IL-6 levels are usually pg/ml whereas levels of the soluble receptor are ng/ml.

*Lotem:* Do you know anything about the mechanism of production of the soluble IL-6 receptor, how it gets into the serum in such high amounts? Is it simply because of shedding of the receptor from cells?

*Kishimoto:* I do not know. We find only a single messenger RNA for the IL-6 receptor.

*Heinrich:* When we cross-linked radioactively labelled IL-6 to the receptor on HepG2 cells, we observed three IL-6-containing complexes, which ran on our gels with molecular masses of 100, 120 and 200 kDa. The 100 kDa complex corresponds to the 80 kDa IL-6-binding protein plus IL-6; the 120 kDa complex can be immunoprecipitated with an antiserum against gp130, and thus contains gp130 (or, more correctly, its equivalent in HepG2 cells) plus ligand. Wouldn't this finding indicate that IL-6 might also bind to gp130? Does IL-6 bind to cells transfected with gp130 cDNA?

*Kishimoto:* We cannot detect binding of IL-6 to gp130. It is possible that a certain portion of gp130 interacts with IL-6 after IL-6 binds to IL-6R, and this may be involved in the formation of the high affinity binding site.

*Ciliberto:* You think that binding of IL-6 triggers a change in the conformation of the receptor, but that it is the receptor itself which binds to gp130. Do you know of any mutations in the receptor which alter conformation such that direct binding to gp130 in the absence of IL-6 is possible?

*Kishimoto:* We have introduced many point mutations into the IL-6 receptor. Certain mutations result in loss of the IL-6-binding activity and others prevent the association with gp130, but we have not found a mutant form of the receptor that binds to gp130 without IL-6.

*Ciliberto:* You have shown that gp130 is expressed by a wide array of cells whereas the receptor is present in only a limited number of cell types. Do you think there could be a response to IL-6 in cells which express only gp130 but which are in close contact with cells expressing the receptor?

*Kishimoto:* At the moment we have no evidence for that kind of interaction.

*Nicola:* Professor Kishimoto, your work provides the most clear-cut evidence I know of to suggest that the mechanism of heterodimeric receptor activation involves the binding of the ligand to one subunit triggering association with a second subunit, the β subunit or gp130. That raises the question of what gp130 'sees'. It seems to me that there are three possibilities. (1) Binding of IL-6 to the α chain (IL-6R) induces a change in the IL-6 molecule such that gp130 can see an altered IL-6 molecule; (2) the binding of IL-6 induces a change in the α chain that allows gp130 to 'see' the altered α chain; or (3) what gp130 actually sees is a combined epitope, part of the IL-6 molecule and part of the IL-6 receptor, in the way that the T cell receptor recognizes the major histocompatibility complex in association with antigen. Do you have any feeling about which of those is the most likely? Your model of interactions between the α chains of the IL-3, IL-5 and GM-CSF receptors and a common β chain must limit the possible interactions of the β chain to those involving something common to all those systems.

*Kishimoto:* There may be structural changes in the receptor molecule after interaction with the ligand. However, this would not explain how different receptors can utilize the same transducer. Common molecular changes in the receptor molecules might induce association with the same transducer.

*Heath:* One could imagine that the IL-6–soluble receptor complex is the ligand for gp130. That means that three things must be present at the same time for a signal to be triggered. That sort of design is rather like the fail-safe mechanism of U.S.A. nuclear warheads, in that at least two separate events have to be controlled for a signal to be generated. In a way, focusing on the receptor as being part of a signalling molecule may be incorrect—the IL-6 receptor may in fact be a presentation receptor. The biological activities of a pleiotropic molecule obviously must be kept under control, and this would provide a neat mechanism for doing that.

*Bazan:* I tend to favour Dr Nicola's third option. What is happening probably does not involve a conformational change of the α receptor on binding of IL-6 allowing gp130 to bind; rather, the α receptor is tethering IL-6 in such a way that gp130 now recognizes *both* molecules simultaneously. Modelling studies indicate that IL-6 will primarily bind to selected loops that converge between the two immunoglobulin-like β-barrel halves of the receptor. In contrast, the interaction between α and β receptor subunits involves residues that are displayed on opposing β-sheets of only the C-terminal β-barrels of paired receptors. This association is similar to the side-by-side packing observed between immunoglobulin domains in antibody structures or in the major histocompatibility complex. We don't yet have enough detailed structural information (that is, a high resolution X-ray fold) to predict which sets of α and β receptors are compatible with each other (regardless of cytokine ligands), but I imagine that a comparison of the GM-CSF, IL-3 and IL-5 α receptors will reveal a select number of amino acids that are sufficient to induce association wiht a common β subunit.

*Gauldie:* I was intrigued by Professor Kishimoto's finding of a 1000-fold excess of the soluble receptor over IL-6 in normal circulation. With a 1000-fold excess of an active ligand-binding receptor, how is there ever any free IL-6?

*Nicola:* Low affinity receptors are characterized by fast on and fast off rates, whereas high affinity receptors always have much slower off rates than on rates; even with excess ligand, given enough time the cell surface high affinity receptor will always win over the low affinity soluble receptor. We have several antibodies to LIF which work well in receptor competition experiments, but they don't inhibit the biological response because of the receptor's kinetic characteristics.

You cannot be sure that an excess of soluble receptor will inhibit biological activity, or that it will necessarily help IL-6 to act on cells which express gp130.

*Gauldie:* I suppose that if the affinity of interaction is low enough the ligand can be separated from the soluble receptor even if it is in excess.

*Nicola:* That would depend on how fast the gel filtration is relative to the dissociation rate. With such experiments you need to do fast gel filtration, with short columns and retention times.

*Williams:* Professor Kishimoto, what controls the level of gp130 in cells? It is clearly one of the key molecules; is its expression controlled by cytokines? Will erythropoietin or IL-3 increase its expression?

*Kishimoto:* I don't know. Injection of IL-6 can augment expression of gp130 for a short time. Two hours after injection the level of gp130 expression returns to normal (Saito et al 1992).

*Gauldie:* In HepG2 cells IL-1 and IL-6 can up-regulate the mRNA for the 80 kDa and 130 kDa proteins (M. Geisterfer, unpublished work). In combination, particularly in the presence of corticosteroids, the up-regulation is enhanced. Several cytokines may well increase expression of the receptor genes. Expression of the 130 kDa protein is not as greatly affected as that of the 80 kDa subunit.

*Kishimoto:* *In vivo* in hepatocytes the expression of gp130 and IL-6R is increased equally.

*Gough:* In those experiments, how much IL-6 did you inject, and how does that amount compare with the levels of endogenous IL-6 in the serum? Is it clear that this was not actually an effect of LPS?

*Kishimoto:* We injected 10 μg IL-6, which was completely free of LPS.

*Heinrich:* When you transfected Jurkat cells with gp130 cDNA you saw an increase in high affinity binding sites (Hibi et al 1990). When we transfected cDNA encoding the 80 kDa IL-6-binding receptor subunit we observed an increase in low affinity binding sites (D. Zohlnhöfer, L. Graeve, S. Rose-John, H. Schooltink, E. Dittrich & P. C. Heinrich, unpublished work). I think this finding complements your results.

*Hirano:* Professor Kishimoto, as you know, the cytoplasmic domain of the IL-2 receptor β chain also contains a serine-rich sequence, which has been shown to be essential for IL-2-mediated signal transduction. Is the serine-rich sequence present in the cytoplasmic domain of gp130 also required for IL-6-mediated signal transduction?

*Kishimoto:* In the IL-2 receptor β chain the serine-rich portion is not as rich in serines as that of gp130; in the cytoplasmic domain of gp130 as well as a 'serine-rich portion' of the IL-2 receptor β chain there are leucine residues in box 2, and these are important.

*Baumann:* Is the initial intracellular event after IL-6 binding identical in liver cells and B cells? Is the same cascade of kinases activated in both cell types?

*Kishimoto:* The sequences of gp130 cloned from plasmacytoma cells, hepatocytes and placenta are exactly the same—there appears to be no heterogeneity amongst different tissues. Differences may occur further downstream, but we do not yet know at which stage the signal-transducing pathways diverge.

*Baumann:* You mentioned that the initial event after IL-6 binding is phosphorylation of gp130. Does the same type of phosphorylation also occur in B cells?

*Kishimoto:* In plasmacytoma cells, myeloma cells, hepatoma cells and the IL-3-dependent transfectant cells there is a tyrosine-specific phosphorylation of gp130.

*Aarden:* What happens to IL-6 after it binds to the receptor? Is it internalized? We have found that in hybridoma cells in culture there is no consumption of IL-6 (Aarden et al 1985). If you start with a few picograms of human natural or recombinant IL-6 in a bottle containing one cell, when the bottle is full of cells there is still the same amount of IL-6 in the supernatant. Recently we found that when we kept cells in medium containing mouse IL-6, either recombinant or natural, the cells became IL-6-independent, and mouse IL-6 is consumed.

*Kishimoto:* We have not studied that.

*Heinrich:* Professor Kishimoto's group reported that $\alpha_2$-macroglobulin is an IL-6-binding protein (Matsuda et al 1989). We have been unable to confirm this (Heinrich et al 1990).

*Kishimoto:* It does bind IL-6.

*Heinrich:* If you incubate serum with radioactively labelled IL-6, a neglible amount of IL-6 is covalently linked to $\alpha_2$-macroglobulin. Obviously, if you activate the $\alpha_2$-macroglobulin with proteases, more or less any protein will bind to it. It is well known that $\alpha_2$-macroglobulin acts as a scavenger for TGF-$\beta$ and IL-1 (Borth & Luger 1989).

*Hirano:* It is likely that there are several IL-6-binding molecules present in serum; $\alpha_2$-macroglobulin is one of them.

*Heinrich:* How is it activated?

*Hirano:* Without any special treament, $\alpha_2$-macroglobulin binds IL-6 at 37 °C.

*Fey:* Is the IL-6 which is bound to $\alpha_2$-macroglobulin still biologically active?

*Kishimoto:* Yes.

*Fey:* $\alpha$-Macroglobulins have an internal thioester in common with complement C3 and C4, and, as Professor Heinrich pointed out, have the propensity to bind non-specifically to more or less any molecule in the neighbourhood at the time when the thioester is hydrolysed. That does not mean they are specific carriers.

## References

Aarden LA, Landsdorp PM, De Groot ER 1985 A growth factor for B cell hybridomas produced by human monocytes. Lymphokines 10:175–185

Borth W, Luger TA 1989 Identification of $\alpha_2$-macroglobulin as a cytokine binding plasma protein. J Biol Chem 264:5818–5825

Fukunaga R, Ishizaka E, Negata S et al 1990 Purification and characterization of the receptor for murine granulocyte colony-stimulating factor. J Biol Chem 265:4008–4015

Fukunaga R, Ishizakaikeda E, Pan CX, Seto Y, Nagata S 1991 Functional domains of the granulocyte colony-stimulating factor receptor. EMBO (Eur Mol Biol Organ) J 10:2855–2865

Heinrich PC, Castell JV, Andus T 1990 Interleukin-6 and the acute-phase response. Biochem J 265:621–636

Hibi M, Murakami M, Saito M, Hirano T, Taga T, Kishimoto T 1990 Molecular cloning and expression of an IL-6 signal transducer, gp130. Cell 63:1149–1157

Matsuda T, Hirano T, Nagasawa S, Kishimoto T 1989 Identification of $\alpha_2$-macroglobulin as a carrier protein for IL-6. J Immunol 142:148–152

Novick D, Engelmann H, Wallach D, Rubenstein M 1989 Soluble cytokine receptors are present in normal human urine. J Exp Med 170:1409–1414

Saito M, Yoshida K, Hibi M, Taga T, Kishimoto T 1992 Molecular cloning of a murine IL-6 signal transducer, gp130, and its regulated expression in vivo. J Immunol, in press

Taga T, Kawanishi Y, Hardy RR, Hirano T, Kishimoto T 1987 Receptors for B cell stimulatory factor 2 (BSF-2): quantitation, specificity, distribution, and regulation of their expression. J Exp Med 166:967– 981

Tovey MG, Content J, Gresser I et al 1988 Genes for IFN-β2 (IL-6), tumor necrosis factor and IL-1 are expressed at high levels in the organs of normal individuals. J Immunol 141:3106–3110

# Molecular biology of the leukaemia inhibitory factor gene

Nicholas M. Gough, Tracy A. Willson, Jürgen Stahl and Melissa A. Brown

*Cancer Research Unit, The Walter and Eliza Hall Institute of Medical Research, Post Office, Royal Melbourne Hospital, Parkville, Melbourne, Victoria 3050, Australia*

*Abstract.* Leukaemia inhibitory factor (LIF) is a polyfunctional cytokine that has been identified and characterized in several laboratories by virtue of a number of different biological activities. LIF is encoded by a unique gene located at 11A1 in the mouse and at 22q12 in man. However, loci related to sequences in the 3′ untranslated region of the mRNA have been detected and located elsewhere in the genome. The LIF gene from four mammalian species has been cloned and sequenced; the sequences are highly conserved within the coding regions and largely non-conserved within the non-coding regions. However, a number of non-coding segments displaying high interspecies similarity are evident; these are candidate control regions. Intriguingly, an exon corresponding to the 5′ end of a variant LIF transcript in the mouse that encodes a potentially matrix-associated form of LIF is not conserved in the human, ovine and porcine genes. The promoter region of the LIF gene contains four well-conserved TATA elements, and two start sites of transcription have been identified. Three regions within the 5′ flanking region have been identified as important for the function of the LIF promoter, including a candidate repressor sequence. The LIF gene is transcribed at only very low levels in normal tissues, but its expression can be increased by various stimuli.

*1992 Polyfunctional cytokines: IL-6 and LIF. Wiley, Chichester (Ciba Foundation Symposium 167) p 24–46*

Leukaemia inhibitory factor (LIF) is a polyfunctional cytokine that has been identified and characterized in several laboratories by virtue of a number of different activities (Table 1). It was identified, characterized, purified and cloned in our laboratory on the basis of its ability to rapidly and potently induce the differentiation of a model mouse myeloid leukaemic cell line, M1 (Gearing et al 1987, Gough et al 1988, Hilton et al 1988, Metcalf et al 1988). Since then, diverse biological activities have been attributed to LIF (see Hilton & Gough 1991 for review). Many of these actions, which are similar to activities displayed by interleukin 6 (IL-6) (such as induction of M1 cell differentiation), are listed in Table 1 and discussed in detail elsewhere in this volume. Like LIF, IL-6 is a polyfunctional cytokine. IL-6 was first cloned in 1980 (as β2-interferon) as a fibroblast product inducible with poly(rI-rC) (Weissenbach et al 1980); it was subsequently re-identified and cloned during 1986–1988 and shown to display

**TABLE 1** LIF and its synonyms

| Acronym | Name | Defining action | References |
|---|---|---|---|
| LIF | Leukaemia inhibitory factor | Induction of M1 differentiation | Gearing et al 1987, Hilton et al 1988, Metcalf et al 1988 |
| D-factor | Differentiation-stimulating factor | Induction of M1 differentiation | Tomida et al 1984, Lowe et al 1989 |
| DIF | Differentiation-inducing factor | Induction of M1 differentiation | Abe et al 1989 |
| DIA | Differentiation inhibitory activity | Inhibition of embryonic stem (ES) cell differentiation | Smith & Hooper 1987, Smith et al 1988, Williams et al 1988 |
| DRF | Differentiation-retarding factor | Inhibition of ES cell differentiation | Koopman & Cotton 1984, Williams et al 1988 |
| HSF-III | Hepatocyte-stimulating factor III | Stimulation of acute-phase protein synthesis | Baumann & Wong 1989 |
| CNDF | Cholinergic neuronal differentiation factor | Induction of cholinergic neuronal differentiation | Yamamori et al 1989 |
| HILDA | Human interleukin for DA cells | Stimulation of murine DA1-1a cell proliferation | Moreau et al 1987, 1988 |
| MLPLI | Melanoma-derived lipoprotein lipase inhibitor | Inhibition of lipoprotein lipase activity | Mori et al 1989 |
| OAF | Osteoclast-activating factor | Stimulation of bone resorption *in vitro* | Abe et al 1986 |

a variety of biological activities. Here, we discuss the molecular biology and molecular genetics of the LIF gene, its expression and the control of its expression, and highlight areas of current interest.

## The LIF locus

Southern blots of mouse or human DNA, probed with DNA probes spanning the LIF-coding region, over a range of hybridization stringencies, all reveal uniquely hybridizing DNA fragments, suggesting that there are not kindred genes with significant similarity to the LIF gene in the mouse or human genome (Gearing et al 1987, Stahl et al 1990, Sutherland et al 1989). This, however, is a relatively insensitive test for related genes; for example, a murine granulocyte-macrophage colony-stimulating factor (GM-CSF) probe will not detect on a genomic Southern blot its human counterpart, which has about 50% nucleotide sequence identity (N. M. Gough, unpublished work). Thus, genes with $\leqslant 50\%$ sequence identity to the LIF-coding region could well exist but remain undetected.

Interestingly, when a mouse cDNA fragment which includes about 800 bp of 3′ untranslated region is used as a probe, a complex hybridization pattern is evident, containing two or three additional bands not seen with the coding region-specific probe (Stahl et al 1990). These extra bands are detected at high stringency (65 °C, 0.2 × standard saline citrate). Whether these additional sequences represent loci which have fortuitous sequence similarity to the LIF 3′ untranslated region, or whether they are evolutionarily related to the LIF gene, is unclear at this stage.

The LIF gene has been localized to human chromosome 22q12.1–12.2 (Budarf et al 1989, Sutherland et al 1989) and to a region of mouse chromosome 11 (11A1) which is known to contain genes equivalent to those on human 22q (Kola et al 1990). These results were derived using human or mouse LIF probes encompassing only the coding regions of the respective genes. Interestingly, Justice et al (1990) reported the localization of the mouse LIF gene (under the synonym HILDA) to chromosome 13. In these experiments, however, a full-length cDNA probe was used, and it is evident that the locus mapped in these studies in fact corresponded to one of the loci related to the 3′ untranslated region.

The human LIF gene is localized to the same cytogenetic band as the t(11;22) (q24;q12) translocation frequently occurring in Ewing sarcoma (a tumour of cells displaying pluripotentiality) and peripheral neuroepithelioma. However, it appears that at the molecular level the LIF gene is unaltered by, and is at least 650 kbp distal to, the Ewing sarcoma break point (Sutherland et al 1989, Budarf et al 1989, Selleri et al 1991).

## The arrangement of the LIF gene

LIF is encoded by a 4.8 kb mRNA which is transcribed from a three-exon gene spanning about 8 kbp of DNA (see Fig. 1) (Stahl et al 1990). In the prototype

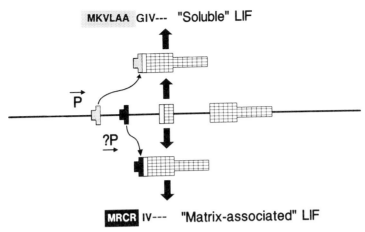

FIG. 1.   Alternative transcripts from the mouse LIF gene. Expression of the prototype LIF mRNA, in which exon 1 encodes the N-terminal six residues (grey block) of the hydrophobic leader of secreted LIF, is shown above the line. A variant mRNA for LIF, possibly transcribed from an alternative promoter (?P) and including an alternative exon 1 which encodes four different N-terminal residues of the hydrophobic leader (black block), is shown below the line.

LIF gene, exon 1 specifies the first six amino acids of the hydrophobic leader sequence, the remainder being encoded by exon 2. Comparison of the LIF gene sequences from four mammalian species (mouse, human, sheep and pig) reveals that the exons are highly conserved, whereas the intronic and flanking sequences, with the exception of about 300 bp of the 5' flanking region, are largely non-conserved (see below).

Recently Rathjen et al (1990) described an alternative transcript of the mouse LIF gene which differs at the 5' end of the mRNA by having a different exon 1. Inspection of the mouse LIF gene sequence places this alternative exon 1 (exon 1b) within intron 1, at a position about 500 bp downstream of exon 1 (Fig. 1). This change alters the N-terminus of the hydrophobic leader, which in turn apparently changes the extracellular fate of the encoded LIF; instead of being soluble, the LIF produced is associated with the extracellular matrix (Rathjen et al 1990). The sequence of exon 1b, however, presents two problems. Firstly, this sequence is not well conserved between species (see Fig. 2): in the human and porcine LIF genes, although sequences clearly similar to this region may be found at a similar location within intron 1, these sequences probably do not constitute a functional exon because they both lack initiation codons, and in the case of the human gene the putative exon 1b equivalent also lacks a splice donor site. In the sheep gene no obvious homologue to this mouse region can be found; only a very poor match containing no splice site or initiation codon, and located at a very different position in the intron (position 1275), is found in intron 1

```
ALTERNATIVE                    MetArgCysArgIleVal---
   MOUSE cDNA:         ---CGCGAGATGAGATGCAGGATTGTG---


                               MetArgCysArg
MOUSE GENE (515):        ---CGCGAGATGAGATGCAGGTATGGG---


                               ValArgTyrArg
HUMAN GENE (532):        ---CGCtAGgTGAGATatAGGggTGtG---


                               ValSerProGly
  PIG GENE (503):        ---tGCtAGgTGAGcccggGGTgTGGG---


                               LysAlaTrpGlu
 SHEEP GENE (1275):      ---gGaGAtAaGgccTGggaGgggttG---
```

FIG. 2.   A cross-species comparison of the alternative exon 1 of the mouse LIF gene. The nucleotide and amino acid sequences of the 5′ end of a variant LIF cDNA clone (Rathjen et al 1990) are given on the top line. The nucleotide and deduced amino acid sequences of the DNA sequence in intron 1 of the mouse gene corresponding to this alternative exon are given below. Regions within intron 1 of the human, porcine and ovine genes which have similarity to this exon are shown. Non-conserved residues are in lower-case letters. The nucleotide position given in parenthesis corresponds to the position within intron 1 of the sequence shown.

(see Fig. 2). The lack of conservation of this alternative exon perhaps calls into question its general biological significance, suggesting that it is either of importance only in the mouse or is present and expressed in the mouse adventitiously. Moreover, as illustrated in Fig. 3, the 5′ splice site (GT) of exon 1b is in a different translation-reading frame of the alternative LIF transcript from the corresponding splice site of the prototype transcript. Indeed, to generate the sequence of the alternative transcript presented by Rathjen et al (1990) it would be necessary to invoke the removal of an extra nucleotide (either the G residue 5′ of the GT, or the G residue 3′ of the CAG) during this RNA-processing event.

## Evolutionary comparison of the LIF gene and protein

LIF is one of the more highly conserved cytokines active within the haemopoietic system, and LIF gene homologues can be detected not only on genomic Southern blots of DNA from all eutherian mammalian species examined (Willson et al 1992), but also in DNA from several marsupial species. The nucleotide sequence of the LIF gene and the inferred amino acid sequence from four mammalian species (mouse, human, sheep and pig) (Stahl et al 1990, Willson et al 1992) and the amino acid sequence of rat LIF (Yamamori et al 1989) have been determined.

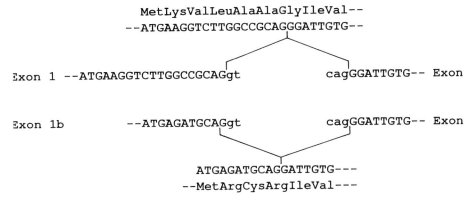

FIG. 3.   Splicing of the prototype and alternative 5′ exons of the mouse LIF gene. The nucleotide sequence of the prototype LIF mRNA and the deduced amino acid sequence are given at the top, and the sequence of a variant LIF cDNA clone (Rathjen et al 1990) is given at the bottom. The nucleotide sequences of the pertinent portions of exon 1, exon 1b and exon 2 are shown in between. Conventional 5′ and 3′ splice sites (GT and CAG) are indicated by lower-case letters.

Comparison of these sequences reveals that at the amino acid level, LIF proteins are highly conserved (Table 2). All six cysteine residues found in mouse LIF are conserved at identical positions in the four other species, suggesting that intramolecular disulphide bonds are vital to the integrity and activity of the LIF molecule. Interestingly, rat LIF has an additional cysteine residue. Six of seven potential N-linked glycosylation sites are absolutely maintained (the seventh being conserved in all species except pig), suggesting that glycosylation of the molecule is important. Intriguingly, although native LIF is heavily glycosylated, glycosylation is not actually required for biological activity, because E. coli-derived unglycosylated recombinant human and mouse LIF, and yeast and baculovirus-derived recombinant LIF, which have different patterns of glycosylation from native LIF, are all biologically active, with similar specific biological activities, at least in vitro. The disposition of charged residues is quite

**TABLE 2   Cross-species similarity of LIF proteins[a]**

|          | Mouse | Rat | Human | Ovine | Porcine |
|----------|-------|-----|-------|-------|---------|
| Mouse    | 100   | 92  | 79    | 74    | 78      |
| Rat      |       | 100 | 81    | 75    | 78      |
| Human    |       |     | 100   | 88    | 87      |
| Ovine    |       |     |       | 100   | 84      |
| Porcine  |       |     |       |       | 100     |

[a]The percent identity of amino acid sequences for the mature LIF proteins from five species are given in pair-wise fashion.

similar in the five sequences. These results collectively suggest that the LIF mole-cules from these five different mammalian species share many structural features.

A comparison of the LIF gene nucleotide sequences indicates that sequence conservation is largely a characteristic of the coding regions (Stahl et al 1990, Willson et al 1992). The non-coding regions of the gene are largely non-conserved, with the exception of the 5′ flanking/promoter region of the gene, which is highly conserved. Alignment of the sequence at the 5′ end of the mouse, human, ovine and porcine LIF genes revealed an 84% sequence identity over a region of approximately 270 bp 5′ of the major transcriptional start site (see below). The significant structural and functional features of the 5′ flanking region, and their cross-species conservation, are discussed below.

## Structural and functional elements in the LIF gene promoter

Nucleotide sequence analysis of the mouse LIF gene revealed four TATA-like elements within a region of about 340 bp 5′ of the translational start codon

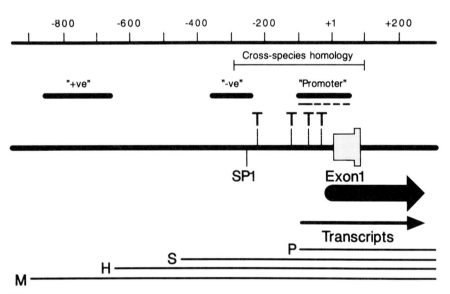

FIG. 4. Structural and functional features within the 5′ flanking region of the LIF gene. The location of exon 1, four conserved TATA boxes (T) and a candidate Sp1-binding site (SP1) are shown along the centre line. Position +1 corresponds to the major start site of transcription. Arrows beneath the centre line indicate transcripts initiated at the major and minor transcriptional start sites. Three regions implicated in the regulation of LIF promoter function (+ ve, − ve and Promoter) are indicated above the line (see text for details). The extent of high cross-species nucleotide sequence identity (>75%) is indicated by the bar. The extent of nucleotide sequence determined within the 5′ flanking region from each of four mammalian species (P, pig; S, sheep; H, human; M, mouse) is shown at the bottom.

within exon 1 (Stahl et al 1990) (see Fig. 4). A transcriptional start site adjacent to the most proximal of these TATA boxes was determined by S1 nuclease protection (Fig. 4) (Stahl et al 1990). In addition, a second transcriptional start site adjacent to the second most distal element was located using a polymerase chain reaction (PCR)-based modification of the primer-extension procedure (Stahl et al 1990). Given the non-quantitative nature of the PCR procedure, however, the relative abundance of transcripts initiating at the more distant site is unclear. Transcripts initiating at the more distal start site could not be detected by the S1 nuclease procedure, suggesting that they are less abundant than those initiating at the more proximal site. No transcripts initiating at the most distal TATA box have been detected, despite the strong cross-species conservation of this region and the high level of similarity with the consensus TATA box sequence.

To delineate regions within the 5′ flanking region of the LIF gene that may constitute functional elements of the promoter, we placed a nested set of DNA fragments, extending from within the 5′ untranslated region up to 900 bp 5′-wards, upstream of the bacterial chloramphenicol acetyltransferase (CAT) reporter gene, and assayed their ability to direct CAT expression in LIF-producing STO-fibroblasts (J. Stahl & N. M. Gough, unpublished work).

The minimal essential region of the LIF promoter was found to consist of the major start site of transcription ($+1$), the proximal TATA box ($-31$) and between 50 and 72 additional 5′ nucleotides ($-32$ to $-103$). Constructs extending 5′-wards to only position $-81$ were inactive, implying that sequences within the interval $-82$ to $-103$ are necessary for promoter function.

A negative regulatory element was identified between positions $-249$ and $-360$. The presence of the $-249$ to $-360$ region (and up to 300 nucleotides further 5′-wards) abolished CAT activity induced through the LIF promoter. However, promoter activity in constructs containing this negative element could be restored either by inclusion of the SV40 viral enhancer element in the construct, or, more interestingly, by inclusion of the sequence $-660$ to $-860$ of the LIF 5′ flanking region ($+$ve in Fig. 4) suggesting the existence of an enhancer-like element within this region.

Cross-species comparison of the 5′ flanking region revealed a region showing high interspecies sequence identity that extended to 270 bp 5′ of the major transcriptional start site. All four TATA boxes are conserved and there are large blocks of conserved nucleotides surrounding the major and minor transcriptional start sites. Indeed, the region defined as the minimal essential promoter region ($-103$ to $+1$) is almost totally conserved. The region harbouring the putative negative element ($-249$ to $-360$) falls immediately 5′ of the conserved region and is only very poorly conserved. Within this region, however, a stretch of 16 conserved nucleotides including a potential Sp1 transcription factor-binding site is found; whether this sequence plays a role in the negative element is

**TABLE 3  Cell types expressing LIF**

| Cell type | Species | Stimuli[a] | Detection method | References |
|---|---|---|---|---|
| T lymphocyte clones and lines, peripheral blood lymphocytes, spleen cells | M,H | Antigen, lectins, PHA, TPA | B,N,P | Moreau et al 1987, 1988, Abe et al 1986, Gearing et al 1987, Le et al 1990, Lübbert et al 1991 |
| Peripheral blood monocytes Various monocytic cell lines | M,H | LPS, VD$_3$, TPA | B,N,P | Abe et al 1989, Anegon et al 1990, Stahl et al 1990 |
| Megakaryocytic cell line | M | — | P | Sasaki et al 1991 |
| Thymic epithelial cells | H | — | N | Le et al 1990 |
| Bone marrow stromal cell lines | M,H | — | P | M. A. Brown, unpublished |
| Fibroblasts Various fibroblastic cell lines | M,H | TNF-$\alpha$, IL-1$\beta$, TPA | B,N | Miyagi et al 1991, Koopman & Cotton 1984, Lübbert et al 1991, Tomida et al 1984 |
| Umbilical vein endothelial cells | H | — | N | Lübbert et al 1991 |
| Osteoblasts Various osteoblastic cell lines | R | TNF-$\alpha$, RA, TGF-$\beta$ | B,N,P | Allan et al 1990, Shiina-Ishimi et al 1986 |

| Cell type/tissue | Species | Stimulus | Assay | Reference |
|---|---|---|---|---|
| Astrocytes | M | LPS, CMV | P | Wesselingh et al 1990 |
| Neuroepithelial cell lines | M | | P | M. A. Brown, unpublished |
| Blastocysts | M | — | P | Murray et al 1990 |
| ES and EC cell lines | M | — | P | M. A. Brown, unpublished |
| Granulated metrial gland cells | M | — | B,P | Croy et al 1991 |
| Buffalo rat liver cells | R | — | B,N | Moreau et al 1988, Smith & Hooper 1987 |
| Heart myoblasts | R | — | B | Yamamori et al 1989 |
| Various tumours: 5637 (bladder), SEKI (melanoma), COLO16 (squamous cell), MiaPaca (pancreatic) | H | TPA | B,N | Williams et al 1988, Gascan et al 1990, Mori et al 1989, Baumann & Wong 1989 |
| Krebs II ascites | M | LPS | B,N | Stahl et al 1990, Hilton et al 1988 |
| Ehrlichs ascites | M | — | B | Lowe et al 1989 |

[a] In most instances the stimulus tested enhanced LIF expression over a basal level of constitutive expression.

M, mouse; H, human; R, rat.

B, bioassay; N, Northern blot; P, polymerase chain reaction-based RNA detection.

PHA, phytohaemagglutinin; TPA, 12-$O$-tetradecanoylphorbol 13-acetate (phorbol 12-myristate 13-acetate, PMA); LPS, lipopolysaccharide; $VD_3$, 1,25-dihydroxyvitamin $D_3$; TNF-$\alpha$, tumour necrosis factor $\alpha$; IL-1$\beta$, interleukin 1$\beta$; RA, retinoic acid; TGF-$\beta$, transforming growth factor $\beta$; CMV, cytomegalovirus.

unclear. Sequence information encompassing the putative enhancer element ( $-660$ to $-860$ ) is so far available only for the mouse gene.

**Expression of the LIF gene**

Table 3 summarizes the various murine and human primary cells and cell lines in which LIF expression has been reported, and indicates the procedure by which expression was detected. LIF transcripts, at low abundance, have been detected by Northern blotting in a number of murine and human cell lines and primary cells *in vitro*, including Krebs II ascites cells—the source from which LIF was purified—primary peripheral blood monocytes and monocytic cell lines (such as WEHI-265 and HL-60), osteoblasts (primary and cell lines), fibroblasts, umbilical vein endothelial cells, peripheral blood lymphocytes and T lymphoid cell lines, thymic epithelial cells and a variety of different tumours. In many of these cell types the abundance of LIF transcripts was increased by various stimuli, such as lipopolysaccharide (LPS), T cell receptor stimuli, TPA (12-*O*-tetradecanoylphorbol 13-acetate), TNF-$\alpha$ and IL-1$\beta$. Expression is often too low to be detected by conventional Northern blot analysis, but is revealed using sensitive polymerase chain reaction-based detection procedures. Such techniques enable detection of LIF transcripts in a broader range of cell lines, including embryonic stem (ES) and embryonal carcinoma (EC) cells, neuroepithelial cell lines and a megakaryocytic line, and also in a number of primary cell populations *in vitro*, including blastocysts, LPS-stimulated astrocytes and granulated metrial gland cells.

In a survey of normal mouse tissues and organs, LIF transcripts were undetectable by Northern blotting. Even by the PCR, LIF transcripts could be detected only sporadically at very low levels, and only by using a high cycle number. In contrast, after injection of LPS LIF transcripts were readily and consistently detectable in a number of organs, including heart, lung and liver (see Fig. 5). LPS induction of LIF expression occurs in parallel with induction of expression of a number of other cytokines, including IL-6, GM-CSF and G-CSF, and is consistent with a role for LIF in mediating the acute inflammatory response.

**Concluding remarks**

The LIF gene is constitutively expressed at low levels in a number of different cell types *in vitro*, and the level of expression can be increased by a variety of stimuli. Basal expression *in vivo* appears to be extremely low, but again the level of expression can be enhanced. The activity of the LIF promoter region in reconstruction experiments involving the CAT reporter gene is weak. Interestingly, a dissection of the LIF 5' flanking region revealed at least three elements that influence the activity of the promoter, one of which appears to

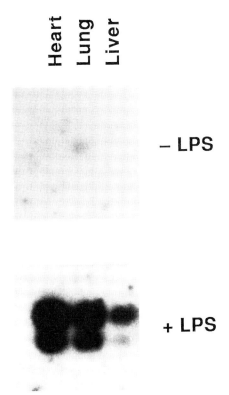

FIG. 5. The induction of LIF expression *in vivo*. Poly(A) RNA was extracted from heart, lung and liver removed from three untreated control mice (– LPS) and three mice injected 3 h previously with 5 μg LPS (+ LPS). First-strand cDNA was synthesized and used as a template for amplification by the polymerase chain reaction, as previously described (Allan et al 1990). PCR products were analysed by Southern blotting.

be a negative control region which may contribute to the low level of basal LIF expression. The precise sequences within these regions that represent functional elements, the interplay between these regions, and their role in regulating the basal and induced expression of the LIF gene remain to be elucidated.

*Acknowledgements*

We are grateful to many of our colleagues for their comments and interest in this work. Work from the authors' laboratory was supported by The National Health and Medical Research Council (Canberra), National Institutes of Health (Bethesda) grant CA22556, the Anti-Cancer Council of Victoria and AMRAD Corporation (Melbourne).

## References

Abe E, Tanaka H, Ishimi Y et al 1986 Differentiation-inducing factor produced from conditional-medium of mitogen-treated spleen cell cultures stimulates bone resorption. Proc Natl Acad Sci USA 83:5958–5962

Abe T, Murakami M, Sato T, Kajiki M, Ohno M, Kodaira R 1989 Macrophage differentiation inducing factor from human monocytic cells is equivalent to murine leukemia inhibitory factor. J Biol Chem 264:8941–8945

Allan EH, Hilton DJ, Brown MA et al 1990 Osteoblasts display receptors for and responses to leukemia inhibitory factor. J Cell Physiol 145:110–119

Anegon I, Moreau J-F, Godard A et al 1990 Production of human interleukin for DA cells (HILDA)/leukemia inhibitory factor (LIF) by activated monocytes. Cell Immunol 130:50–65

Baumann H, Wong GG 1989 Hepatocyte-stimulating factor III shares structural and functional identity with leukemia-inhibitory factor. J Immunol 143:1163–1167

Budarf M, Emanuel BS, Mohandas T, Goeddel DV, Lowe DG 1989 Human differentiation-stimulating factor (leukaemia inhibitory factor, human interleukin DA) gene maps distal to the Ewing sarcoma breakpoint on 22q. Cytogenet Cell Genet 52:19–22

Croy BA, Guilbert LJ, Brown MA et al 1991 Characterization of cytokine production by the metrial gland and granulated metrial gland cells. J Reprod Immunol 19:149–166

Gascan H, Anegon I, Praloran V et al 1990 Constitutive production of human interleukin for DA cells/leukemia inhibitory factor by human tumor cell lines derived from various tissues. J Immunol 144:2592–2598

Gearing DP, Gough NM, King JA et al 1987 Molecular cloning and expression of cDNA encoding a murine myeloid leukaemia inhibitory factor (LIF). EMBO (Eur Mol Biol Organ) J 6:3995–4002

Gough NM, Gearing DP, King JA et al 1988 Molecular cloning and expression of the human gene homologous to murine myeloid leukaemia inhibitory factor (LIF). Proc Natl Acad Sci USA 85:2623–2627

Hilton DJ, Gough NM 1991 Leukemia inhibitory factor: a biological perspective. J Cell Biochem 46:21–26

Hilton DJ, Nicola NA, Gough NM, Metcalf D 1988 Resolution and purification of three distinct factors produced by Krebs ascites cells which have differentiation-inducing activity on murine myeloid leukemic cell lines. J Biol Chem 263:9238–9243

Justice MJ, Silan CM, Ceci JD, Buchberg AM, Copeland NG, Jenkins NA 1990 A molecular genetic linkage map of mouse chromosome 13 anchored by the beige (bg) and satin (sa) loci. Genomics 6:341–351

Kola I, Davey A, Gough NM 1990 Localization of the murine leukaemia inhibitory factor gene near the centromeres on chromosome 11. Growth Factors 2:235–240

Koopman P, Cotton RGH 1984 A factor produced by feeder cells which inhibits embryonal carcinoma cell differentiation: characterization and partial purification. Exp Cell Res 154:233–242

Le PT, Lazorick S, Whichard LP et al 1990 Human thymic epithelial cells produce IL-6, granulocyte-monocyte CSF and leukemia inhibitory factor. J Immunol 145:3310–3315

Lowe DG, Nunes W, Bombara M et al 1989 Genomic cloning and heterologous expression of human differentiation-stimulating factor. DNA (NY) 8:351–359

Lübbert M, Mantovani L, Lindemann A, Mertelsmann R, Herrmann F 1991 Expression of leukemia inhibitory factor is regulated in human mesenchymal cells. Leukemia (Baltimore) 5:361–365

Metcalf D, Hilton DJ, Nicola NA 1988 Clonal analysis of the action of the murine leukemia inhibitory factor on leukemic and normal murine hemopoietic cells. Leukemia (Baltimore) 2:216–221

Miyagi T, Akashi M, Yamato K, Miyoshi I, Koeffler HP 1991 D-factor: modulation of expression in fibroblasts. Leuk Res 15:441–451

Moreau J-F, Bonneville M, Godard A et al 1987 Characterization of a factor produced by human T cell clones exhibiting eosinophil-activating and burst promoting activities. J Immunol 138:3844–3849

Moreau J-F, Donaldson DD, Bennett F, Witek-Giannotti J, Clark SC, Wong GG 1988 Leukaemia inhibitory factor is identical to the myeloid growth factor human interleukin for DA cells. Nature (Lond) 336:690–692

Mori M, Yamaguchi K, Abe K 1989 Purification of a lipoprotein lipase-inhibiting protein produced by a melanoma cell line associated with cancer cachexia. Biochem Biophys Res Commun 160:1085–1092

Murray R, Lee F, Chiu C-P 1990 The genes for leukemia inhibitory factor and interleukin-6 are expressed in mouse blastocysts prior to the onset of hemopoiesis. Mol Cell Biol 10:4953–4956

Rathjen DP, Toth S, Willis A, Heath JK, Smith AG 1990 Differentiation inhibiting activity is produced in matrix-associated and diffusible forms that are generated by alternate promoter usage. Cell 62:1105–1114

Sasaki H, Kajigaya Y, Hirabayashi Y et al 1991 Establishment and characterization of a murine megakaryoblastic cell line growing in protein-free culture (L8057Y5). Leukemia (Baltimore) 5:408–415

Selleri L, Hermanson GG, Eubanks JH, Lewis KA, Evans GA 1991 Molecular localization of the t(11;22) (q24;q12) translocation of Ewing sarcoma by chromosomal in situ suppression hybridization. Proc Natl Acad Sci USA 88:887–891

Shiina-Ishimi Y, Abe E, Tanaka H, Suda T 1986 Synthesis of colony-stimulating factor (CSF) and differentiation-inducing factor (D-factor) by osteoblastic cells, clone MC3T3-E1. Biochem Biophys Res Commun 134:400–406

Smith AG, Hooper ML 1987 Buffalo rat liver cells produce a diffusible activity which inhibits the differentiation of murine embryonal carcinoma and embryonic stem cells. Dev Biol 121:1–9

Smith AG, Heath JK, Donaldson DD et al 1988 Inhibition of pluripotential embryonic stem cell differentiation by purified polypeptides. Nature (Lond) 336:688–690

Stahl J, Gearing DP, Willson TA, Brown MA, King JA, Gough NM 1990 Structural organization of the genes for murine and human leukemia inhibitory factor. J Biol Chem 265:8833–8841

Sutherland GR, Baker E, Hyland VJ, Callen DF, Stahl J, Gough NM 1989 The gene for human leukemia inhibitory factor (LIF) maps to 22q12. Leukemia (Baltimore) 3:9–13

Tomida M, Yamamoto-Yamaguchi Y, Hozumi M 1984 Purification of a factor inducing differentiation of mouse myeloid leukaemic M1 cells from conditional medium of mouse fibroblast L929 cells. J Biol Chem 259:10978–10982

Weissenbach J, Chernajovsky Y, Zeevi M et al 1980 Two interferon mRNAs in human fibroblasts. *In vitro* translation and *Escherichia coli* cloning studies. Proc Natl Acad Sci USA 77:7152–7156

Wesselingh SL, Gough NM, Finlay-Jones JJ, MacDonald PJ 1990 Detection of cytokine mRNA in astrocyte cultures using the polymerase chain reaction. Lymphokine Res 9:177–185

Williams RL, Hilton DJ, Pease S et al 1988 Myeloid leukaemia inhibitory factor maintains the developmental potential of embryonic stem cells. Nature (Lond) 336:684–687

Willson TA, Metcalf D, Gough NM 1992 Cross-species comparison of the sequence of the leukaemia inhibitory factor gene and its protein. Eur J Biochem 204:21–30

Yamamori T, Fukada K, Aebersold R, Korsching S, Fann M-J, Patterson PH 1989 The cholinergic neuronal differentiation factor from heart cells is identical to leukemia inhibitory factor. Science (Wash DC) 246:1412–1416

## DISCUSSION

*Dexter:* Does the increased LIF expression you see after treatment with LPS or TNF result from an effect on transcription rates or from post-transcriptional modification?

*Gough:* We do not know.

*Dexter:* In tissues in which you see expression of LIF, do the tissue cells themselves express LIF or are carrier cells such as monocytes or T cells responsible?

*Gough:* We don't know. However, for the purposes of this discussion I don't think it actually matters greatly, because for the most part I have described essentially the 'lack' of expression. Various organs do express tiny amounts of LIF but I am not sure whether or not it's in carrier haemopoietic cells.

*Fey:* We have recently cloned and sequenced the rat LIF gene and have obtained results similar to yours (G. Baffet, R. Fletcher, M.-Z. Cui, W. Northemann & G. H. Fey, unpublished work). The overall identity with the mouse and human LIF genes was 91.5% and 80%, respectively. Conservation with the mouse gene was particularly strong ($>98\%$) in the 300 bp 5' flanking region. The rat gene also contained sequences resembling exon 1b that has been described in the mouse, but so far there is no evidence that it is actually used in rats. With high resolution Northern blots we detected RNA for the alternative species in the mouse but not in the rat. This alternative species seems to be mouse-specific rather than rodent-specific, and therefore its biological importance is questionable.

*Gough:* In the rat gene, is the structure of the exon 1b region conserved?

*Fey:* There is an upstream stop signal and a methionine initiation codon at positions equivalent to those in the mouse gene, a few nucleotides upstream of the exon–intron boundary. It could potentially be used as an exon but we don't have evidence that it actually is.

*Heath:* What tissues did you look at?

*Fey:* We have worked with primary rat macrophage cultures, unstimulated or treated with LPS, and with regenerating rat liver after partial hepatectomy.

*Heath:* Those are situations where you would expect the 'diffusible' form to predominate. Have you tried probe protection methods?

*Fey:* We have not tried other detection methods.

*Heath:* The difference in size between the two transcripts is fairly small (15 nucleotides) so it is important to use probe protection techniques to accurately visualize the two transcripts.

*Gough:* Do you find the alternative transcript in BRL cells?

*Heath:* We haven't looked, because we don't have the right probe.

Dr Gough, I don't quite understand your point about the splice sequences, because the published results are from protection experiments using probes which span the putative splice junctions and there's absolutely no evidence for the incorrect splicing you suggested.

*Gough:* Those experiments actually don't address the nature of the genomic sequence at the point corresponding to the splice site. I drew the structure of the gene at the intron–exon boundary and compared it with the structure of the cDNA sequence, and that raised a problem. We checked our gene sequence and found it to be correct across that region, so the problem cannot be resolved by the gene sequence being incorrect.

*Heath:* I was interested in your cross-species comparisons. We haven't looked at pigs or cows, and have looked at only a few human examples. With humans, by using probes which span the putative exon 1, the first exon that you described, we can find evidence for two transcripts which differ, but at present we have no idea what the origins of these two transcripts are.

*Gough:* In a way, that's negative evidence. While it suggests that there is an alternative splice at the exon 1–2 boundary, it doesn't tell you what has been spliced in.

*Heath:* That's correct and will remain so until the appropriate cDNA clones have been isolated. It's interesting that the region including exon 1b is fairly conserved.

*Gough:* Actually, exon 1b is not well conserved, but lies adjacent to a conserved region in intron 1. The important feature is not the overall similarity, but the sequence identity at the nucleotide level, and that's where you run into the problem.

*Lee:* Dr Heath, is there any evidence for differential regulation of this alternative splicing in cell types in which you might expect to see the matrix-bound form rather than the soluble form?

*Heath:* The published evidence is based on the induction of LIF expression by various growth factors. These experiments, which use probes that distinguish between the two forms, suggest that the diffusible form is influenced by external stimuli more than the matrix-bound form; that's consistent with the sequence upstream of exon 1a that Nick Gough described which contains various kinds of consensus binding sequences for PEA-3 and AP1, whereas the sequences upstream of the putative exon 1b are GC-rich and more resemble a housekeeping promoter.

In ES cells there are low, but detectable, levels of transcription, predominantly of the matrix form (Rathjen et al 1990), and when the cells differentiate there is massive induction of the diffusible form, although the matrix form is still present. Probe protection assays on dissected material show that in the extra-embryonic ectoderm both forms are expressed at Days 6–7, but at Day 8 the diffusible form predominates. The two forms therefore do appear to be differentially expressed in at least some situations.

*Gough:* What do you know about how the alternative form of LIF actually associates with the extracellular matrix?

*Heath:* We got into this area through analysis of cells which had LIF-like effects on ES cells but apparently didn't secrete LIF into the culture medium. The most striking example was the $10t_{1/2}$ cell line. We found that the biological activity was all associated with the cell culture dish—that is, after we removed the cells we found biological activity when we replated ES cells into the dish from which the cells had been removed. *In vitro*, in particular situations, there is therefore a LIF-like bioactivity in the 'solid phase'. After isolating the cDNA clones we found a close correlation between the form which is expressed and the location of the biological activity. In unstimulated $10t_{1/2}$ cells, for example, the M (matrix) form predominates, whereas in early Ehrlich ascites cells the D (diffusible) form predominates. If you transfect these cDNA clones into cells which express very low levels of LIF under particular kinds of promoters you can show that the biological activity follows the appropriate 5′ end. If you use high level expression promoters, as we initially did, you find activity associated with the matrix but also substantial activity in the medium; if you use weaker promoters the matrix-associated activity predominates. It seems that after transfection of the matrix cDNA, the matrix compartment fills up and then LIF starts spilling out into the medium.

More recently, Agnes Mereau in my lab has shown that if you detach cells, such as UMR-K6 osteoblast cell lines or PYS2, radiolabelled pro-karyotically expressed (i.e., non-glycosylated) LIF binds to specific high affinity sites associated with the matrix—that is, the material which is left behind when you remove the cells. The affinity of these sites is about 500 pM, which is lower than the 50–100 pM affinity of the sites associated with the cell. LIF therefore has an inherent ability to associate with the extracellular matrix, and it appears to do that by association with a specific binding protein. That leaves us with the question, what is the biological function of the amino terminus? We have to conclude that it acts as some kind of intracellular routing or trafficking signal, because it's removed before secretion of the molecule from the cell yet clearly seems to influence the localization of the protein. We have not yet attached that sequence to a heterologous protein to see whether it will target the protein. However, that experiment might not work, because it appears that part of the mature LIF molecule is also required for matrix association.

*Dexter:* Have you used enzymes to digest various components of the matrix, to see which species are binding the LIF?

*Heath:* The matrix-associated activity is insensitive to heparinase and chondroitinase, so the mechanism of association is not analogous to those of fibroblast growth factor or colony-stimulating factor.

*Stewart:* Because LIF inhibits the differentiation of embryonic stem (ES) cells *in vitro* we have been interested in determining what its function is in regulating the growth and development of mouse embryos *in vivo*, as ES cells are derived from the inner cell mass of the blastocyst. We began by looking for expression of LIF in various tissues from adult mice as well as in the developing embryo. Using Northern analysis and RNAse protection analysis, we found low levels of expression in many adult mouse tissues and our results were very similar to those Dr Gough described. In postimplantation mouse embryos weak signals were detected in the embryo itself, with slightly higher levels being found in the yolk sac. We did, however, detect a much stronger signal in the uterus of adults. Consequently, we analysed LIF expression in this tissue during the entire period of pregnancy.

LIF transcripts were not detected in the uterus during oestrus, but on the first day of pregnancy (Day 1 = day of plug) a weak signal was observed, with the levels declining over the next two days. On the fourth day of pregnancy a clear burst of LIF expression occurred, with a 5–10-fold increase in transcript levels. These levels were maintained for only one day, and by the fifth day had declined to those seen on the second and third days. Throughout the rest of pregnancy LIF expression remained at this basal level. Histological and *in situ* hybridization analysis revealed that the site of LIF expression on the fourth day was the secretory endometrial glands present in the uterine endometrium, with a fainter signal detectable in the endometrial epithelium. By the fifth day of pregnancy, when the blastocysts had begun to implant with concurrent formation of the decidua, the endometrial glands were disappearing and also were ceasing to express LIF. These results showed that LIF was synthesized for one day in the endometrial glands of the uterus and, presumably, was secreted into the uterine lumen along with other proteins produced by these glands.

The coincidence between the burst of LIF expression and the presence of the embryos as blastocysts in the uterus was intriguing. Does the blastocyst itself stimulate expression, or is expression under maternal control? To investigate this question we looked at LIF expression in pseudopregnant mice and in females undergoing delayed implantation.

Pseudopregnancy results after females mate with vasectomized males. These mice experience the same physiological changes as in normal pregnancy except that the uterus does not contain viable embryos. LIF expression during the first five days of pseudopregnancy showed the same pattern, with a burst occurring on the fourth day, as in normal pregnant females. These results showed that LIF expression was not dependent on the presence of viable embryos and suggested that expression was under maternal control.

Delayed implantation can be induced either experimentally or naturally. Naturally induced delay occurs when a female has just given birth. She mates immediately and supports two groups of offspring. While the first litter is being suckled the second group develops to the blastocyst stage in the reproductive tract. Rather than implanting, the blastocysts remain suspended in the uterine lumen with cellular proliferation halted. Implantation can be delayed for 6–10 days beyond the normal time. Delay can be induced experimentally by ovariectomizing females on the third day of pregnancy. Removal of the ovaries prevents the surge of oestrogen expression that would occur on Day 3–4 and which is apparently necessary for implantation. If the mice are maintained on progesterone the blastocysts remain viable in the uterine lumen but do not implant. Implantation can be induced by removal of the suckling litter, or by intraperitoneal injection of oestrogen in the experimentally induced state.

In both cases, while delay was maintained no LIF expression could be detected, even though the blastocysts were present in the uterine lumen and endometrial glands were seen in the endometrium. However, after interruption of delay a clear burst of LIF expression was observed some 12–24 hours later, coinciding with the beginning of blastocyst implantation.

Our results indicated that the burst of LIF expression always preceded implantation of the blastocysts, suggesting that LIF secreted into the uterine lumen might stimulate implantation; whether the stimulation is direct or indirect is not yet known. The results also show that this expression is under maternal control (Bhatt et al 1991).

*Metcalf:* What controls the development of the endometrial gland, if it's not the blastocyst? Is it ovarian regulated?

*Stewart:* It's apparently regulated maternally by changes in the levels of various hormones produced during oestrous and pregnancy. In delayed implantation the endometrial glands are still present in the endometrium, though they're not expressing LIF, and they disappear with the implantation of the embryo and the formation of the decidua.

*Martin:* Can you detect a LIF message by *in situ* hybridization in either the endometrial gland or the endometrium of non-pregnant mice treated with oestrogen?

*Stewart:* We haven't looked at that. It is unlikely to be expressed because we found no evidence for LIF transcripts during oestrous and the endometrial glands are not present. There is a suggestion of a signal in the endometrial epithelium, which lines the uterine lumen.

*Romero:* Do you have any information on the expression of LIF around the time of parturition?

*Stewart:* Expression on Day 19, which is the day before parturition, is low. We have never looked at the uterus immediately after parturition.

*Lee:* We have seen expression of both LIF mRNA and protein in blastocysts just before implantation (Murray et al 1990). The LIF could be coming from either direction. We don't yet know whether it's the trophoblast or the endosome mast cells that produce it.

*Stewart:* When you looked for the expression of protein you incubated the blastocysts in 20% fetal calf serum, and I would be concerned that the serum might be stimulating LIF expression.

*Lee:* That doesn't apply to the mRNA though, which was detected in freshly isolated blastocysts.

*Heath:* It's curious that no one has detected LIF in the blastocyst by *in situ* hybridization.

*Stewart:* We can see a LIF signal in the trophectoderm of blastocysts that are allowed to attach and outgrow *in vitro*, but again that could be because they were grown in medium containing fetal calf serum, which might actually induce LIF. That result tells you only that the cells are capable of producing LIF and not whether they actually produce it *in vivo*.

*Gough:* Without 'over-culturing', we have also detected LIF using the polymerase chain reaction in blastocysts, but there is about an hour of *in vitro* manipulation which could be the stimulus for induction.

*Gauldie:* Cells actively synthesizing several growth factors can be detected in tissues by *in situ* hybridization and if you stain the same sections with antibodies against the growth factors you find that many of them, TGF-β for example, are associated with the matrix around the cells where they are produced. Have you done any immunohistochemistry on your sections?

*Stewart:* We have an antiserum that recognizes recombinant LIF. With this, we saw no staining of the endometrial glands, but that might be because the cells could be secreting it very rapidly into the lumen after synthesis.

*Gauldie:* Was there any LIF bound to the matrix around the glands? If it's matrix-associated it might stick to the matrix and present the cytokine signal in that way.

*Stewart:* We didn't see any evidence for that, but we don't really know how good our antiserum is.

*Heath:* We have confirmed the matrix localization results using an antibody against mouse LIF, not by histochemical methods but by a combination of biochemical fractionation, immunoprecipitation and Western blotting (L. Grey and J. K. Heath, unpublished work).

*Lotem:* I have heard a report from Dr Brûlet that his group has transgenic mice expressing these two forms of LIF separately. They found that embryos expressing the soluble form developed to full term and their offspring were normal. With the matrix form, the embryos stopped developing before gastrulation (P. Brûlet, personal communication). LIF is expressed on Day 4 of pregnancy, so what do you think the role of these two forms of LIF is in the normal development of the embryo, Dr Stewart?

*Stewart:* I can't really address that issue. Doug Hilton and Lindsay Williams's work (personal communication) suggests that there are LIF receptors on the trophoblast of the blastocyst. I think that the natural tendency of the mouse embryo during development is to go into a state of delayed implantation. Perhaps LIF, if it is in fact secreted into the uterine lumen, in some way stops the blastocyst from going into this state of delay, and stimulates implantation. In our manipulations of mouse pregnancy, LIF expression always precedes implantation.

*Metcalf:* Dr Lotem, did either of these two transgenic mice develop proliferation of osteoblasts or new bone formation? Was the viable type perfectly normal?

*Lotem:* The viable type was apparently normal but the animals were not characterized further; there was no assessment of proliferation of osteoblasts or new bone formation.

*Heath:* The effect is apparently during gastrulation.

*Gearing:* What promoter was used?

*Lotem:* I don't know.

IL-6 is expressed in mice in newly formed blood vessels (during angiogenesis) both in the developing ovarian follicles and in the maternal decidua after embryonic implantation (Motro et al 1990). Is this also true of LIF?

*Stewart:* We have not detected LIF expression after implantation by *in situ* hybridization.

Professor Kishimoto, you have shown high levels of IL-6 receptor expression in Day 8, Day 9 and Day 10 mouse embryos (Saito et al 1992). Have you localized that expression of the receptor by *in situ* hybridization or by immuno-histochemistry? In those embryos, is IL-6 itself also found?

*Kishimoto:* We do not yet know which embryonic tissues express the receptor. IL-6 mRNA is also expressed between Day 8 and Day 12.

*Sehgal:* Dr Stewart, what is the basis of your belief that the glandular cells secrete LIF into the lumen? Could the secretion be towards the basal side, the stromal side? Is there any evidence for a vectorial secretion?

*Stewart:* That's possible, but you would need to do immuno-electron microscopy to be sure. We have found LIF-like activity in Day 4 uterine washings using the M1 differentiation assay. I realize that this approach is open to a variety of criticisms, because IL-6 can also induce M1 cell differentiation, but we failed to detect IL-6 mRNA in the uterus during pregnancy using Northern analysis.

*Sehgal:* You suggested that LIF affects implantation. Has anybody verified this in laminin substrate experiments or shown that it affects embryonic outgrowth in such experiments?

*Stewart:* Bob Seamark and his colleagues have some results suggesting that mouse blastocysts grown in the presence of LIF hatched from the zona at a higher frequency and attached more rapidly to the surface of petri dishes than those grown in its absence (Robertson et al 1990).

*Sehgal:* Dr Daniel Carson and Dr Andrew Jacobs at the MD Anderson Cancer Center in Houston, Texas, in collaboration with my laboratory, have found that IL-6 is secreted in a vectorial fashion from the endometrial epithelium in the mouse. By growing endometrial epithelial cells on membrane filters to generate a three-dimensional structured epithelial cell system they observed preferential vectorial secretion of IL-6 towards the apical side.

*Metcalf:* Does anyone have any information on IL-6 expression in the blastocyst?

*Lee:* Essentially, it is co-expressed with LIF.

*Gearing:* In mice producing excess LIF spermatogenesis is completely inhibited and the corpora lutea are defective.

Dr Matsui in Brigid Hogan's lab has shown that LIF and the c-*kit* ligand cause the proliferation of primordial germ cells in culture (Matsui et al 1991). This suggest another early action of LIF. LIF appears to cause the proliferation of such cells but only in the presence of the c-*kit* ligand.

*Sehgal:* Has the effect of glucocorticoids on LIF expression been investigated?

*Fey:* We have studied that in hepatoma cells and in macrophages (Baffet et al 1992). The LIF gene can be induced by LPS, and glucocorticoids prevent the induction, as they do with IL-6, in rat and mouse macrophages and in both human and rat hepatoma cells.

*Martin:* Are there recognizable oestrogen-responsive elements in the 5′ regions of the IL-6 or LIF genes? Have there been any studies of promoter regulation by oestrogen?

*Gough:* In the LIF gene we have looked for various sequence elements and motifs that might be expected to be present, but have found nothing convincing.

*Lotem:* Both IL-6 and LIF seem to be able to induce a cascade of other inducers, other cytokines. A toxic effect induced by a high concentration of LIF may actually result from production of TNF or β-interferon.

*Gauldie:* We have talked about the expression of IL-6 and LIF in gestation, and about the inducible expression of the two molecules in the adult system. We should not expect the same roles to be played in these two situations. Both cytokines may act as growth factors or inhibitors in the gestational stage, and as intracellular signalling molecules, with totally different roles, in the adult. Thus, you might expect to see expression of the genes in one or two particular cell types in gestation and never see it again in that particular cell type in the adult. We should bear this in mind when trying to produce transgenic mice. The reason why some die, or are never born, may have nothing to do with the adult role of the cytokine that is thought of initially.

## References

Bhatt H, Brunet L, Stewart CL 1991 Uterine expression of leukemia inhibitory factor coincides with the onset of blastocyst implantation. Proc Natl Acad Sci USA 88:11408–11412.

Matsui Y, Toksoz D, Nishikawa S et al 1991 Effect of Steel factor and leukemia inhibitory factor on murine primordial germ cells in culture. Nature (Lond) 353:750–752

Motro B, Itin A, Sachs L, Keshet E 1990 Pattern of interleukin 6 gene expression *in vivo* suggests a role for this cytokine in angiogenesis. Proc Natl Acad Sci USA 87:3092–3096

Murray R, Lee F, Chiu C-P 1990 The genes for leukemia inhibitory factor and interleukin 6 are expressed in mouse blastocysts prior to the onset of hemopoiesis. Mol Cell Biol 10:4953–4956

Rathjen DP, Nichols J, Edwards D, Heath JK, Smith AG 1990 Developmentally
    programmed induction of differentiation inhibiting activity and the control of stem
    cell populations. Genes & Dev 4:2308–2318
Robertson SA, Lavvanos TC, Seamark RF 1990 In vitro models of the maternal–fetal
    interface. In: Wegmann TG, Nisbet-Brown E, Gill TJ III (eds) The molecular and
    cellular immunobiology of the maternal–fetal interface. Oxford University Press,
    Oxford, p 191–206
Saito M, Yoshida K, Hibi M, Taga T, Kishimoto T 1992 Molecular cloning of a murine
    IL-6 signal transducer, gp130, and its regulated expression *in vivo*. J Immunol, in press

# Regulation of expression of the interleukin 6 gene: structure and function of the transcription factor NF-IL6

Shizuo Akira, Hiroshi Isshiki, Toshihiro Nakajima, Shigemi Kinoshita, Yukihiro Nishio, Shunji Natsuka and Tadamitsu Kishimoto

*Institute for Molecular and Cellular Biology, Osaka University, 1-3 Yamada-oka, Suita, Osaka 565, Japan*

*Abstract.* The interleukin 6 (IL-6) promoter is rapidly and transiently activated by other cytokines, including IL-1 and tumour necrosis factor (TNF), as well as by phorbol esters and cyclic AMP agonists. Studies using promoter mutants suggested that an IL-1-responsive element mapped within the $-180$ to $-123$ region of the IL-6 promoter. A nuclear factor (NF-IL6) that recognized a unique sequence containing an inverted repeat, ACATTGCACAATCT, was identified within the region. Direct cloning of the human NF-IL6 revealed its similarity to C/EBP, a liver- and adipose tissue-specific transcription factor. C/EBP and NF-IL6 recognize the same nucleotide sequence, but exhibit distinct patterns of expression. NF-IL6 is expressed at a low level in normal tissues, but is rapidly and drastically induced by bacterial lipopolysaccharide (LPS) or inflammatory cytokines such as IL-1, TNF and IL-6. Recently, NF-IL6 has been shown to be identical to IL-6DBP, the DNA-binding protein which is responsible for IL-6-mediated induction of several acute-phase proteins. Evidence that NF-IL6 DNA-binding activity is increased after IL-6 stimulation without increased NF-IL6 protein synthesis demonstrates the importance of post-translational modification. There are some results indicating that phosphorylation is involved in transcriptional and binding activities of NF-IL6. Taken together, these findings indicate that NF-IL6 may be an important transcription factor on the signal transduction pathways of IL-1 and IL-6.

*1992 Polyfunctional cytokines: IL-6 and LIF. Wiley, Chichester (Ciba Foundation Symposium 167) p 47–67*

Interleukin 6 (IL-6) is a polyfunctional cytokine that is produced by a wide range of cells and plays a central role in host defence mechanisms (Kishimoto & Hirano 1988, Akira et al 1990a, Hirano et al 1990). Expression of the IL-6 gene is induced by a variety of cytokines including IL-1, tumour necrosis factor (TNF), platelet-derived growth factor (PDGF) and β-interferon, as well as by serum,

47

bacterial lipopolysaccharide (LPS), phorbol esters and agents that increase intracellular cyclic AMP. Infection with various viruses is also a potent stimulus for IL-6 production. IL-6 is not produced under normal circumstances, but when invasion, such as bacterial and viral infection, or tissue injury takes place in the host IL-6 is rapidly and transiently induced. There is growing evidence that deregulation of IL-6 gene expression is involved in the pathogenesis of autoimmune diseases and certain lymphoid malignancies including multiple myelomas. Therefore, unravelling the molecular mechanisms underlying abnormal IL-6 gene expression may give insight into the pathophysiology of these IL-6-associated diseases. In recent years considerable information has been accumulated on the role of *cis*-acting elements and *trans*-acting factors that are essential for IL-6 gene expression. Here, the structure and function of the transcription factor NF-IL6, originally identified as a nuclear factor that binds to the IL-1-responsive element of the IL-6 gene, will be discussed.

### Identification of an IL-1-responsive element and nuclear factors that bind to it

Comparison of the human (Yasukawa et al 1987) and the mouse IL-6 genes (Tanabe et al 1988) revealed highly similar regions extending about 350 bp upstream of the transcriptional start site. Potential transcriptional control elements were identified within the conserved region of the IL-6 promoter, as indicated in Fig. 1.

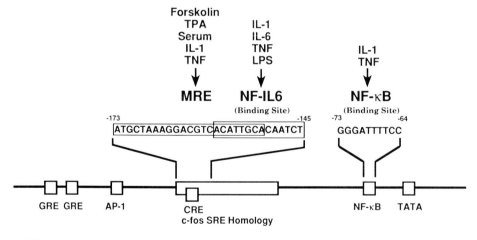

FIG. 1. Transcriptional control motifs identified in the IL-6 promoter. c-fos SRE homology represents a 70% nucleotide sequence identity across a 50-nucleotide stretch of the c-*fos* enhancer including an SRE (serum response element). CRE, cyclic AMP response element; GRE, glucocorticoid response element; AP-1, activation protein 1; MRE, multi-response element; TNF, tumour necrosis factor; TPA, 12-*O*-tetradecanoyl-phorbol-13-acetate; LPS, bacterial lipopolysaccharide.

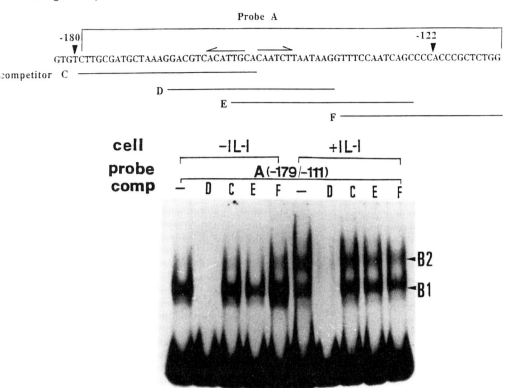

FIG. 2. Identification of constitutively expressed and IL-1-inducible complexes in glioblastoma SK-MG-4 cells. End-labelled fragment A (−179 to −111) was incubated with crude glioblastoma SK-MG-4 nuclear extracts (from cells grown with or without IL-1) either with no competitor (−) or in the presence of the indicated competitor (comp) at a 100-fold molar excess. Competitors used are shown at the top. The complexes were separated by electrophoresis on a 5% native polyacrylamide gel. Specific protein–DNA complexes (B1 and B2) are indicated by arrows.

To analyse the sequences involved in the response to IL-1, we transfected a series of 5′ end deletion mutants of the human IL-6 gene into mouse fibroblasts, L cells, whose endogenous IL-6 gene was inducible by IL-1, obtaining stable transformants. Northern blot analysis of the exogenous human IL-6 gene in the transformants before and after stimulation with IL-1 indicated that the −180 to −123 region of the IL-6 promoter was involved in the IL-1 inducibility (Isshiki et al 1990). Next, nuclear factors binding to the IL-1-responsive element were examined by a gel retardation assay. Nuclear extracts from cells of a human glioblastoma cell line, SK-MG-4, cultured with or without IL-1, were incubated with an end-labelled 70 bp fragment (−179 to −111) corresponding to the IL-1-responsive element of the human IL-6 gene either alone or with competitor. As shown in Fig. 2, a specific protein–probe complex, B1, was identified when the probe was incubated with the untreated nuclear extract. In addition to this complex, a new more slowly migrating complex, B2, was detected with the nuclear extract from IL-1-treated cells. The formation of these labelled complexes was inhibited only by adding unlabelled oligo-nucleotide D, suggesting that the protein-binding sites of the complexes were within the − 164 to − 139 region. A methylation interference assay showed that the DNA-binding protein involved in the constitutive complex and that in the inducible complex bound to a 14 bp nucleotide (ACATTGCACAATCT) containing an inverted repeat (palindrome). Because these two DNA-binding proteins exhibited indistinguishable DNA-binding characteristics in the competition gel retardation assay and in the methylation interference assay, they were termed, collectively, NF-IL6 (Isshiki et al 1990). Furthermore, it was ascertained by CAT assay (in which the activity of the chloramphenicol acetyltransferase 'reporter' gene, under the control of the IL-6 promoter, is assayed) that the 14 bp NF-IL6-binding sequence is an IL-1-responsive element in glioblastoma SK-MG-4 cells.

Ray et al (1989) showed that a 23 bp IL-6 multi-response element (MRE) (− 173 to − 151, see Fig. 1) is responsible for induction of IL-6 gene expression by IL-1, TNF and serum, as well as by activators of protein kinase A (forskolin) and protein kinase C (phorbol ester). They also identified several sequence-specific complexes found in increased amounts in HeLa cell nuclear extracts after stimulation. This region includes a CRE (cyclic AMP response element) motif and the upper half of the 14 bp NF-IL6-binding sequence.

There is growing evidence for the involvement of NF-$\varkappa$B in the pathways of signal transduction for several cytokines such as IL-1, TNF and interferons. In fact, the possibility that NF-$\varkappa$B mediates IL-6 induction has been suggested by several investigators (Shimizu et al 1990, Liberman & Baltimore 1990, Zhang et al 1990). Thus, as shown in Fig. 1, three regions within the IL-6 promoter have been shown to be involved in the regulation of the IL-6 gene.

## Molecular cloning of NF-IL6

To clone the gene encoding NF-IL6, we screened a λgt 11 phage library generated from the cDNA of LPS-stimulated human peripheral monocytes with a $^{32}$P-labelled probe consisting of four copies of the 14 bp NF-IL6-binding sequence according to the method developed by Singh et al (1988). The cloned NF-IL6 encoded a single open reading frame 345 amino acids long (Akira et al 1990b). A search of a protein database indicated that NF-IL6 contained a region highly similar to the C-terminal portion of C/EBP (CCAAT/enhancer binding protein), a rat liver nuclear factor containing the leucine zipper structure (Landschulz et al 1988a,b). C/EBP has been shown to be responsible for the expression of several liver-specific genes, including the albumin gene. The highly

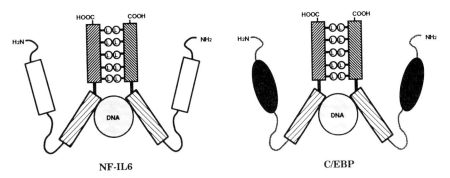

FIG. 3. Structural comparison of NF-IL6 and C/EBP. The leucine zipper and basic DNA-binding domains of these two proteins are almost identical.

conserved region occurs between residues 261 and 345 of NF-IL6, where there is 73% sequence identity, extending to 95% when conservative changes are included. This region of C/EBP has been shown to contain a basic domain and a leucine zipper structure that are essential for DNA binding and dimerization, respectively. NF-IL6 also contains a potential leucine zipper, consisting of five leucines in every seventh position (Fig. 3).

A genomic clone corresponding to human NF-IL6 cDNA was also isolated. The sequence of the genomic clone was found to be contiguous with the sequence of the NF-IL6 cDNA, indicating that the NF-IL6 gene lacks introns, like the C/EBP gene.

Like C/EBP, NF-IL6 binds avidly to sequences containing CCAAT and viral enhancer core (TGTGGAAA) sequences. Competition analysis of the binding of NF-IL6 with the sequences to which C/EBP-like proteins have been reported to bind revealed that NF-IL6 and C/EBP recognized the same nucleotide sequences. The 'best fit' consensus sequence for NF-IL6 was found to be T(T/G)NNGNAA(T/G) (Fig. 4).

When a combination of the NF-IL6 expression vector and the reporter plasmid was transfected into a T cell line (Jurkat cells) which did not express the NF-IL6 gene, the promoter harbouring the NF-IL6-binding sites could be activated by co-transfection of the correctly expressed NF-IL6 cDNA, but not by co-transfection of the antisense cDNA, demonstrating that NF-IL6 is a positive transcription factor.

**NF-IL6 is an inducible transcription factor**

NF-IL6 expression was examined in various mouse tissues. As shown in Fig. 5, mRNA for NF-IL6 was undetectable or expressed at a low level in all normal tissues examined. However, NF-IL6 mRNA was drastically induced by stimulation with LPS, IL-1, TNF or IL-6. LPS induced high levels of NF-IL6

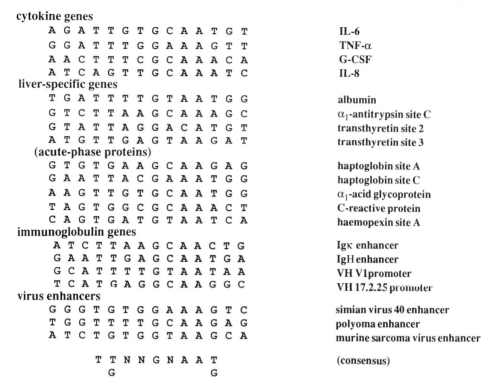

cytokine genes
```
A  G  A  T  T  G  T  G  C  A  A  T  G  T        IL-6
G  G  A  T  T  T  G  G  A  A  A  G  T  T        TNF-α
A  A  C  T  T  T  C  G  C  A  A  A  C  A        G-CSF
A  T  C  A  G  T  T  G  C  A  A  A  T  C        IL-8
```
liver-specific genes
```
T  G  A  T  T  T  T  G  T  A  A  T  G  G        albumin
G  T  C  T  T  A  A  G  C  A  A  A  G  C        α₁-antitrypsin site C
G  T  A  T  T  A  G  G  A  C  A  T  G  T        transthyretin site 2
A  T  G  T  T  G  A  G  T  A  A  G  A  T        transthyretin site 3
```
(acute-phase proteins)
```
G  T  G  T  G  A  A  G  C  A  A  G  A  G        haptoglobin site A
G  A  A  T  T  A  C  G  A  A  A  T  G  G        haptoglobin site C
A  A  G  T  T  G  T  G  C  A  A  T  G  G        α₁-acid glycoprotein
T  A  G  T  G  G  C  G  C  A  A  A  C  T        C-reactive protein
C  A  G  T  G  A  T  G  T  A  A  T  C  A        haemopexin site A
```
immunoglobulin genes
```
A  T  C  T  T  A  A  G  C  A  A  C  T  G        Igκ enhancer
G  A  A  T  T  G  A  G  C  A  A  T  G  A        IgH enhancer
G  C  A  T  T  T  T  G  T  A  A  T  A  A        VH V1 promoter
T  C  A  T  G  A  G  G  C  A  A  G  G  C        VII 17.2.25 promoter
```
virus enhancers
```
G  G  G  T  G  T  G  G  A  A  A  G  T  C        simian virus 40 enhancer
T  G  G  T  T  T  T  G  C  A  A  G  A  G        polyoma enhancer
A  T  C  T  G  T  G  G  T  A  A  G  C  A        murine sarcoma virus enhancer

         T  T  N  N  G  N  A  A  T              (consensus)
         G                    G
```

FIG. 4.   Summary of sequences to which NF-IL6 binds.

mRNA in all tissues except for the brain. IL-1 and TNF increased NF-IL6 mRNA in the lung, liver and kidney, whereas the predominant effect of IL-6 was in the liver. Thus, the pattern of induction of NF-IL6 mRNA differed according to the stimulus. Rehybridization of the filter with the mouse IL-6 probe showed that the pattern of NF-IL6 induction correlated quantitatively with that of IL-6 induction, except in the liver, where very low levels of IL-6 mRNA induction followed expression of NF-IL6. This suggests that NF-IL6 may be involved in the regulation of genes other than the IL-6 gene in the liver.

## The involvement of NF-IL6 in the expression of acute-phase proteins

Inflammation is a physiological response to a variety of stimuli, including infections and tissue injury. It is accompanied by the acute-phase response, which involves significant alterations in the serum levels of several plasma proteins known as acute-phase proteins. These proteins are synthesized mainly by the liver. During the acute-phase response synthesis of proteins such as haptoglobin, haemopexin, C-reactive protein (CRP) and $\alpha_1$-acid glycoprotein increases (so-called positive acute-phase proteins), while there is an accompanying decrease in the synthesis of liver-specific proteins including albumin and transthyretin (negative acute-phase proteins). IL-6 is a major hepatocyte-stimulating factor that regulates the production of a wide spectrum of acute-phase proteins (Andus et al 1987, Gauldie et al 1987). Several acute-phase protein genes are transcriptionally regulated by IL-6 (Morrone et al 1988). However, the precise mechanism of the coordinated expression of the liver-specific acute-phase protein genes during the acute-phase response is currently unknown. Cortese and his colleagues (Oliviero & Cortese 1989, Poli & Cortese 1989) have identified the IL-6-responsive elements (IL-6-REs) in the haptoglobin, haemopexin and CRP genes, and the IL-6-stimulated DNA-binding protein (IL-6DBP), in the human hepatoma cell line Hep3B using site-specific mutagenesis, CAT assay and gel retardation assays. Competition experiments indicated that the same nuclear factor bound to these particular sequences. We noticed that the IL-6-REs identified by Cortese and his colleagues were quite similar to the recognition sequence of NF-IL6 (Fig. 4). Actually, recombinant NF-IL6 was able to bind to the IL-6-REs of several acute-phase protein genes. Furthermore, formation of all the DNA–protein complexes between the haptoglobin site C IL-6-RE and proteins in extracts from IL-6-treated hepatoma cells was competitively inhibited by antibody against recombinant NF-IL6 (Isshiki et al 1991). The 14 bp palindromic NF-IL6-binding motif was also shown, by CAT assay, to work as an IL-6-RE in the hepatoma cells.

Recently, Cortese and his colleagues (Poli et al 1990) have cloned and characterized a cDNA coding for IL-6DBP. The sequence of rat IL-6DBP is very similar to that of human NF-IL6, suggesting that IL-6DBP is the rat homologue of NF-IL6.

The time course of NF-IL6 mRNA induction by IL-6 was studied in the mouse liver. mRNAs from the liver were analysed by Northern blotting at different times after intravenous injection of recombinant human IL-6. NF-IL6 mRNA was detected at a low level in the control mouse liver, but its expression was drastically increased by 15 minutes after IL-6 stimulation, reaching a maximum level between one and two hours. After this time the level of NF-IL6 mRNA decreased, reaching an almost basal level by six hours. The time course of haptoglobin mRNA expression was studied in parallel. The increase in haptoglobin gene expression was detected as early as 30 minutes after IL-6 injection. Haptoglobin mRNA reached a maximum, moderately increased level at three hours, which persisted for up to six hours. Thus, the increase in haptoglobin mRNA was preceded by NF-IL6 mRNA induction.

In contrast, mRNA for C/EBP was constitutively expressed before IL-6 injection, but a dramatic decrease in its level was observed by one hour later; this low level was maintained until beginning to increase at four hours after injection. These results indicated that C/EBP mRNA is down-regulated in the liver in

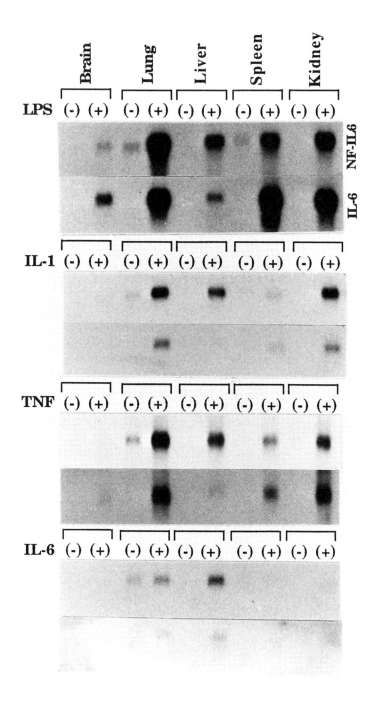

response to IL-6. Such reciprocal expression of NF-IL6 and C/EBP in response to IL-6 was also demonstrated in Eμ-IL6 transgenic mice (Suematsu et al 1989) in which the serum IL-6 level was continuously high and the serum albumin level was depressed. These results suggest that the positive acute-phase protein genes are regulated by NF-IL6, rather than by C/EBP, even though these factors recognize the same DNA sequences. In fact, Natsuka et al (1991) have demonstrated that high expression of the stably transfected NF-IL6 gene results in induction of haptoglobin protein in Hep 3B cells.

Thus, the following scenario may be deduced from these findings. C/EBP regulates a broad spectrum of liver-specific genes such as albumin and transthyretin in normal physiological conditions, under which NF-IL6 expression is suppressed. Once the acute-phase reaction is initiated, NF-IL6 is rapidly induced, and it may be involved in the induction of acute-phase proteins. C/EBP expression is suppressed by IL-6. At the same time, NF-IL6 may suppress the expression of negative acute-phase proteins, including albumin and transthyretin, by competing with C/EBP for DNA binding.

Several studies have shown that C/EBP-like factors may be involved in the expression of several other acute-phase proteins, including $\alpha_1$-acid glycoprotein (Chang et al 1990), angiotensinogen (Brasier et al 1990), complement C3 (Wilson et al 1990) and serum amyloid A (Li et al 1990).

## The involvement of NF-IL6 in macrophage-specific gene expression

Mouse myeloid leukaemia M1 cells are myeloblastic in morphology and can be induced to differentiate into macrophages by exposure to IL-6. Human monocytic leukaemia U-937 cells also differentiate into cells with macrophage characteristics after exposure to LPS plus TPA (12-O-tetradecanoyl-phorbol-13-acetate). Both cell lines expressed a low level of NF-IL6 and C/EBP mRNAs before stimulation. When these cell lines were stimulated to differentiate into macrophages, mRNA for NF-IL6 was increased markedly two days later. The increase in NF-IL6 mRNA coincided with the morphological change. The expression of IL-6 mRNA was transiently increased, reaching a maximum level at two days then returning to a basal level. The level of C/EBP mRNA did not change during differentiation. In human HL-60 promyelocytic cells, which undergo differentiation to granulocytes or monocyte/macrophages, depending on the inducer, NF-IL6 mRNA was up-regulated during macrophage differentiation but not during neutrophil differentiation.

---

FIG. 5. Tissue distribution of NF-IL6 expression in mice. mRNA from various tissues was obtained from control mice ( − ) and from mice intravenously injected four hours previously ( + ) with LPS (20 μg), IL-1 ($2 \times 10^4$ units), TNF ($6 \times 10^4$ units) or IL-6 ($5 \times 10^4$ units). Poly(A) RNA (2 μg per lane) was hybridized with an NF-IL6-specific probe. The same filter was subsequently rehybridized with the mouse IL-6 probe.

On stimulation with a variety of agents, including LPS, macrophages produce inflammatory cytokines such as IL-1, TNF and IL-6. NF-IL6-binding motifs were identified in the promoter regions of these genes (Fig. 4) and the binding of recombinant NF-IL6 to these sequences was established by a competition gel retardation assay. Granulocyte colony-stimulating factor (G-CSF) is produced in macrophages after stimulation with LPS. Nishizawa & Nagata (1990) found three regulatory elements (G-CSF proximal regulatory elements GPE1, GPE2 and GPE3) in the promoter of the G-CSF gene by analysing a set of linker-scanning mutants and internal deletion mutants. They showed that GPE1 played an important role in LPS-induced expression of the G-CSF gene in macrophages. Specific binding of recombinant NF-IL6 to GPE1 was shown by a methylation interference assay. The lysozyme gene is also activated during differentiation of cells to macrophages. Recombinant NF-IL6 was shown to bind to the 10 bp sequence ATATTGCAAC found at position −208 to −199, a functionally important region within the lysozyme gene (Steiner et al 1987).

Taken together, these results indicated that NF-IL6 mRNA increased markedly during macrophage differentiation and that it might play an important role in the regulation of several genes associated with macrophage differentiation and activation (Natsuka et al 1992). In this respect, noteworthy is the finding that C/EBP mRNA increases markedly during differentiation of 3T3-L1 pre-adipocytes to adipocytes and that C/EBP can *trans* activate the promoters of the adipocyte-specific genes 422 (aP2) and stearoyl-CoA desaturase (SCD) 1 (Christy et al 1989) (Fig. 6).

## Post-translational modification of NF-IL6

There is some evidence indicating that NF-IL6 is also regulated at a post-translational level. Although IL-6 was rapidly and drastically induced by IL-1 in the glioblastoma cell line SK-MG-4, there was little increase of NF-IL6 mRNA or protein. However, a gel retardation assay showed that a new slower-migrating complex is formed after IL-1 stimulation in addition to the already existing complex, as shown in Fig. 2. In the hepatoma cell line HepG2 a gel retardation assay revealed a large increase in NF-IL6-binding activity after treatment with IL-6, as shown in Fig. 7. Northern blotting showed that, in contrast to normal hepatocytes, HepG2 cells expressed NF-IL6 mRNA constitutively and that there was only a slight increase in expression at 40 hours after stimulation with IL-6. Furthermore, Western blot analysis did not reveal any changes in the amount of protein recognized by the NF-IL6-specific antibody after IL-6 stimulation. This indicates that the increased DNA-binding activity may be due to a modification of pre-existing NF-IL6 protein in HepG2 cells. Phosphorylation of NF-IL6 is probably important for its interaction with specific DNA sequences because phosphatase treatment abolished NF-IL6's DNA-binding activity in

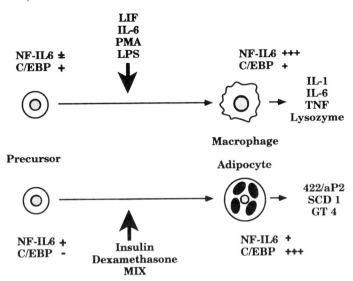

FIG. 6.   Differential expression of NF-IL6 and C/EBP during differentiation to macrophages and adipocytes. LIF, leukaemia inhibitory factor; PMA, phorbol 12-myristate 13-acetate (TPA); MIX, 1-methyl-3-isobutyl xanthine; SCD 1, stearoyl-CoA desaturase; GT 4, glucose transporter 4; 422/aP2 and SDC 1 are adipocyte-specific genes.

HepG2 cells. Our preliminary results indicate that IL-6 stimulation increases phosphorylation at specific sites on NF-IL6 in Hep3B cells.

Poli et al (1990) also demonstrated that post-translational modifications play an important role in the activation of transcription through IL-6DBP (the rat homologue of NF-IL6). By CAT co-transfection experiments, they showed that IL-6DBP is not capable of significantly activating promoters in the absence of IL-6, but increases transcription more than 20-fold in the presence of IL-6. They measured the amount of IL-6DBP–DNA complexes by immunoprecipitation with an IL-6DBP-specific antiserum. The amount of IL-6-RE probe precipitated by the anti-IL-6DBP antibodies was at least three-fold higher in extracts from IL-6-treated cells than in extracts from untreated cells, even in the presence of cycloheximide, suggesting that IL-6 increases the DNA-binding activity of NF-IL6.

## Heterodimer formation among the C/EBP family members

There are at least four members of the C/EBP family, as shown schematically in Fig. 8. These include C/EBP, NF-IL6 (also known as liver-enriched transcriptional activator protein, LAP, [Descombes et al 1990], IL-6DBP and acid glycoprotein enhancer binding protein, AGP/EBP [Chang et al 1990] ), Ig/EBP-1 (also known as GPR1-BP, Nishizawa et al 1991) and NF-IL6β

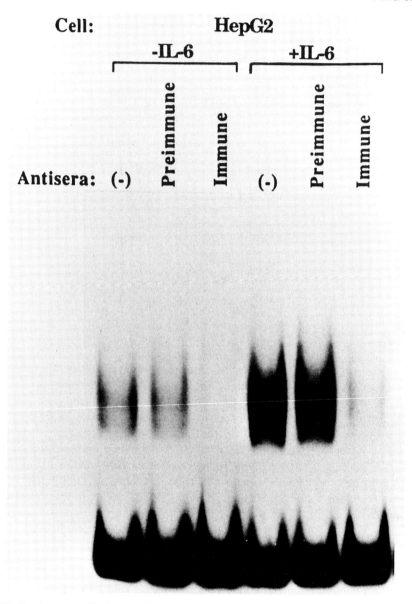

FIG. 7.   Increase of NF-IL6-binding activity after treatment of HepG2 cells with IL-6.
Nuclear extracts were prepared from untreated and IL-6-treated (500 units, 40 h) HepG2
cells. A gel retardation assay was performed using an end-labelled fragment of DNA
containing the haptoglobin site C IL-6-responsive element. An anti-peptide antibody
raised against the sequence of the putative DNA-binding domain of NF-IL6 (immune)
or preimmune serum was added to the binding reactions where indicated. The antibody
abolished formation of labelled protein–DNA complexes.

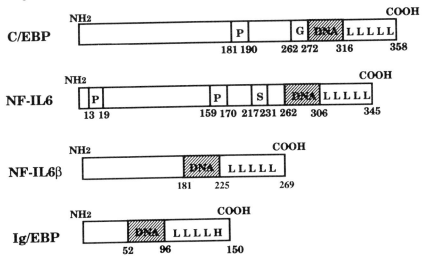

FIG. 8.   Schematic illustration of members of the C/EBP family. DNA, DNA-binding domain; L, leucine; H, histidine; P, proline-rich region; G, glycine-rich region; S, serine-rich region.

(our unpublished data). These members of the C/EBP family recognize the same nucleotide sequences, but they exhibit distinct patterns of expression and may be functionally different. C/EBP mRNA is found in adipose tissue, liver and placenta, tissues which play a vital role in energy metabolism (Birkenmeier et al 1989, Xanthopoulos et al 1989). The promoters of several liver- and adipose tissue-specific genes have been found to be *trans* activated by C/EBP. These include the genes that encode serum albumin, phosphoenolpyruvate phosphocarboxykinase, 422/aP2 protein, stearoyl acyl-CoA desaturase 1 and the insulin-responsive glucose transporter. The expression of NF-IL6 mRNA is normally suppressed but is drastically and rapidly induced in many tissues by LPS or inflammatory cytokines such as IL-1, TNF and IL-6, indicating that NF-IL6 may be involved in the regulation of the genes activated in the inflammatory reaction. NF-IL6β was recently isolated using a DNA probe from the DNA-binding portion of NF-IL6 and antisera raised against its DNA-binding domain (our unpublished results). This gene, like that for NF-IL6, is expressed at undetectable or minor levels in normal tissues, but is rapidly and drastically induced by LPS or inflammatory cytokines including IL-1 and IL-6. In contrast, Ig/EBP-1 mRNA is ubiquitously expressed and does not appear to be influenced by LPS treatment (Roman et al 1990, Nishizawa et al 1991). There are functional differences between the C/EBP family members, but the demonstration of the formation of heterodimers between the members makes the situation more complicated. In some cell types two, or even all, of the members of the C/EBP family are co-expressed; in such cells, they may exist in the form of heterodimers, as well as in homodimeric form, and the heterodimers may possess different transcriptional and binding activities from the homodimers. A change in the concentration of a single member would alter the proportions of homo- and heterodimers and would affect the expression of the genes controlled by the C/EBP family members. Differential utilization of homo- and heterodimers of these proteins may modulate the expression of the various genes which are involved in signal transduction and cell differentiation.

*Acknowledgements*

We would like to thank Ms K. Ono and Ms K. Kubota for secretarial assistance. This work was supported in part by grants from the Ministry of Education, Science and Culture, Japan.

**References**

Akira S, Hirano T, Taga T, Kishimoto T 1990a Biology of multifunctional cytokines: IL 6 and related molecules (IL 1 and TNF). FASEB (Fed Am Soc Exp Biol) J 4:2860–2867

Akira S, Isshiki H, Sugita T et al 1990b A nuclear factor for IL-6 expression (NF-IL6) is a member of a C/EBP family. EMBO (Eur Mol Biol Organ) J 9:1897–1906

Andus T, Geiger T, Hirano T et al 1987 Recombinant human B cell stimulatory factor 2 (BSF-2/IFNβ$_2$) regulates β-fibrinogen and albumin mRNA levels in Fao-9 cell. FEBS (Fed Eur Biochem Soc) Lett 221:18–22

Birkenmeier EH, Gwynn B, Howard S et al 1989 Tissue-specific expression, developmental regulation, and genetic mapping of the gene encoding CCAAT/enhancer binding protein. Genes & Dev 3:1146–1156

Brasier AR, Ron D, Tate JE, Habener JF 1990 A family of constitutive C/EBP-like DNA binding proteins attenuate the IL-1α induced, NFϰB mediated transactivation of the angiotensinogen gene acute-phase response element. EMBO (Eur Mol Biol Organ) J 10:6181–6191

Chang C-J, Chen T-T, Lei H-Y, Chen D-S, Lee S-C 1990 Molecular cloning of a transcription factor, AGP/EBP, that belongs to members of the C/EBP family. Mol Cell Biol 10:6642–6653

Christy RJ, Yang VW, Ntambi JM et al 1989 Differentiation-induced gene expression in 3T3-L1 preadipocytes: CCAAT/enhancer binding protein interacts with and activates the promoters of two adipocyte-specific genes. Genes & Dev 3:1323–1335

Descombes P, Chojkier M, Lichtsteiner S, Falvey E, Schibler U 1990 LAP, a novel member of the C/EBP gene family, encodes a liver-enriched transcriptional activator protein. Genes & Dev 4:1541–1551

Gauldie J, Richards C, Harnish D, Lansdorp P, Baumann H 1987 Interferon β$_2$/B cell stimulatory factor type 2 shares identity with monocyte-derived hepatocyte-stimulating factor and regulates the major acute phase protein response in liver cells. Proc Natl Acad Sci USA 84:7251–7255

Hirano T, Akira S, Taga T, Kishimoto T 1990 Biological and clinical aspects of interleukin 6. Immunol Today 11:443–449

Isshiki H, Akira S, Tanabe O et al 1990 Constitutive and IL-1 inducible factors interact with the IL-1 responsive element in the IL-6 gene. Mol Cell Biol 10:2757–2764

Isshiki H, Akira S, Sugita T et al 1991 Reciprocal expression of NF-IL6 and C/EBP in hepatocytes: possible involvement of NF-IL6 in acute phase protein gene expression. New Biol 3:63–70

Kishimoto T, Hirano T 1988 Molecular regulation of B lymphocyte response. Annu Rev Immunol 6:485–512

Landschulz WH, Johnson PF, Adashi EY, Graves BJ, McKnight SL 1988a Isolation of a recombinant copy of the gene encoding C/EBP. Genes & Dev 2:786–800

Landschulz WH, Johnson PF, McKnight SL 1988b The leucine zipper: a hypothetical structure common to a new class of DNA binding proteins. Science (Wash DC) 240:1759–1764

Li X, Huang JH, Rienhoff Jr HY, Liao WS-L 1990 Two adjacent C/EBP binding sequences that participate in the cell-specific expression of the mouse serum amyloid A3 gene. Mol Cell Biol 10:6624–6631

Liberman TA, Baltimore D 1990 Activation of interleukin-6 gene expression through the NF-$\varkappa$B transcription factor. Mol Cell Biol 10:2327–2334

Morrone G, Ciliberto G, Oliviero S et al 1988 Recombinant interleukin 6 regulates the transcriptional activation of a set of human acute phase genes. J Biol Chem 263:12554–12558

Natsuka S, Isshiki H, Akira S, Kishimoto T 1991 Augmentation of the haptoglobin production in Hep3B cell line by a nuclear factor NF-IL6. FEBS (Fed Eur Biochem Soc) Lett 291:58–62

Natsuka S, Akira S, Nishio Y et al 1992 Macrophage differentiation specific expression of NF-IL6, a transcription factor for IL-6. Blood 79:460–466

Nishizawa M, Nagata S 1990 Regulatory elements responsible for inducible expression of the granulocyte colony-stimulating factor gene in macrophages. Mol Cell Biol 10:2002–2011

Nishizawa M, Wakabayashi-Ito N, Nagata S 1991 Molecular cloning of cDNA and a chromosomal gene encoding GRE1-BP, a nuclear protein which binds to granulocyte colony-stimulating factor promoter element 1. FEBS (Fed Eur Biochem Soc) Lett 282:95–97

Oliviero S, Cortese R 1989 The human haptoglobin gene promoter: interleukin-6-responsive elements interact with a DNA-binding protein induced by interleukin-6. EMBO (Eur Mol Biol Organ) J 8:1145–1151

Poli V, Cortese R 1989 Interleukin 6 induces a liver-specific nuclear protein that binds to the promoter of acute-phase genes. Proc Natl Acad Sci USA 86: 7547–7551

Poli V, Mancini FP, Cortese R 1990 IL-6DBP, a nuclear protein involved in interleukin-6 signal transduction, defines a new family of leucine zipper proteins related to C/EBP. Cell 63:643–653

Ray A, Sassone-Corsi P, Sehgal PB 1989 A multiple cytokine- and second messenger-responsive element in the enhancer of the human interleukin-6 gene: similarities with c-*fos* gene regulation. Mol Cell Biol 9:5537–5547

Roman C, Platero JS, Shuman J, Calame K 1990 Ig/EBP-1: a ubiquitously expressed immunoglobulin enhancer binding protein that is similar to C/EBP and heterodimerizes with C/EBP. Genes & Dev 4:1404–1415

Shimizu H, Mitomo K, Watanabe T, Okamoto S, Yamamoto K-I 1990 Involvement of a NF-$\varkappa$B-like transcription factor in the activation of the interleukin-6 gene by inflammatory lymphokines. Mol Cell Biol 10:561–568

Singh H, LeBowitz JH, Baldwin AS, Sharp PA 1988 Molecular cloning of an enhancer binding protein: isolation by screening of an expression library with a recognition site D. Cell 52:415–423

Steiner C, Muller M, Baniahmad A, Renkawitz R 1987 Lysozyme gene activity in chicken macrophages is controlled by positive and negative regulatory elements. Nucleic Acids Res 15:4163–4178

Suematsu S, Matsuda T, Aozasa K et al 1989 IgG1 plasmacytosis in interleukin 6
    transgenic mice. Proc Natl Acad Sci USA 86:7547–7551
Tanabe O, Akira S, Kamiya T, Wong GG, Hirano T, Kishimoto T 1988 Genomic
    structure of the murine IL-6 gene: High degree conservation of potential regulatory
    sequences between mouse and human. J Immunol 141:3875–3881
Wilson DR, Juan TS-C, Wilde MD, Fey GH, Darlington GJ 1990 A 58-base-pair region
    of the human C3 gene confers synergistic inducibility by interleukin-1 and interleukin-6.
    Mol Cell Biol 10:6181–6191
Xanthopoulos KG, Mirkevitch J, Decker T, Kuo CF, Darnell JE 1989 Cell-specific
    transcriptional control of the mouse DNA-binding protein mC/EBP. Proc Natl Acad
    Sci USA 86:4117–4121
Yasukawa K, Hirano T, Watanabe Y, Muratani K, Matsuda T, Kishimoto T 1987
    Structure and expression of human B cell stimulatory factor 2 (BSF-2) gene. EMBO
    (Eur Mol Biol Organ) J 6:2939–2945
Zhang Y, Lin J-X, Vilcek J 1990 Interleukin-6 induction by tumor necrosis factor and
    interleukin-1 in human fibroblasts involves activation of a nuclear factor binding to
    a $\kappa$B-like sequence. Mol Cell Biol 10:3818–3823


## DISCUSSION

*Sehgal:* Have you investigated the proportion of NF-IL6 in the cytoplasm?
Is it predominantly a cytoplasmic protein? Can it be swept into the nucleus on
addition of IL-6?

*Akira:* There is no movement from the cytoplasm to the nucleus in hepatoma
cells.

*Sehgal:* We have found that in normal keratinocytes most of the staining
for NF-IL6, *in situ*, is cytoplasmic, not nuclear (J. Krueger, A. Gottlieb &
P. B. Sehgal, unpublished results). In some experiments we see a shift
in NF-IL6 immunostaining from the cytoplasm to the nucleus on addition
of IL-6.

*Kishimoto:* NF-IL6 moves from the cytoplasm to the nucleus in PC12 cells
after stimulation with forskolin.

*Baumann:* It appears that all isoforms of C/EBP share the same DNA-binding
specificity. The DNA sequence recognized by C/EBPs is not different from the
sequence Dr Akira defined as the binding site for NF-IL6 (or IL-6DBP). Every
isoform of C/EBP, when bound to a given regulatory element, has the potential
to activate transcription of a linked gene. Did you compare in your system the
*trans* activation by NF-IL6 and by C/EBP (or C/EBPα)? If so, did you get
the same level of *trans* activation?

*Akira:* Yes; C/EBP bound to the NF-IL6-binding sequence and caused the
same amount of *trans* activation as NF-IL6.

*Baumann:* This raises an interesting problem: C/EBP is present at high levels
in the liver and will recognize the same DNA sequence as NF-IL6, yet it does
not seem to *trans* activate as well as NF-IL6.

*Ciliberto:* We have done experiments comparing IL-6DBP, not with NF-IL6, but with C/EBP. C/EBP is a very good *trans* activator, but its activity is not induced by IL-6, whereas NF-IL6 and IL-6DBP are strongly induced by IL-6. Why, then, do natural promoters which carry common binding sites for the two factors behave differently from synthetic promoters carrying multiple copies of the binding site? If you do the experiment with a complex natural promoter you get completely different results, probably because there are protein–protein interactions which attract a specific factor to one promoter—for example, C/EBP to the albumin promoter, and NF-IL6 to the C-reactive protein promoter. As a consequence, the albumin promoter would be down-regulated and the C-reactive promoter would be up-regulated by IL-6.

*Kishimoto:* NF-IL6 may be necessary but not sufficient for acute-phase protein gene expression; certain other nuclear factors may be necessary. NF-IL6 can cooperate with other nuclear factors and the acute-phase gene promoters.

*Fey:* Dr Akira, you said that in normal mouse livers levels of NF-IL6 were very low unless you stimulated with LPS. That same protein, cloned not only by Poli et al (1990) but also by Descombes et al (1990), was found at high levels in normal rat livers. Actually, it is one of the factors that maintains the normal expression of the albumin gene. Could you clarify this discrepancy?

*Akira:* In older mice, or mice maintained in poor conditions, NF-IL6 is induced in the liver. The expression of NF-IL6 is modulated by circumstances. For example, even fasting induces NF-IL6.

*Gearing:* Do you have any evidence for formation of a heterodimer between NF-IL6 and NF-IL6β?

*Akira:* Yes, we have observed this *in vitro*. When recombinant NF-IL6 and NF-IL6β are preincubated for 15 min at 37 °C a heterodimer is formed which can be detected by a gel shift assay.

*Sehgal:* When you add IL-6 to hepatocyte cultures, I assume that dexamethasone is also present. Is that correct?

*Akira:* No.

*Sehgal:* So you see maximal levels of haptoglobin expression by the addition of IL-6 only to HepG2 cells. That's different from what is customary.

*Heinrich:* Primary cultures of hepatocytes need glucocorticoids for acute-phase protein expression; HepG2 cells don't necessarily.

*Fey:* Some HepG2 sublines show little or no synergism between IL-6 and glucocorticoids, because they have low concentrations of the glucocorticoid receptor (Baumann et al 1990).

*Gauldie:* This is a general question which we should think about. The hepatoma cell line is used as a model system, without cortisteroid, but the normal hepatocyte, in the liver, will be responding in the presence of corticosteroids. I wonder whether the extent of induction of NF-IL6 that Dr Akira saw could occur in hepatoma cells if dexamethasone were present. Such regulation may be totally different from that seen in normal hepatocytes. In intact liver the regulation you see is the correct regulation, that of acute-phase protein genes, and not the expression of the IL-6 gene. The hepatocyte does not express the IL-6 gene under normal physiological conditions, whereas primary hepatocytes or hepatoma cells in culture in the absence of corticosteroid will express IL-6.

*Ciliberto:* Does IL-6 induce NF-IL6β mRNA and protein in Hep3B cells?

*Akira:* IL-6 induces mRNA for NF-IL6β, but I don't know about the expression of the protein because I don't have a specific antibody.

*Ciliberto:* So you cannot assess the relative contributions of NF-IL6 and NF-IL6β in acute-phase protein induction in Hep3B cells.

*Akira:* That's correct.

*Baumann:* When you treat hepatoma cells or liver cells with IL-1 there is an increase both in mRNA for NF-IL6 and in the protein. Does IL-1 stimulate haptoglobin in Hep3B cells?

*Akira:* I don't have any results on that.

*Baumann:* You showed an induction of NF-IL6 in animals injected with IL-1. The NF-IL6 in turn should activate the haptoglobin gene because this gene contains two binding sites for NF-IL6 (or IL-6DBP) (Oliviero & Cortese 1989). In reality, haptoglobin expression is not turned on by IL-1 in Hep3B cells or in mouse liver cells.

*Akira:* I am interested in modification of NF-IL6 by IL-1 or IL-6. In this case, modification by IL-6 stimulation may be different from modification by IL-1. IL-1 may not activate haptoglobin.

*Stewart:* Does NF-IL6 also stimulate expression of gp130?

*Akira:* I have no idea.

*Fey:* In rats, the family of C/EBP-like proteins has been extensively characterized, and at least three different species have been identified—C/EBP, NF-IL6 (also called IL-6DBP or LAP in rats) and DBP. Is NF-IL6β the human equivalent of the rat DBP member of this C/EBP family?

*Ciliberto:* I don't think so, because DBP doesn't have a leucine zipper.

*Gearing:* I understand that LIF upregulates NF-IL6 (Natsuka et al 1992). Does NF-IL6 have any effect on LIF expression? There is an NF-IL6-like element upstream of the LIF gene.

*Akira:* We have examined the promoter regions of IL-6 and LIF, but found no apparent similarities.

*Sehgal:* In Fig. 1 you showed GRE elements upstream of $-225$ in the human IL-6 promoter. Do you have any results to indicate that these are functionally relevant?

*Akira:* No.

*Sehgal:* So these are 'theoretical' GREs predicted by computer searching for GRE motifs. Our experiments indicate that the glucocorticoid receptor binds elsewhere, across the multiple response element (MRE, $-141$ to $-173$), the TATA box and the transcription start site in this promoter (Ray et al 1990). In functional assays the wild-type glucorticoid receptor represses the IL-6 promoter in a dexamethasone-responsive manner. We have done additional experiments that suggest that some mutations in the first zinc finger of the glucorticoid receptor enable hormone-induced up-regulation of IL-6 transcription without an apparent need for direct binding of the glucocorticoid receptor mutant to the target DNA (Ray et al 1991). In our system we produce the wild-type or mutant receptor in HeLa cells using a constitutive expression plasmid. This allows us to co-transfect various IL-6 promoter–chloramphenicol acetyltransferase (CAT) constructs into such cells, and to see whether the addition of dexamethasone causes repression or up-regulation, or has no effect. With the wild-type glucocorticoid receptor in HeLa cells, dexamethasone plus IL-1 gives a good, solid repression of IL-6 promoter function. With mutant receptors containing complete deletions of the first and second zinc fingers, or only second finger deletions, dexamethasone has no effects. A point mutation (G-421) in the zinc coordination site of the first zinc finger or a deletion of the first finger (420–451) results in dexamethasone-responsive activation, rather than repression, of IL-6 promoter constructs.

This result is essentially heretical. Thus, we had to verify that in our hands, in our lab, with the HeLa cells that we used, that the glucocorticoid receptor mutants had the previously described inactive phenotypes with respect to activation of mammary tumour virus (MTV)–CAT promoter constructs (Hollenberg & Evans 1988). In our hands G-421 and Δ420–451 were inactive on the MTV–CAT promoter, yet these two mutants mediated dexamethasone-responsive up-regulation of the IL-6 promoter. Standard DNA-binding–immunoprecipitation assays verified that the wild-type glucocorticoid receptor bound the GRE in the MTV promoter, whereas the G-421 and Δ420–451 mutants did not. Similarly, the wild-type receptor binds to the IL-6 promoter, whereas these two mutants did not. Thus, at least in HeLa cells, these mutated forms of the receptor, which show no detectable direct binding to the IL-6 promoter DNA, can transcriptionally up-regulate the IL-6 gene. We are not dealing with a general up-regulation of transcription because these mutants have no effect on the c-*fos* promoter. With the complete IL-6 promoter the up-regulation can be as much as 10-fold. The phenotypic switch is actually a factor of 100 or more because the wild-type glucocorticoid receptor causes 10–50-fold repression. Our working hypothesis is that there is direct interaction of these mutant glucocorticoid receptor proteins with transcription factors such as NF-IL6. A second transcription factor might indirectly tether the receptor onto the promoter.

*Heath:* Can you rule out effects on the stability of the CAT message?

*Sehgal:* In our experiments the working end of the RNA is exactly the same in all of these constructs. The RNA start site, the polyadenylation signals and so on are the same. We have not done formal kinetic analysis to verify that there are no effects on stability.

*Heath:* One of the problems with CAT assays is that the levels of CAT activity are extraordinarily sensitive to small changes in the amount of transcript.

*Sehgal:* Everything I have discussed so far has been concerned with the inducible expression of the IL-6 promoter. A number of us, including Professor Hirano and Professor Kishimoto, are also interested in the dysregulated up-regulation of IL-6 expression in tumour cells. We have done some work with p53 and the *RB* (retinoblastoma) gene product (Santhanam et al 1991). Wild-type or mutant p53 was expressed in transfected HeLa cells using constitutive expression plasmids. In a functional assay involving IL-6–CAT reporter plasmids mouse wild-type p53 strongly repressed the IL-6 promoter. The mutant Glu-168→Gly + Met-234→Ile, which is a transforming mutant, is somewhat less active as a repressor. Human wild-type p53 also repressed the IL-6 promoter; the mutant Val-143→Ala was just as active as the wild-type in repressing the IL-6 promoter when serum was the inducer, but with IL-1 as the inducer the mutant was less active than the wild-type. RB also repressed the IL-6 promoter. This raises a whole range of questions about nuclear 'anti-oncogenes' as transcriptional repressors of the IL-6 gene, and whether mutations in these proteins relieve the repression, to allow a permissive state for the up-regulation of IL-6 in tumour cells.

*Kishimoto:* Is this really specific suppression or is it non-specific? Anything can be suppressed by transfection of p53.

*Sehgal:* Some promoters can be up-regulated, not repressed, by p53. There has been a report (Weintraub et al 1991) that p53 can act as a transcriptional activator. Moshe Arens' group also have results suggesting that p53 acts as a transcriptional repressor (personal communication).

*Gough:* What's the genetic background of the cells in terms of the endogenous p53?

*Sehgal:* Fortunately, in HeLa cells there is intact, low level expression of wild-type p53.

*Gough:* Do you know for certain that this is the case in your clones?

*Sehgal:* Yes, from the literature.

*Heath:* Could I ask Professor Kishimoto and Dr Akira about the signal transduction between the receptor and the nucleus? There have been a variety of reports recently showing tyrosine phosphorylation of various substrates occurring rapidly after stimulation with IL-6.

*Kishimoto:* At the moment, we do not know what kind of tyrosine kinases can interact with gp130. Antibodies against Src, Fyn and Lyn cannot detect the tyrosine kinase. I would like to know if the LIF receptor can activate tyrosine kinases.

*Gearing:* One of the early responses of M1 cells to LIF or IL-6 is the phosphorylation of a 160 kDa protein on tyrosine residues (Lord et al 1991).

*Kishimoto:* Nakajima & Wall (1991) also reported phosphorylation of a 160 kDa protein in response to IL-6 stimulation.

# References

Baumann H, Jahreis GP, Morella KK 1990 Interaction of cytokine- and glucocorticoid-response elements of acute-phase plasma protein genes. J Biol Chem 265:22275–22281

Descombes P, Chojkier M, Lichtsteiner S, Falvey E, Schibler U 1990 LAP, a novel member of the C/EBP gene family, encodes a liver-enriched transcriptional activator protein. Genes & Dev 4:1541–1551

Hollenberg SM, Evans RM 1988 Multiple and cooperative trans-activation domains of the human glucocorticoid receptor. Cell 55:899–906

Lord KA, Abdollahi A, Thomas SM et al 1991 Leukemia inhibitory factor and interleukin-6 trigger the same immediate early response. including tyrosine phosphorylation upon induction of myeloid leukemic differentiation. Mol Cell Biol 11:4371–4379

Nakajima K, Wall R 1991 Interleukin-6 signals activating junB and TIS11 gene transcription in a B-cell hybridoma. Mol Cell Biol 11:1409–1418

Natsuka S, Akira S, Nishio Y et al 1992 Macrophage differentiation-specific expression of NF-IL6, a transcription factor for interleukin-6. Blood 79:460–466

Oliviero S, Cortese R 1989 The human haptoglobin gene promoter: interleukin-6-responsive elements interact with a DNA-binding protein induced by interleukin-6. EMBO (Eur Mol Biol Organ) J 9:1145–1151

Poli V, Mancini FP, Cortese R 1990 IL-6DBP, a nuclear protein involved in interleukin-6 signal transduction, defines a new family of leucine zipper proteins related to C/EBP. Cell 63:643–653

Ray A, LaForge KS, Sehgal PB 1990 On the mechanism for efficient repression of the interleukin-6 promoter by glucocorticoids: enhancer, TATA box, and RNA start site (Inr motif) occlusion. Mol Cell Biol 10:5736–5746

Ray A, LaForge KS, Sehgal PB 1991 Repressor to activator switch by mutations in the first Zn finger of the glucocorticoid receptor—is direct binding necessary? Proc Natl Acad Sci USA 88:7086–7090

Santhanam U, Ray A, Sehgal PB 1991 Repression of the interleukin-6 gene promoter by p53 and the retinoblastoma susceptibility gene product. Proc Natl Acad Sci USA 88:7605–7609

Weintraub H, Hauschka S, Tapscott J 1991 The MCK enhancer contains a p53 responsive element. Proc Natl Acad Sci USA 88:4570–4571

# The action of interleukin 6 on lymphoid populations

Lucien A. Aarden and Cees van Kooten

*Central Laboratory of the Netherlands Red Cross Bloodtransfusion Service, PO Box 9190, 1006 AD Amsterdam, The Netherlands*

*Abstract.* We have analysed the role of IL-6 in activation of human lymphocytes isolated from peripheral blood. We found little effect of IL-6 on proliferation of T cells. IL-6 did stimulate production of IgM by B cells. However, it did not behave as a terminal differentiation factor and was required only during the first two days of the culture. Under most conventional stimulation conditions T cells themselves produce little or no IL-6. High IL-6 production could be induced by stimulating the cells with a combination of phorbol-12-myristate 13-acetate (PMA) and anti-CD28 antibodies.

*1992 Polyfunctional cytokines: IL-6 and LIF. Wiley, Chichester (Ciba Foundation Symposium 167) p 68–79*

Interleukin 6 (IL-6) has been shown to have pronounced effects on the growth and differentiation of lymphocytes. It supports growth of B cell hybridomas and plasmacytomas. It induces growth in some Epstein–Barr virus (EBV)-transformed human B cell lines and differentiation in others. In freshly isolated mouse B cells, IL-6, acting in synergy with IL-1, induces proliferation and differentiation as well. In human B cells IL-6 induces differentiation with little effect on growth. T cell proliferation is also stimulated by IL-6 and again there is a clear synergy with IL-1. Similar effects have been observed for the induction of cytotoxic T lymphocytes in accessory cell-depleted allogenic mixed lymphocyte reactions. The role of IL-6 in the activation of B cells and in T cells has been reviewed recently by Kishimoto & Hirano (1988) and by Van Snick (1990), respectively.

Here, we describe our own experiments on the effects of IL-6 on human T and B cells. We shall show that we have met considerable difficulties in reproducing the above-mentioned results. Using lymphocytes isolated from the peripheral blood of healthy individuals we found that IL-6 had little influence on T cell growth. In B cells IL-6 did stimulate immunoglobulin production. However, it acted early in culture and not as a terminal differentiation factor. Finally, we address the production of IL-6 by T cells. We observed that T cells can make considerable amounts of IL-6 after co-stimulation with anti-CD28 and phorbol 12-myristate 13-acetate (PMA).

68

## IL-6 and T cell proliferation

Cultured human monocytes produce large amounts of IL-6 (Aarden et al 1987). This production is dependent on stimulation by substances such as lipopolysaccharide (LPS). Picogram/ml quantities of LPS are sufficient and in our experience it is extremely difficult to prevent activation of monocytes by stimuli present in conventional culture media and fetal calf sera (Aarden et al 1991). Therefore, we have studied the role of IL-6 in lymphocyte populations, depleted of monocytes by counterflow elutriation to eliminate endogenous IL-6. As an alternative, we have used the human T cell line Jurkat. Stimulation was carried out by addition of a monoclonal antibody to CD3. In general, T cells are activated by anti-CD3 only when the antibodies are cross-linked. This can be achieved by interaction with Fc receptors present on accessory cells such as monocytes, or by coating the antibodies to the culture well (Van Lier et al 1989). Another elegant solution of this problem is to make use of an agonistic isotype switch variant of anti-CD3 of the IgE class (Van Lier et al 1987). Because of its restricted flexibility in the hinge region this antibody can cross-link T cells by itself, rendering the cells responsive to a second signal—IL-2, IL-4 or anti-CD28 (Fig. 1). In view of the reported effects of IL-6 and IL-1 on T cell proliferation, we tested whether these cytokines, either alone or in combination, could supply the second signal for proliferation or for the initiation of IL-2 production.

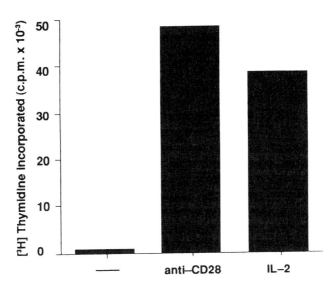

FIG. 1.   T cell proliferation induced by anti-CD3 IgE. 40 000 purified human T cells were incubated in 0.2 ml flat-bottomed wells with anti-CD3 IgE (100 ng/ml) in combination with anti-CD28 (15E8, 5 µg/ml), or with IL-2 (3 ng/ml), or without an additional stimulus (−). After four days, incorporation of [³H]thymidine was measured. Details are described in Van Lier et al 1987.

**TABLE 1   Induction of T cell proliferation by IL-6**

| | [$^3H$] Thymidine incorporated (c.p.m.)[a] | |
| --- | --- | --- |
| Treatment | − anti-CD3 | + anti-CD3 (100 ng/ml) |
| Control | 20 | 600 |
| IL-1 (0.5 ng/ml) | 20 | 2500 |
| IL-6 (50 ng/ml) | 30 | 3200 |
| IL-1 + IL-6 | 30 | 7200 |
| IL-2 (3 ng/ml) | 2000 | 36 000 |

[a]40 000 purified human T cells were incubated with combinations of cytokines with or without anti-CD3.

Table 1 shows an example typical of numerous experiments; we found very little effect of IL-6 or IL-1 on T cell activation, particularly in comparison with the effect of IL-2. As we were convinced that something was wrong with the system we then tried a number of variations. We replaced the anti-CD3 IgE by phytohaemagglutinin, concanavalin A, combinations of anti-CD2 antibodies, anti-CD28, PMA and 21 other monoclonal antibodies to CD3; all these gave the same result—IL-6 and IL-1 had little or no effect. We then tested cells from a variety of donors at various cell concentrations. We used T cell subsets [CD4$^+$, CD8$^+$, CD45RA$^+$ (naive T cells), CD45RO$^+$ (memory T cells)] and we used titrations of coated anti-CD3, all without success. Finally, we tested the effect of saturating amounts of neutralizing antibodies to IL-6 under conditions where T cell activation took place, but we never saw any effects. We have to conclude that with these target cells IL-6 plays a minor role in T cell activation.

## IL-6 and B cell activation

IL-6 has been found to enhance production of immunoglobulins by primary human B cells in various systems. Muraguchi et al (1988) used peripheral blood mononuclear cells stimulated with pokeweed mitogen and found that IL-6 was required late in culture only. That led to the notion that IL-6 acts as a terminal differentiation factor, in agreement with the results obtained with lymphoblastoid cell lines such as CESS (Hirano et al 1986). However, Emilie et al (1988), using *Staphylococcus aureus* Cowan I to stimulate purified B cells in the presence of glucocorticoids, reported that IL-6 is required early in culture. We have not been able to repeat the experiments of Muraguchi and co-workers. This was no surprise, because in our hands the endogenous production of IL-6 by mononuclear cells was high, which is why we used a monocyte-independent activation system in which anti-CD3 was coated onto the well (Van Lier et al 1989). In this system, IL-6 strongly enhances production of IgM by lymphocytes

**TABLE 2    Role of IL-6 in production of IgM by human B cells**

| Treatment[a] | IgM produced ($\mu$g/ml $\pm$ SEM, n = 4) |
|---|---|
| Control | 1.4 $\pm$ 0.3 |
| Anti-IL-6 (1 $\mu$g/ml)[b] | 0.3 $\pm$ 0.1 |
| IL-6 (0.5 ng/ml) | 6.1 $\pm$ 0.5 |
| IL-6 (0.5 ng/ml) + anti-IL-6 | 0.1 $\pm$ 0.1 |
| IL-6 (50 ng/ml) + anti-IL-6 | 5.1 $\pm$ 0.5 |

[a]40 000 lymphocytes were incubated in 0.2 ml round-bottomed wells coated with anti-CD3 IgA (0.68 $\mu$g/ml). Culture conditions were as described in van Kooten et al 1991.
[b]Affinity-purified goat anti-IL-6 antibodies were used.

and the effect of IL-6 can be blocked by anti-IL-6 (Table 2). When the addition of IL-6 was delayed until Day 2 of these 8-day cultures it no longer stimulated IgM production. In line with this, addition of anti-IL-6 at Day 2 does not eliminate the stimulatory effect of IL-6 (van Kooten et al 1990). Therefore, we conclude that in this system IL-6 is involved in an early activation event, before the B cells have begun to produce IgM. Using purified B and T cells pre-incubated with IL-6 we found that T cells were most probably the targets for IL-6 (van Kooten et al 1990).

**TABLE 3    Proliferation of and IL-6 production by human T cells**

| Treatment[a] | IL-6 produced[b] (pg/ml) | [$^3$H] Thymidine incorporated (c.p.m.) |
|---|---|---|
| Control | 1 | 30 |
| PMA (1 ng/ml) | 3 | 10 400 |
| IL-1 (0.5 ng/ml) | 1 | 70 |
| PHA (1 $\mu$g/ml) | 2 | 27 400 |
| Anti-CD3 (3.4 $\mu$g/ml)[c] | 7 | 64 900 |
| IL-2 | 1 | 200 |
| PHA + IL-1 | 12.5 | 45 300 |
| Anti-CD3 + PMA | 22 | 64 500 |
| PHA + IL-2 (3 ng/ml) | 15 | 48 400 |

[a]40 000 purified human T cells or monocytes were incubated with the indicated stimuli. After four days proliferation was assessed by measuring incorporation of [$^3$H]thymidine and IL-6 was measured in the supernatant using the B9 bioassay (Aarden et al 1987).
[b]Under these conditions stimulated monocytes produce 4000 pg IL-6/ml.
[c]Anti-CD3 was coated onto the culture wells.
PHA, phytohaemagglutinin.
PMA, phorbol 12-myristate 13-acetate.

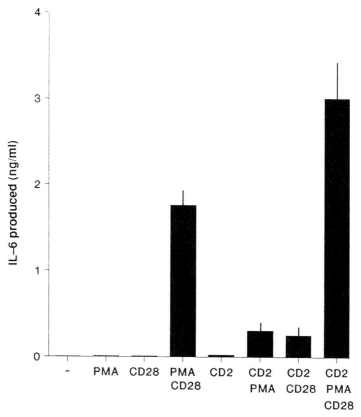

FIG. 2.   Production of IL-6 by purified T cells. 200 000 purified T cells in 0.5 ml wells were incubated for 65 h with the indicated stimulants, and IL-6 in the supernatant was measured using the B9 cell proliferation assay. Phorbol 12-myristate 13-acetate (PMA) was used at 2 ng/ml, anti-CD28 (CD28) at 5 µg/ml and anti-CD2 (CD2) (a combination of CLB-T11.1/2 and CLB-T11.2/1) at 1/2000 diluted ascites. Stimulation with agonistic anti-CD3 antibodies gave results similar to those obtained with anti-CD2. Error bars represent SEM, $n = 5$. Further details are described in van Kooten et al 1991.

## Production of IL-6 by T cells

For some time it has been a matter of debate whether freshly isolated T cells can be stimulated to produce IL-6. Using the sensitive B9 cell proliferation bioassay we were able to measure production of IL-6 by a single monocyte in 0.2 ml (Aarden et al 1987) and yet production of IL-6 by 1000 purified T cells in 0.2 ml could barely be detected. We then analysed production of IL-6 by T cells in bulk cultures incubated with a variety of stimuli (Table 3). The maximal amount of IL-6 produced was about 400-fold less than that produced by monocytes. Despite this, IL-6 was identified as a product of a human T-cell lymphotropic virus (HTLV-1)-transformed T cell line and was cloned from this line (Hirano et al 1986). Soon afterwards it was reported that primary T cells could produce IL-6 if monocytes were present (Horii et al 1988). However, the presence of (IL-6-producing) monocytes made analyses of the production cumbersome. Then antibodies to CD28 were discovered. These antibodies are potent co-stimulators of anti-CD3-, anti-CD2- or PMA-stimulated T cells (June et al 1990, Van Lier et al 1988). The CD28 molecule is a 44 kDa homodimeric glycoprotein which is present on most mature T cells and which is involved in cell–cell interactions through its association with the B cell activation antigen B7/BB1 (Linsley et al 1990). It turned out that anti-CD28 in combination with PMA stimulates levels of IL-6 production by T cells approaching those observed with monocytes (Fig. 2). Furthermore, after this stimulation T cells become potent producers of IL-1$\alpha$, TNF-$\alpha$ (van Kooten et al 1991) and IL-8 (M. Hart, unpublished work). The production of IL-6 by T cells is much slower than that by monocytes—IL-6 becomes detectable only after two days of culturing and increases up to Day 7, whereas monocytes stop producing IL-6 within 20 hours. Memory (CD45RO) and naive (CD45RA) T cells produce IL-6 equally well.

## Concluding remarks

In general, the effect of IL-6 on human peripheral T and B cells has been rather disappointing in our hands, and our results seem to be at odds with a number of published reports. Obviously, methodological differences might be responsible, but we have tried hard to test as many variables as possible. A clue might be offered by an observation made by Houssiau et al (1989). They studied the role of IL-6 in the activation of human tonsillar T cells and found the response to IL-6 and IL-1 to be much less reproducible than the response to IL-2. In their experiments only T cells with low buoyant density were able to respond to IL-6 and IL-1. We are currently investigating whether the same is true for T cells from peripheral blood.

## References

Aarden LA, De Groot ER, Schaap OL, Lansdorp PM 1987 Production of hybridoma growth factor by monocytes. Eur J Immunol 17:1411–1416

Aarden L, Helle M, Boeije L, Pascual-Salcedo D, De Groot E 1991 Differential induction of interleukin 6 production by monocytes, endothelial cells and smooth muscle cells. Eur Cytokine Net 2:115–120

Emilie D, Crevon MC, Auffredou MT, Galanaud P 1988 Glucocorticosteroid-dependent synergy between interleukin 1 and interleukin 6 for human B lymphocyte differentiation. Eur J Immunol 18:2043–2047

Hirano T, Yasukawa K, Harada H et al 1986 Complementary DNA for a novel human interleukin (BSF-2) that induces B lymphocytes to produce immunoglobulin. Nature (Lond) 324:73–76

Horii Y, Muraguchi A, Suematsu S et al 1988 Regulation of BSF-2/IL-6 production by human mononuclear cells. Macrophage-dependent synthesis of BSF-2/IL-6 by T cells. J Immunol 141:1529–1535

Houssiau FA, Coulie PG, Van Snick J 1989 Distinct roles of interleukin 1 and interleukin 6 in human T cell activation. J Immunol 3:2520–2524

June CH, Ledbetter JA, Linsley PS, Thompson CB 1990 Role of CD28 receptor in T-cell activation. Immunol Today 11:211–216

Kishimoto T, Hirano T 1988 Molecular regulation of B lymphocyte response. Annu Rev Immunol 6:485–512

Linsley PS, Clark EA, Ledbetter JA 1990 The cell antigen, CD28, mediates adhesion with B cells by interacting with the activation antigen B7/BB1. Proc Natl Acad Sci USA 87:5031–5035

Muraguchi A, Hirano T, Tang B et al 1988 The essential role of B cell stimulatory factor 2 (BSF-2/IL6) for the terminal differentiation of B cells. J Exp Med 167:332–344

van Kooten C, van Oers MHJ, Aarden LA 1990 Interleukin-6 enhances human Ig production, but not as a terminal differentiation factor for B lymphocytes. Res Immunol 141:341–356

van Kooten C, Rensink I, Pascual-Salcedo D, van Oers R, Aarden LA 1991 Monokine production by human T cells: Interleukin-α production restricted to memory T cells. J Immunol 146:2654–2658

Van Lier RAW, Boot JHA, Verhoeven AJ, De Groot ER, Brouwer M, Aarden LA 1987 Functional studies with anti-CD3 heavy chain isotype switch-variant monoclonal antibodies. Accessory cell-independent induction of interleukin 2 responsiveness in T cells by ε-anti-CD3. J Immunol 139:2873–2879

Van Lier RAW, Brouwer M, Aarden LA 1988 Signals involved in T cell activation. T cell proliferation induced through the synergistic action of anti-CD28 and anti-CD2 monoclonal antibodies. Eur J Immunol 18:167–172

Van Lier RAW, Brouwer M, Rebel VI, Van Noesel CJM, Aarden LA 1989 Immobilized anti-CD3 monoclonal antibodies induce accessory cell-independent lymphokine production, proliferation and helper activity in human T lymphocytes. Immunology 68:45–50

Van Snick J 1990 Interleukin-6: an overview. Annu Rev Immunol 8:253–278

## DISCUSSION

*Lotem:* Where the addition of IL-6 to B cell cultures on Day 2 did not have an effect, could that result from a loss, because of the absence of IL-6 during the first two days, of a certain population of cells that should produce the antibody?

*Aarden:* That would be in line with your finding that IL-6 affects cell survival, but production of IgG starts after Day 5 in these kinds of culture, so there would have to be a selective survival of cells which are involved five days later in immunoglobulin production—that of course could be true. Our experiments on separated T and B cells, in which we incubated cells with IL-6 for two days, added anti-IL-6, mixed the cells and assayed eight days later for immunoglobulin production (van Kooten et al 1990), suggested that T cells are the primary targets for IL-6.

*Dexter:* I was quite relieved to hear your experiences because we have had tremendous problems reproducing some of the claimed effects of IL-6 on primary myeloma cells. What we suspect, with some supporting data, is that the response of haemopoietic cells to IL-6 depends, at least in part, on the serum used in the assays. Many of the reported effects of IL-6 on haemopoietic myeloid cells cannot be reproduced in serum-free conditions. I wonder if this could account for some of the differences that you and other people have observed.

*Aarden:* We tried different human sera and fetal calf sera. We always include β-mercaptoethanol in our medium; I have seen reports that IL-1 activity is evident only if β-mercaptoethanol is omitted (Howard et al 1983).

*Dexter:* Have you tried serum-free or serum-deprived conditions?

*Aarden:* No. At the low cell densities that we use, 40 000 cells per 200 μl well, the only serum-free system that works is the Iscove system in which there is still so much albumin that I don't think it can be fairly called serum-free.

*Hirano:* The experimental system by which you showed that IL-6 may not be a late-acting factor for B cells seems to be complicated, because it contains not only B cells but also T cells. The requirement for IL-6 in an early phase in your experimental system may reflect the effect of IL-6 on T cells but not B cells. Did you examine the kinetics of the expression of the IL-6 receptor in B cells in your system? In my experience, expression of the receptor is delayed, and requires many signals.

*Aarden:* We haven't investigated that. It would be a good thing to look at, because if there are no receptors in the first few days it's difficult to see how IL-6 could act, though it certainly does. It's true that we used T cells, but that's also true for the pokeweed system. Emilie et al (1988) have used purified B cells stimulated with *Staphylococcus aureus* in the absence of T cells but in the presence of glucocorticoids, which are essential. They found that IL-6 induces immunoglobulin production but is required only for the first two days, as we found.

*Hirano:* That experiment may indicate that IL-6 acts on B cells at an early phase. However, it fails to demonstrate that IL-6 is not necessary in the late phase of B cell differentiation, because *S. aureus*-stimulated B cells can produce IL-6.

*Aarden:* You need IL-6 for only two days, then later on you see immunoglobulin production. We tried the pokeweed system, but this system requires the use of mononuclear cells, and monocytes produce a lot of IL-6. Therefore, the only sensible experiment is to add neutralizing antibodies to IL-6—this had no effect.

*Gough:* If I understood correctly, if you used PMA and anti-CD2 or anti-CD3 as the stimulus you saw no IL-6 production—you needed to stimulate CD28 to get IL-6 production. This caused a small increase in IL-1, TNF-α etc. On the figure where you showed IL-6 production there appeared to be clear induction of IL-6 with anti-CD2 + PMA.

*Aarden:* There is an effect but it's less than 10% of the maximal response.

*Gough:* Have you tried using a calcium ionophore with any of those treatments?

*Aarden:* We have tried an ionophore in the absence of anti-CD28, but we haven't combined anti-CD28 with a calcium ionophore and looked at IL-6. We tried a calcium ionophore with PMA, but that didn't work.

*Gauldie:* Given that accessory cells, and other cells, are capable of making IL-6, how important do you think T cell-derived IL-6 is?

*Aarden:* The conditions of stimulation are completely different. Production of IL-6 by monocytes can be stimulated with IL-1, TNF or LPS, but T cell IL-6 production cannot be stimulated by any of those treatments. In an antigen-driven, sterile situation, I imagine that T cells could play a role in chronic infections or parasitic diseases.

*Gauldie:* From what you said about CD28 and its counter-ligand on B cells, I assumed you were alluding to the fact that there may be a tight junction between these two cells, yet antibodies to IL-6 are capable of inhibiting the reaction. It is possible that there are other cells that could provide that signal more easily than T cells.

*Metcalf:* What worries me about IL-6 is that it is reported to be a proliferative stimulus for plasmacytoma cells, for cell lines cultured under certain conditions, yet it seems to be a differentiation-inducing stimulus, and not a proliferative stimulus, for normal B cells. I can accept that there may be a difference in the responses shown by neoplastic and normal cells, but in IL-6 transgenic animals there is clearly an accumulation of non-neoplastic plasma cells. So, how does this occur in the transgenic animals? Is the accumulation of plasma cells due not to proliferative stimulation but to an increased lifespan of the cells, or is the response the outcome of a complex network interaction occurring *in vivo*?

*Aarden:* In mouse spleen B cells the combination of IL-6 and IL-1 is a potent stimulus for proliferation (Vink et al 1988). That might explain the B cell growth and the differentiation. What decides whether growth or differentiation is induced, I don't know.

*Kishimoto:* In IL-6 transgenic mice of the B6 strain there are no malignant tumours—only proliferation of normal plasma cells. This indicates that IL-6 is a growth factor not only for plasmacytoma cells but also for normal plasma cells.

*Metcalf:* In the photographs of these tissues I have seen there seem to be few mitotic figures—that is, there is not much mitotic activity.

*Hirano:* We have never examined the mitotic index of plasma cells in IL-6 transgenic mice. Several plasma cells were multinucleated, indicating that they were proliferating.

*Metcalf:* Are you comfortable with the conclusion that, in combination with certain factors, IL-6 can act as a proliferative stimulus?

*Hirano:* It's very difficult to induce the proliferation of human B cells *in vitro* with IL-6. However, Vink et al (1988) demonstrated that IL-6, in the presence of IL-1, can induce mouse B cell growth.

*Heinrich:* The fate of IL-6 may differ in homologous and heterologous systems, which means that human IL-6 may be internalized only by human target cells. This could be a problem in the interpretation of the results obtained with transgenic mice in which the human IL-6 gene is expressed.

*Kishimoto:* In Castleman's disease, in humans, there is also massive proliferation of mature plasma cells, not tumours.

*Dexter:* I am not aware of any results indicating that normal mature plasma cells undergo division. Even in mature myeloma cells you don't see mitotic figures.

*Kishimoto:* The point is whether the plasma cell proliferation is monoclonal or polyclonal. In B6 transgenic mice, as well as in Castleman's disease, plasma cell growth is polyclonal. When we prepare IL-6 transgenic BALB/c strain mice we can get monoclonal transplantable plasmacytomas. In certain cases of Castleman's disease we can detect a chromosomal translocation including the *bcl*-2 gene. During proliferation of plasma cells in the presence of IL-6 a certain cell undergoes oncogene rearrangements, which result in monoclonal growth.

*Aarden:* In the first myeloma patient treated with an anti-IL-6 monoclonal antibody that we developed, within four days the percentage of myeloma cells in S phase went from over 20% to 0 (Aarden 1991). This suggests that under those conditions IL-6 is a proliferative factor for B cells.

*Hirano:* Jourdan et al (1990) demonstrated that there are several proliferating plasma cells in the bone marrow of patients with Castleman's disease (Jourdan et al 1990). This again suggests that IL-6 may cause proliferation of normal plasma cells.

*Aarden:* Cees van Kooten is looking at a number of B cell lymphomas. The general picture is that if there is an effect of IL-6, it is an inhibition of proliferation of B-CLL cells without an induction of differentiation. IL-6 has been observed to induce differentiation or proliferation, or to inhibit proliferation of B cells.

*Kishimoto:* Several studies have revealed a difference in the requirement for IL-6 between naive T cells and memory T cells. You said that they produce similar amounts of IL-6, but what is your experience with their IL-6 requirements?

*Aarden:* Using separated cells and stimulating under different conditions we find minor effects of IL-6 on proliferation of both naive and memory cells. In one of our donors, IL-6 induced proliferation of both naive and memory T cells. In general, we see little effect of IL-6 or anti-IL-6.

*Metcalf:* What is known about the level of expression of mRNA for IL-6 in the developing thymus, and in the pre-adult thymus, where the lymphoid population is increasing? Is there any suggestion that IL-6 has an important function within the thymus itself?

*Aarden:* I haven't seen any data on that.

*Metcalf:* It seems to me that this is a poorly understood area. Both IL-2 and the IL-2 receptor can be expressed in cells in the thymus (McGuire & Rothenberg 1987, Ceredig et al 1985), but mice in which the IL-2 gene has been deleted appear to develop a normal thymus (Schorle et al 1991), so IL-2 seems to have been ruled out as the factor controlling the enormous proliferation of lymphocytes in the thymus. Is IL-6 a possible alternative?

*Aarden:* Not that I am aware of. The presence of the counter structure of CD28 has now been demonstrated in the thymus (Turka et al 1991). Activation of T cells via anti-CD2 combined with anti-CD28 is an antigen receptor-independent mechanism for stimulating T cells. That might be an interesting candidate mechanism for control of thymocyte proliferation.

*Metcalf:* If CD28 is also present on liver cells or toe-nail cells, then it's not an interesting candidate. Is CD28 limited in its expression?

*Aarden:* I don't know. The only published result that I am aware of is that the B7 antigen is expressed on activated B cells and on thymic epithelial cells.

*Gough:* Is IL-6 produced by thymic cells?

*Lee:* It has been reported to be (Murray et al 1989, Helle et al 1988).

*Gauldie:* Is there any information about the activity of LIF on a lymphoid population? Is LIF involved in the same way as IL-6 in the generation of the immune response?

*Metcalf:* You will appreciate from the paper you have just heard that there are many ways to test both T and B lymphocytes, and I am not satisfied that LIF has been tested exhaustively. There is a small subset of lymphocyte-like cells in the marrow, the spleen and the thymus, 1–2% of the total, that seem to have receptors for LIF. Most standard tests for T cell proliferation in mice have been done, and the results are negative. The same sorts of tests on B lymphocytes suggested that LIF might have effects that are similar to those of IL-6.

*Heath:* LIF is produced by a wide variety of B cell- and T cell-derived tumours.

*Gearing:* It is also produced by the thymic epithelium (Le et al 1990).

*Metcalf:* There are reasons to hope that it has some actions, but, in my opinion, the work that has been done so far has not been exhaustive. You have seen today how one needs to examine these responses in fine detail before a clear answer emerges, and, even then, you don't know what the relevance of the response is to the *in vivo* situation.

## References

Aarden LA 1991 Plasmacytomas. In: Klein B (ed) Mechanisms of B-cell neoplasia 1991. Workshop at the Basel Institute for Immunology. Editions Roche, Basel, p 429–433
Ceredig R, Lowenthal JW, Nabholz E, MacDonald HR 1985 Expression of interleukin-2 receptors as a differentiation marker on intrathymic stem cells. Nature (Lond) 314:98–100

Emilie D, Crevon MC, Affredou MT, Galanaud P 1988 Glucocorticosteroid-dependent synergy between interleukin 1 and interleukin 6 for human B lymphocyte differentiation. Eur J Immunol 18:2043–2047

Helle M, Brakenhoff JPJ, De Groot ER, Aarden LA 1988 Interleukin 6 is involved in interleukin 1-induced activities. Eur J Immunol 18:957–959

Howard M, Mizel SB, Lachman L, Ansel J, Johnson B, Paul WE 1983 Role of interleukin 1 in anti-immunoglobulin-induced B-cell proliferation. J Exp Med 157:1529–1543

Jourdan M, Bataille R, Seguin J, Zhang XG, Chaptal PA, Klein B et al 1990 Constitutive production of interleukin-6 and immunologic features in cardiac myxomas. Arthritis Rheum 33:398–402

Le PT, Lazorick S, Whichard LP et al 1990 Human thymic epithelial cells produce IL-6, granulocyte-monocyte CSF, and leukemia inhibitory factor. J Immunol 145:3310–3315

McGuire KL, Rothenberg EU 1987 Inducibility of interleukin-2 RNA expression in individual mature and immature T lymphocytes. EMBO (Eur Mol Biol Organ) J 6:939–946

Murray R, Suda T, Wrighton N, Lee F, Zlotnik A 1989 IL-7 is a growth and maintenance factor for mature and immature thymocyte subsets. Int Immunol 1:526–531

Schorle H, Hottschke T, Hienig T, Schimpl A, Horat I 1991 Development and function of T cells in mice rendered interleukin-2 deficient by gene targetting. Nature (Lond) 352:621–624

Turka LA, Linsley PS, Paine R III, Schieven GL, Thompson CB, Ledbetter JA 1991 Signal transduction via CD4, CD8, and CD28 in mature and immature thymocytes. Implications for thymic selection. J Immunol 146:1428–1436

van Kooten C, van Oers MHJ, Aarden LA 1990 Interleukin 6 enhances human Ig production, but not as a terminal differentiation factor for B lymphocytes. Res Immunol 141:341–356

Vink A, Coulie PG, Wauters P, Nordan RP, Van Snick J 1988 B-cell growth and differentiation activity of interleukin-HP1 and related murine plasmacytoma growth factors: synergy with interleukin-1. Eur J Immunol 18:607–612

# Regulation of leukaemic cells by interleukin 6 and leukaemia inhibitory factor

Joseph Lotem and Leo Sachs

*Department of Molecular Genetics and Virology, Weizmann Institute of Science, Rehovot 76100, Israel*

*Abstract.* Interleukin 6 (IL-6) and leukaemia inhibitory factor (LIF) can have pleiotropic effects on different cell types. M1 myeloid leukaemic cells respond to IL-6 with activation of a terminal differentiation programme which includes activation of genes for certain haemopoietic regulatory proteins (IL-6, IL-1α, IL-1β, granulocyte-macrophage colony-stimulating factor [GM-CSF], M-CSF, tumour necrosis factor and transforming growth factor [TGF] β1) and for receptors for some of these proteins, thus establishing a network of positive and negative regulatory cytokines. IL-6 and some other cytokines also induce during differentiation sustained levels of transciption factors that can regulate and maintain gene expression in the differentiation programme. M1 leukaemic cells induced to differentiate with IL-6 undergo programmed cell death (apoptosis) on withdrawal of IL-6, and can be rescued from apoptosis by IL-6, IL-3, M-CSF, G-CSF or IL-1, but not by GM-CSF. These differentiating leukaemic cells can also be rescued from apoptosis by the tumour promoter TPA (12-*O*-tetradecanoylphorbol-13-acetate) but not by the non-tumour-promoting isomer 4-α-TPA, and rescue from apoptosis can be achieved by different pathways. Apoptosis can also be induced in undifferentiated M1 leukaemic cells by expression of the wild-type form of the tumour suppressor p53 protein and IL-6 can rescue the cells from this wild-type p53-mediated apoptosis. There are clones of M1 cells that differentiate with IL-6 but not with LIF and another M1 clone that differentiates with either IL-6 or LIF. Differentiation induced by IL-6 or LIF is inhibited by TGF-β1. The pleiotropic effects of LIF, like those of IL-6, are presumably also in a network of interacting regulatory proteins.

*1992 Polyfunctional cytokines: IL-6 and LIF. Wiley, Chichester (Ciba Foundation Symposium 167) p 80–99*

Interleukin 6 (IL-6) and leukaemia inhibitory factor (LIF) can have pleiotropic effects on different cell types including certain leukaemic cells, hepatocytes and neuronal cells (for reviews see Hirano et al 1990 for IL-6 and Metcalf 1991 for LIF). Although these cytokines show some similar activities in certain cell types (see above reviews), they have distinct actions on other cells such as normal embryonic stem cells, in which LIF inhibits differentiation but IL-6 does not, normal myeloid precursor cells, in which IL-6 has viability-promoting and differentiation-inducing activity (including differentiation to megakaryocytes) (Lotem et al 1989a) whereas LIF does not, and normal B cells, in which IL-6 induces differentiation (Hirano et al 1990) but LIF does not. Although IL-6 and LIF differ in their effects on normal embryonic stem cells, both can inhibit the differentiation of F9 teratocarcinoma cells (Hirayoshi et al 1991). Even among cell types that respond similarly to IL-6 and LIF, such as M1 leukaemic cells, there are clonal differences in responsiveness (Lotem et al 1989b). The physiological consequences of sustained high levels of IL-6 and LIF *in vivo* also differ (Suematsu et al 1989, Metcalf & Gearing 1989). Here, we analyse the regulation of myeloid leukaemic cells induced to undergo a terminal differentiation programme by IL-6 or LIF.

## Regulation of a cytokine network during differentiation of myeloid leukaemic cells

Incubation of different clones (Fibach et al 1973) of M1 myeloid leukaemic cells (Ichikawa 1969) with IL-6 induces the cells to differentiate along the macrophage and/or granulocyte pathway, so that they express typical differentiation markers and characteristics such as cell adherence, Fc and C3 rosettes, synthesis and secretion of lysozyme, phagocytic activity, cell migration in agar and chemotaxis, and adopt the morphology of mature macrophages or granulocytes (reviewed in Sachs 1982, 1987a,b, 1990). The IL-6-responsive clones of M1 cells isolated by us in Rehovot, Israel (Fibach et al 1973) do not differentiate when cultured with LIF (Lotem et al 1989b), whereas another M1 clone (clone T22) isolated in Saitama, Japan, differentiates with either IL-6 or LIF (Lotem et al 1989b, Hozumi et al 1990, Metcalf 1991). The differentiation of M1 cells induced with IL-6 or LIF was inhibited by transforming growth factor (TGF) β1 (Lotem & Sachs 1990). The LIF-sensitive and LIF-resistant clones of M1 cells displayed a similar weak responsiveness to granulocyte colony-stimulating factor (G-CSF) and IL-1, but differed in their susceptibility to induction of differentiation with bacterial lipopolysaccharide (LPS) (Lotem et al 1989b). These clonal differences can explain the isolation from the same source, Krebs ascites cell conditioned medium, of two different differentiation-inducing proteins for M1 cells (IL-6 and LIF) in two laboratories using different M1 subclones (Shabo et al 1988, Hilton et al 1988).

Induction of differentiation in M1 leukaemic cells using IL-6 induced mRNA for several cytokines including IL-6 itself and granulocyte-macrophage (GM) CSF (Shabo et al 1989a), IL-1α, IL-1β, M-CSF, tumour necrosis factor (TNF) and TGF-β1 (Lotem et al 1991b). mRNA for all these cytokines was also induced during GM-CSF-stimulated differentiation in another differentiation-competent clone of myeloid leukaemic cells (7-M12) that was not derived from the M1 leukaemia (Shabo et al 1989a, Lotem et al 1991b). Human U-937 myeloid leukaemic cells induced to differentiate with GM-CSF were also reported to express mRNA for M-CSF, IL-1β and TNF (Lindemann et al 1988). In addition, induction of differentiation in our M1 cells with IL-1α, G-CSF, dexamethasone, LPS or cytosine arabinoside resulted in induction of mRNA for IL-6 and GM-CSF (Shabo et al 1989b), and induction of differentiation in the Saitama M1 clone T22 with LIF or vitamin $D_3$ induced IL-6 (Miyaura et al 1989) and TNF (Michishita et al 1990). LIF caused the Saitama M1 clone T22 cells (data not shown), like the IL-6-treated Rehovot M1 clone 11 cells (Lotem & Sachs 1986), to produce biologically active GM-CSF.

The genes for all these cytokines are expressed by normal cells from the myelomonocytic lineages either without stimulation or after incubation with different cytokines (reviewed in Lotem et al 1991b). However, whereas normal monocytes/macrophages were reported to produce G-CSF when stimulated with different cytokines (Metcalf & Nicola 1985, Lu et al 1988, Motoyoshi et al 1989, Oster et al 1989), the differentiating M1 and 7-M12 leukaemic cells did not express G-CSF mRNA (Lotem et al 1991b). The results indicate (Lotem et al 1991b) (Table 1) that induction of differentiation in different leukaemic clones with IL-6, LIF or GM-CSF is associated with a nearly normal pattern of expression of the network of haemopoietic regulatory proteins. The results with G-CSF, however, indicate that leukaemic clones may have specific defects in

**TABLE 1  Induction of cytokine gene expression during differentiation of myeloid leukaemic cells**

| Cells | Inducer added | mRNA induction[a] | | | | | | | | |
|---|---|---|---|---|---|---|---|---|---|---|
| | | IL-6 | GM-CSF | M-CSF | IL-1α | IL-1β | TNF | TGF-β1 | G-CSF | IL-3 |
| Leukaemic M1 clone 11 | IL-6 | + | + | + | + | + | + | + | − | − |
| Leukaemic M1 clone T22 | LIF | + | + | | | | + | | − | |
| Leukaemic 7-M12 | GM-CSF | + | + | + | + | + | + | + | − | − |
| Leukaemic U-937 | GM-CSF | | | + | + | + | | | | |
| Normal myelo-monocytic[b] | | + | + | + | + | + | + | + | + | − |

[a]Blanks indicate cases where results on induction of a specific mRNA have not been reported. (−) no induction; (+) induction of mRNA.
[b]Normal myelomonocytic cells produce the above cytokines either without induction or when cultured with cytokines.

the normal network; other defects may be expressed in other types of leukaemic cells. In addition to the induction of genes for different cytokines, including M-CSF, IL-6 treatment of M1 leukaemic cells also induces mRNA for M-CSF receptors (c-*fms*) (Lotem et al 1991b) and functional M-CSF and IL-3 cell surface receptors (Lotem & Sachs 1986, 1989). In 7-M12 leukaemic cells, however, differentiation in response to GM-CSF is associated with induction of M-CSF but not of c-*fms* mRNA or functional M-CSF receptors (Lotem & Sachs 1986, 1989, Lotem et al 1991b). This shows that the expression of a cytokine can be dissociated from the expression of its receptor and that there can be different specific defects in the network of cytokines and their receptors in different leukaemic clones.

### Induction of expression of genes for transcription factors during differentiation of myeloid leukaemic cells

Interaction of cell surface receptors with the appropriate ligands initiates intracellular signalling that can lead to various cellular reactions such as mitogenesis or differentiation. This signalling can bring about these changes by controlling the levels or activities of transcription factors that activate specific genes in the mitogenic or differentiation pathways. Induction of differentiation in M1 myeloid leukaemic cells with IL-6 brought about an immediate–early increase in the levels of mRNA for c-*jun*, *jun*B and c-*fos*, followed by high levels of these mRNAs sustained even at four days after addition of IL-6 (Shabo et al 1990). Similar sustained high levels of these mRNAs were also observed during differentiation of 7-M12 myeloid leukaemic cells with GM-CSF (Shabo et al 1990). mRNA for another transcription factor, zif/268 (Egr-1) was induced in GM-CSF-treated 7-M12 cells but not in IL-6-treated M1 leukaemic cells (Shabo et al 1990). These results indicate that differentiation in myeloid leukaemic cells induced by IL-6 and GM-CSF, and presumably also by LIF, is associated with sustained increased levels of mRNA for certain transcription factors and that zif/268 gene expression is not essential for differentiation to mature macrophages. Induction of these and possibly other different transcription factors may thus account for the induction and maintenance of the various differentiation-associated properties, including switching on of the cytokine network, caused by IL-6 or GM-CSF in myeloid leukaemic cells.

### Induction of programmed cell death in myeloid leukaemic cells

Normal myeloid precursor cells require exogenous haemopoietic regulatory proteins to remain viable and to multiply and differentiate (reviewed in Sachs 1982, 1987a,b, 1990; Metcalf 1985). Withdrawal of these proteins leads to cell death by apoptosis (programmed cell death) (Williams et al 1990), which is associated with chromatin condensation and DNA fragmentation (Wyllie 1980). M1 myeloid leukaemic cells are growth factor independent. However, M1 cells induced to differentiate with IL-6 enter a growth factor-dependent state so that the cells undergo apoptosis after withdrawal of IL-6 (Fibach & Sachs 1976, Lotem & Sachs 1982, 1989, Lotem et al 1991a). The programme for cell death is turned on before terminal differentiation and these differentiating cells can be rescued from apoptosis and made to continue to multiply by adding back IL-6, or by treatment with IL-3, M-CSF, G-CSF or IL-1, but not GM-CSF (Lotem & Sachs 1989). On one clone of M1 cells surface receptors for GM-CSF were not detected after induction of differentiation with IL-6 (Lotem & Sachs 1989), whereas another clone was induced by IL-6 to express such receptors (Lotem & Sachs 1986). Neither of these clones could be rescued from apoptosis by GM-CSF (Lotem & Sachs 1989).

Differentiating M1 cells could also be rescued from apoptosis by the tumour-promoting phorbol ester 12-$O$-tetradecanoylphorbol-13-acetate (TPA) but not by its non-tumour-promoting isomer 4-$\alpha$-TPA (Lotem et al 1991a). IL-3, M-CSF, G-CSF and IL-6 gave better protection aginst apoptosis when combined with TPA than when used alone (Table 2). Pre-exposure of M1 cells to a high concentration of TPA during IL-6-induced differentiation did not prevent apoptosis after withdrawal of IL-6 and the cells could still be rescued from apoptosis (after withdrawal of IL-6 and TPA) by IL-3, M-CSF, G-CSF or IL-6, but not by TPA (Lotem et al 1991a) (Table 2). These results indicate that TPA and the cytokines rescued the differentiating cells from apoptosis by different pathways and that rescue by the cytokines does not appear to be mediated by the protein kinase C that is activated by TPA. We have suggested that TPA may act as a tumour promoter by inhibiting apoptosis (Lotem et al 1991a). Although no experiments have yet been reported on the development of LIF-induced apoptosis in M1 clone T22 leukaemic cells, these cells appear to develop M-CSF dependency after induction of differentiation with LIF (Metcalf 1991). This suggests that both IL-6 and LIF can initiate a programme for cell death in M1 myeloid leukaemic

**TABLE 2   Cytokines and TPA rescue differentiating myeloid leukaemic cells from apoptosis by different pathways**

| Pretreat-ment[a] | TPA | No. of viable cells $\times 10^{-4}/ml^c$ | | | | |
|---|---|---|---|---|---|---|
| | | None[b] | IL-3 | IL-6 | M-CSF | G-CSF |
| IL-6 | − | $8\pm2$ | $72\pm6$ | $61\pm6$ | $44\pm5$ | $123\pm10$ |
| | + | $42\pm6$ | $136\pm15$ | $128\pm8$ | $98\pm6$ | $189\pm12$ |
| IL-6+TPA | − | $8\pm3$ | $76\pm8$ | $63\pm10$ | $42\pm4$ | $118\pm16$ |
| | + | $10\pm2$ | $78\pm6$ | $70\pm12$ | $48\pm5$ | $110\pm12$ |

[a]M1 leukaemic cells were incubated for three days with IL-6 without or with 1 μg/ml TPA.
[b]After pretreatment, cells were washed and subcultured for three days at $20\times10^4$ cells/ml with (+) or without (−) 10 ng/ml TPA, with no further additions (none), or with different cytokines.
[c]The number of viable cells was determined three days after subculture as described previously (Lotem & Sachs 1989).

cells. Because this programme is already present in normal myeloid precursor cells and in more differentiated cells including mature granulocytes and macrophages, induction of programmed cell death in myeloid leukaemic cells during differentiation might be an important physiological way of suppressing leukaemia.

Wild-type p53 protein is a product of a tumour suppressor gene, the expression of which is abrogated in many tumour cell types including myeloid leukaemic cells (reviewed in Levine et al 1991). We have used a temperature-sensitive p53 mutant (Ala-135→Val) which behaves like other p53 mutants at 37.5 °C (that is, it lacks anti-proliferative effects) but like wild-type p53 at 32.5 °C (Michalovitz et al 1990). There is a clone of M1 cells that completely lacks expression of p53 mRNA and protein (Yonish-Rouach et al 1991). When these cells were transfected with DNA encoding the temperature-sensitive p53 mutant there was no change in their behaviour at 37.5 °C (at which the Val-135 p53 protein has mutant properties) but at 32.5 °C the wild-type activity of this p53 protein resulted in apoptotic cell death (Yonish-Rouach et al 1991). This induction of apoptosis was not associated with differentiation (Yonish-Rouach et al 1991). Therefore, apoptosis can be initiated in M1 cells not only by a differentiation-associated process induced by IL-6, but also by expression of wild-type p53 in undifferentiated cells. This induction of apoptosis by wild-type p53 was inhibited by IL-6 (Yonish-Rouach et al 1991).

These results indicate that the wild-type p53-mediated apoptosis in M1 myeloid leukaemic cells reflects a physiological process. Wild-type p53 may also be involved in mediating apoptosis in normal myeloid cells deprived of the appropriate growth factors. There are, however, presumably also alternative pathways leading to apoptotic cell death in normal cells of the myeloid lineage as well as in cells from other lineages.

Our results on the induction of differentiation in myeloid leukaemic cells indicate that differentiation can be associated with induction of a network of cytokines and cytokine receptors and with initiation of a programme for cell death, which are expressed by normal cells of the myelomonocytic lineages. The network, which has arisen through evolution (Sachs 1991), includes both positive (IL-6, IL-1, GM-CSF, M-CSF) and negative (TGF-β1, TNF) regulators of viability and growth. TNF and TGF-β1 may function in the cells as switch-off signals, turning off proliferation once the cells have progressed into the differentiation pathway and thus coupling the expression of differentiation to growth arrest. The growth of M1 cells can be arrested by TGF-β1 but not by TNF (Lotem et al 1991b). GM-CSF that is produced during differentiation cannot rescue M1 cells from apoptosis (Lotem & Sachs 1989). The results indicate that there can be specific defects not only in the expression of components of the cytokine network in certain leukaemic cells but also in the responsiveness of leukaemic cells to different components of this network (Lotem et al 1991b).

*Acknowledgements*

This work was supported by the National Foundation for Cancer Research, Bethesda, MD, the Jerome A. & Estelle R. Newman Assistance Fund and the Ebner Foundation for Leukemia Research. We thank R. Kama and N. Dorevitch for excellent technical assistance.

## References

Fibach E, Sachs L 1976 Control of normal differentiation of myeloid leukemic cells. XI. Induction of a specific requirement for cell viability and growth during differentiation of myeloid leukemic cells. J Cell Physiol 89:259–266

Fibach E, Hayashi M, Sachs L 1973 Control of normal differentiation of myeloid leukemic cells to macrophages and granulocytes. Proc Natl Acad Sci USA 70:343–346

Hilton DJ, Nicola NA, Metcalf D 1988 Purification of a murine leukemia inhibitory factor from Krebs ascites cells. Anal Biochem 173:359–367

Hirano T, Akira S, Taga T, Kishimoto T 1990 Biological and clinical aspects of interleukin 6. Immunol Today 11:443–449

Hirayoshi K, Tsuru A, Yamashita M et al 1991 Both D factor/LIF and IL-6 inhibit the differentiation of mouse teratocarcinoma F9 cells. FEBS (Fed Eur Biochem Soc) Lett 282:401–404

Hozumi M, Tomida M, Yamamoto-Yamaguchi Y et al 1990 Protein factors that regulate the growth and differentiation of mouse myeloid leukaemia cells. In: Molecular control of haemopoiesis. Wiley, Chichester (Ciba Found Symp 148) p 25–42

Ichikawa Y 1969 Differentiation of a cell line of myeloid leukemia. J Cell Physiol 74:223–234

Levine AJ, Momand J, Finlay CA 1991 The p53 tumour suppressor gene. Nature (Lond) 351:453–456

Lindemann A, Riedel D, Oster W, Mertelsmann R, Herrmann F 1988 Recombinant human granulocyte-macrophage colony-stimulating factor induces secretion of autoinhibitory cytokines by U-937 cells. J Immunol 18:369–374

Lotem J, Sachs L 1982 Mechanisms that uncouple growth and differentiation in myeloid leukemic cells. Restoration of requirement for normal growth-inducing protein without restoring induction of differentiation-inducing protein. Proc Natl Acad Sci USA 79:4347–4351

Lotem J, Sachs L 1986 Regulation of cell surface receptors for different hematopoietic growth factors on myeloid leukemic cells. EMBO (Eur Mol Biol Organ) J 5:2163–2170

Lotem J, Sachs L 1989 Induction of dependence on hematopoietic proteins for viability and receptor upregulation in differentiating myeloid leukemic cells. Blood 74:579–585

Lotem J, Sachs L 1990 Selective regulation of the activity of different hematopoietic regulatory proteins by transforming growth factor β1 in normal and leukemic myeloid cells. Blood 76:1315–1322

Lotem J, Shabo Y, Sachs L 1989a Regulation of megakaryocyte development by interleukin 6. Blood 74:1545–1551

Lotem J, Shabo Y, Sachs L 1989b Clonal variation in susceptibility to differentiation by different protein inducers in the myeloid leukemia cell line M1. Leukemia 3:804–807

Lotem J, Cragoe EJ Jr, Sachs L 1991a Rescue from programmed cell death in leukemic and normal myeloid cells. Blood 78:953–960

Lotem J, Shabo Y, Sachs L 1991b The network of hematopoietic regulatory proteins in myeloid cell differentiation. Cell Growth & Differ 2:421–427

Lu L, Walker D, Graham CD, Waheed A, Shadduck RK, Broxmeyer HE 1988 Enhancement of release from MHC class II antigen-positive monocytes of hematopoietic colony stimulating factors CSF-1 and G-CSF by recombinant human tumour necrosis factor-alpha: synergism with recombinant human interferon-gamma. Blood 72:34–41

Metcalf D 1985 The granulocyte-macrophage colony-stimulating factors. Science (Wash DC) 229:16–22

Metcalf D 1991 The leukemia inhibitory factor (LIF). Int J Cell Cloning 9:95–108

Metcalf D, Gearing DP 1989 Fatal syndrome in mice engrafted with cells producing high levels of the leukemia inhibitory factor. Proc Natl Acad Sci USA 86:5948–5952

Metcalf D, Nicola NA 1985 Synthesis by mouse peritoneal cells of G-CSF, the differentiation inducer for myeloid leukemia cells: stimulation by endotoxin, M-CSF and multi-CSF. Leuk Res 9:35–50

Michalovitz D, Halevy O, Oren M 1990 Conditional inhibition of transformation and cell proliferation by a temperature-sensitive mutant of p53. Cell 62:671–680

Michishita M, Yoshida Y, Uchino H, Nagata K 1990 Induction of tumour necrosis-$\alpha$ and its receptors during differentiation in myeloid leukemic cells along the monocytic pathway. J Biol Chem 265:8751–8759

Miyaura C, Jin CH, Yamaguchi Y et al 1989 Production of interleukin 6 and its relation to the macrophage differentiation of mouse myeloid leukemia cells (M1) treated with differentiation-inducing factor and $1\alpha$,25-dihydroxyvitamin $D_3$. Biochem Biophys Res Commun 158:660–666

Motoyoshi K, Yoshida K, Hatake K et al 1989 Recombinant and native human urinary colony-stimulating factor directly augments granulocytic and granulocyte macrophage colony-stimulating factor production of human peripheral blood monocytes. Exp Hematol 17:68–71

Oster W, Lindemann A, Mertelsmann R, Herrmann F 1989 Granulocyte-macrophage colony stimulating factor (CSF) and multilineage CSF recruit human monocytes to express granulocyte CSF. Blood 73:64–67

Sachs L 1982 Normal developmental programmes in myeloid leukemia: regulatory proteins in the control of growth and differentiation. Cancer Surv 1:321–342

Sachs L 1987a The molecular regulators of normal and leukaemic blood cells. The Wellcome Foundation Lecture 1986. Proc R Soc Lond B Biol Sci 231:289–312

Sachs L 1987b The molecular control of blood cell development. Science (Wash DC) 238:1374–1379

Sachs L 1990 The control of growth and differentiation in normal and leukemic blood cells. Cancer (Phila) 65:2196–2206

Sachs L 1991 Keynote address for symposium on a visionary assessment of the scientific, clinical and economic implications of hematopoietic growth factors. Cancer (Phila) (suppl) 67:2681–2683

Shabo Y, Lotem J, Rubinstein M et al 1988 The myeloid blood cell differentiation-inducing protein MGI-2A is interleukin 6. Blood 72:2070–2073

Shabo Y, Lotem J, Sachs L 1989a Autoregulation of interleukin 6 and granulocyte-macrophage colony-stimulating factor in the differentiation of myeloid leukemic cells. Mol Cell Biol 9:4109–4112

Shabo Y, Lotem J, Sachs L 1989b Regulation of the genes for interleukin-6 and granulocyte-macrophage colony-stimulating factor by different inducers of differentiation in myeloid leukemic cells. Leukemia (Baltimore) 3:859–865

Shabo Y, Lotem J, Sachs L 1990 Regulation of genes for transcription factors by normal hematopoietic regulatory proteins in the differentiation of myeloid leukemic cells. Leukemia (Baltimore) 4:797–801

Suematsu S, Matsuda T, Aozasa k et al 1989 IgG1 plasmacytosis in interleukin 6 transgenic mice. Proc Natl Acad Sci USA 86:7547–7551

Williams GT, Smith CA, Spooncer E, Dexter TM, Taylor DR 1990 Haemopoietic colony stimulating factors promote cell survival by suppressing apoptosis. Nature (Lond) 343:76–79

Wyllie AH 1980 Glucocorticoid-induced thymocyte apoptosis is associated with endogenous endonuclease activation. Nature (Lond) 284:555–556

Yonish-Rouach E, Resnitzky D, Lotem J, Sachs L, Kimchi A, Oren M 1991 Wild type p53 induces apoptosis of myeloid leukaemic cells that is inhibited by interleukin 6. Nature (Lond) 352:345–347

## DISCUSSION

*Sehgal:* Hoffman-Liebermann & Liebermann (1991a) described M1 cell lines in which *myc* and *myb* were constitutively expressed and observed that these lines were blocked at specific points in the differentiation pathway. Have you had any experience with those kinds of systems?

*Lotem:* We haven't worked with that kind of system but Dr Kimchi in our department has (Resnitzky & Kimchi 1991). I get the impression that there is not really a block. There is a constitutive expression of c-*myc* and continued cell multiplication in the presence of IL-6, but the cells do express differentiation markers. The responsiveness to IL-6 or LIF is not blocked, but the cells don't seem to reach the phase of $G_0/G_1$ growth arrest and terminal cell differentiation.

*Sehgal:* There was a report from that group (Resnitzky et al 1986) that the inhibition of proliferation of M1 cells by interferon was dependent on down-regulation of *myc*. One of the most impressive results from the studies by Hoffman-Liebermann & Liebermann (1991b) was that in M1 clones which have high levels of *myc* expression, interferons still inhibit proliferation. Thus, the inference is that the original conclusion, drawn 4–5 years ago, that *myc* is an obligatory intermediate in the inhibition of proliferation by interferon, is not valid. What is your opinion on this controversial area?

*Lotem:* My feeling is that interferon is not the only answer. Though it's true that antibodies against interferon will keep the cells in the multiplication phase for a longer period, other factors, such as TNF, TGF-β and prostaglandin $E_2$, and possibly other inhibitory molecules, may act together with β-interferon to cause the inhibition of growth which is associated with differentiation. The presence or absence of β-interferon may not be the only important thing.

*Baumann:* Do the M1 cells which are sensitive to both IL-6 and LIF (clone T22) show identical responses to the two cytokines?

*Lotem:* Yes, as far as I can tell. The only difference is that they seem to respond to lower concentrations of LIF than IL-6. Once differentiation has been induced it's very difficult to know whether it was induced by IL-6 or LIF; there are no differences in any molecular markers.

*Heinrich:* What is known about the DNAses which act during apoptosis? Are they of lysosomal origin, or are they activated nuclear enzymes?

*Lotem:* Not much is known. The activity is not cytoplasmic; it is something in the nucleus which is activated. The physiology of the activation of apoptosis is not well understood. You don't see apoptosis until about 12–20 hours after removal of the growth factor, because there's an entire cascade of events leading to activation of an endonuclease. In lymphocytes, addition of a nuclease inhibitor inhibits apoptosis, but that's the end stage, and there may be many different processes leading up to that.

*Dexter:* Your results raise a question that's bothered me for some time—do normal progenitor cells, in response to one cytokine, produce a range of other cytokines such as IL-1, GM-CSF or M-CSF, so that what we are really seeing when we add a single cytokine is a coordinated response to a network of cytokines that are produced by normal progenitor cells? Do you have any results that address this question?

*Lotem:* The only results we have on normal progenitor cells relate to the production of IL-6. Incubation of myeloid precursors with IL-3 or with the other three CSFs induced production of IL-6, which we previously called MGI-2 (Lotem & Sachs 1983a, Lotem et al 1988). It's really the early progenitors, not the accessory cells in the bone marrow, that produce IL-6, because as you purify the bone marrow myeloid precursors and the number of accessory cells decreases, more IL-6 is produced.

*Dexter:* How pure was that cell population?

*Lotem:* In our final population about 90% of the cells looked like early myeloid cells.

*Dexter:* Their appearance is not the important factor— how efficiently do they clone?

*Lotem:* The cloning efficiency was about 20%, compared with a cloning efficiency of about 0.2–0.4% in the original population (Lotem & Sachs 1983a). Recently, three groups have confirmed our results, observing production of IL-6 in in response to IL-3 treatment, with a more enriched cell population (Schneider et al 1991).

*Akira:* Are there any effects of anti-IL-6 oligonucleotide on M1 cell differentiation?

*Lotem:* We haven't done experiments in the M1 system to see whether inhibition of endogenous IL-6 inhibits differentiation. I doubt that it would have such an effect, because the amount of IL-6 produced by M1 cells during differentiation is quite small. I was not trying to suggest that it is the endogenous IL-6 that is induced which causes the cells to differentiate, although it may contribute. The IL-6 produced by the cells is really a differentiation marker, but the amount of IL-6 added externally is sufficient to induce differentiation.

*Lee:* The effects of IL-6 and LIF on M1 cells suggest that they are good inducers of differentiation of myeloid leukaemic cells. What has been done to explore their therapeutic potential?

*Lotem:* Even in M1 cells there is a clonal distribution of responses to these cytokines. There are human and murine leukaemic cells that differentiate with IL-6 but I am not aware of any human myeloid leukaemic cells that differentiate in response to LIF. It is unlikely that these two molecules could be used as general inducers of differentiation in human myeloid leukaemia. In model systems in mice with IL-6-sensitive leukaemic cells we have shown that injection of IL-6 inhibits the development of leukaemia (Lotem & Sachs 1981). With the diffusion chamber model, by which you can measure *in vivo* differentiation, you can see that injection of IL-6 increases the differentiation of M1 cells (Lotem & Sachs 1983b).

*Kishimoto:* Does either α-interferon or β-interferon induce differentiation of M1 cells? A recent study has shown that IL-6 and LIF induce IRF, an interferon-inducing transcription factor.

*Lotem:* α-Interferon or β-interferon will only stop the cells from multiplying. Neither induce differentiation.

*Stewart:* Does the M1 clone which doesn't respond to LIF actually express LIF receptors?

*Lotem:* We have never looked and it is possible that they don't. I mentioned this clone, clone 11, because differentiation-inducing activity for M1 cells is often reported and M1 cells have been regarded as an entity. The differentiation-inducing activity of LIF is supposed to have certain properties, such as a molecular mass of 55 kDa, and presence in certain conditioned media; in our case the inducer of differentiation had a molecular mass of 21–28 kDa, and the presence of the activity in conditioned media was different from that reported for LIF. Don Metcalf and his colleagues sent us some purified LIF and we found it was completely inactive on our M1 clone 11 cells. All these discrepancies can be explained by clonal differences.

*Sehgal:* In this particular M1 clone, clone 11, IL-6 has two effects—it causes a decrease in cell number, or decreased proliferation, and an increase in differentiation markers. Can you dissociate these two effects? Are the effects on proliferation and on differentiation separable?

*Lotem:* The answer to that question depends on how you define differentiation. We can get expression of early differentiation markers with little inhibition of cell multiplication. However, to get mature macrophages there has to be substantial inhibition of cell growth. After addition of IL-6, there is usually one round of cell multiplication and then the cell number remains more or less the same and the cells change their morphology and become true macrophages. With dexamethasone, the cells seem to continue to proliferate at a slower rate, but eventually the percentage of mature cells obtained is rather small.

*Metcalf:* There are leukaemias, such as HL-60, where you can suppress self-generation but the cells retain their blast cell morphology and perhaps express only a few membrane differentiation markers. There are other cell lines with which dramatic morphological maturation occurs. The common action isn't the morphological maturation, but the suppression of self-generation. Suppression of self-generation is not necessarily detectable as changes in the length of the cell cycle. Sometimes, cell proliferation is actually stimulated as the crucial process of suppression of self-generation is occurring.

*Gauldie:* Dr Lotem, you showed that your cells will stop proliferating and will differentiate when exposed to LIF (clone T22) or IL-6 (clones 11 and T22). The level of IL-6 in the circulation will rise and fall. Perhaps naturally existing myeloid leukaemias are those cells which have lost the receptor and will not respond, because there is likely to be enough of the cytokine around to inhibit their proliferation and cause differentiation.

*Lotem:* That could well be true. Dr N. Haran-Ghera in our institute induced myeloid leukaemias by irradiation and injection of dexamethasone and followed the time taken to produce full leukaemia. If the mice were injected with IL-6 early on, the incidence of leukaemias was lower and the leukaemia appeared much later than without IL-6 (personal communication). I don't know whether that is really due to the differentiation of the cells. In most cases you end up with a leukaemic cell population which does not differentiate with IL-6, so it's possible that the cells which respond to IL-6 are first eliminated and another leukaemic clone develops.

*Metcalf:* Could you speculate what the difference might be between a normal embryonic stem cell, in which LIF has the opposite effect, preventing induction of differentiation, and a myeloid leukaemic cell? There was an experiment done some years ago in which leukaemia cells, presumably M1, were injected into embryos. After birth there was persistence of donor-type cells as morphologically mature cells (Cootwine et al 1982). If LIF is an active agent at the blastocyst stage, which is possible, is this an example of two cell types in the same organism responding in different ways?

*Lotem:* We have not looked at embryonic cells. Those experiments involved injection of leukaemic cells into the placenta itself. For a few weeks after birth mature granulocytes or mature macrophages (depending on the clone of leukaemic cells injected) of the offspring mice had the genetic markers of the injected myeloid leukaemic cells (Webb et al 1984); it appears that the injected myeloid leukaemic cells, instead of developing into a leukaemia, are affected somehow by the environment in the embryo and are forced to become normal stem cells, and to participate in normal differentiation. I don't see a contradiction between the fact that LIF can act on M1 cells as a growth-inhibiting molecule and on embryonic stem cells as an inhibitor of differentiation. The effect of TPA on different cell systems also differs; it suppresses growth and induces differentiation of HL-60 cells, whereas it has the opposite effect on Friend erythroleukaemic cells.

*Metcalf:* I suppose that's one difference between the two molecules. No one has suggested that IL-6 has an action on normal embryonic stem cells like that of LIF.

*Heath:* IL-6 has no biological activity towards ES cells, and in fact the receptors are not expressed (A. Mereau & J. K. Heath, unpublished results). Professor Kishimoto has found expression of gp130 in ES cells but we haven't.

*Lotem:* The expression of receptors or lack of expression cannot be the only explanation. GM-CSF cannot rescue M1 clone 12 cells from apoptosis and the cells don't have GM-CSF receptors. However, in M1 clone 11 cells, which express GM-CSF receptors during differentiation, GM-CSF again does not prevent apoptosis. Having the receptor is not enough—the cells must lack something in the steps after receptor binding.

*Heath:* You don't have to look as far as the pluripotential stem cells to find a difference between effects of IL-6 and LIF. The human interleukin for DA (HILDA) assay for LIF represents factor dependency for a myeloid leukaemia cell line. So, even within myeloid leukaemias there can be both inhibition of differentiation and induction of differentiation. I don't find this surprising; as Dr Lotem said, the difference is not simply due to the receptor, but to a particular combination of secondary and tertiary signalling systems within the cell. The ligand, in a sense, is simply the starting point and doesn't have any intrinsic encoded function.

*Lotem:* Even within one M1 subclone you can get different effects. You can start with a clone which is not growth factor-dependent; when you add IL-6 the cells stop growing and start to differentiate but when you remove IL-6 they die. If you add IL-6 again at this stage it functions not only as a differentiation factor but also as a viability factor.

*Metcalf:* I am not surprised either; I just wanted someone to clarify the nature of the problem and explain that the response of a cell depends on the programming of that cell and the pathways for signalling that are open to it.

*Nicola:* There is still a problem. There are cell types which will respond differently to two different signals. How is the integrity of the signal maintained from the receptor to the nucleus if there is essentially just a single type of reaction to different signals? The 32D clone 3 cell is a myeloid cell line which depends on IL-3 for its growth. If you add G-CSF to those cells, differentiation to granulocytes and suppression of growth result (Valtieri et al 1987). Many reports have suggested that G-CSF and IL-6 have similar mechanisms of action, involving common tyrosine phosphorylation signals, for example. There must be mechanisms other than simple cell programming—there has to be coupling between the receptors and the signals which is specific for each type of ligand and its receptor.

*Lee:* That doesn't mean that the signal diverges downstream of tyrosine phosphorylation. There could be many steps at which the two signals could diverge.

*Nicola:* Absolutely; all I'm saying is that firing of the receptor is not *the* signal which the cell responds to as intrinsically programmed.

*Heath:* It isn't essential to think in qualitative terms. There are now a number of reasonably well documented situations where one growth factor elicits different biological effects according to its concentration— for example, FGF in the case of lens fibre differentiation, or activin in mesoderm induction in frogs. Quantitative aspects may well be important also. The efficiency with which a particular receptor interacts with the downstream tyrosine kinase, for example, could in principle play an important role in the programming.

*Nicola:* There can be a quantitative difference in the response to a factor, but in 32D clone 3 cells there is a qualitative difference—there is no concentration of IL-3 that will ever induce granulocyte differentiation in this cell line.

*Heath:* My point was simply that qualitative differences can have quantitative undertones.

*Metcalf:* I think many people might be wondering whether differences in receptor subunits can result in different types of signalling or responses in cells of different lineages. Perhaps the arrangement of subunits in a hepatocyte, for example, is different from that in a leukaemic cell. There is a possibility that mutations in one or other of the receptor subunits would result in a selective loss of functional activity.

*Heinrich:* The 80 kDa human IL-6 receptor from natural killer cells was first cloned by Professor Kishimoto's group. The clone we got from a human liver cDNA library had an identical sequence (Schooltink et al 1991). Therefore, the difference in biological responses to IL-6 of liver cells and natural killer cells cannot be due to a difference between the IL-6-binding proteins.

*Metcalf:* Unless there is a third subunit which has yet to be discovered.

*Williams:* I don't think it is clear at this point whether with all the biological effects that have been described for the two molecules the same range of amino acids sequences are involved. I would like to hear those working with mutant forms of the ligands describe exactly which assays activity has been tested in, to give us an idea about whether the large range of biological effects are attributable to a small number of amino acids.

*Aarden:* We have made a series of deletion mutations in IL-6, starting from the N-terminus. Deletion of the first 28 amino acids has no effect on activity in biological assays (Brakenhoff et al 1989). With deletion of the next four amino acids you see a complete disappearance of biological activity. This is true with the B9 hybridoma growth assay, the T1165 plasmacytoma growth assay, the CESS induction of IgG1 assay and the HepG2 induction of fibrinogen assay.

*Williams:* Those are similar sorts of assays though. What are the effects on biologically more diverse processes, such as neuronal growth?

*Aarden:* The induction of acute-phase proteins, B cell growth and B cell differentiation are quite diverse effects.

*Nicola:* There have been sporadic reports of mutant growth factors in which different activities are dissociated, but very few of them have stood the test of time. You have to be careful about the nature of the assay. If you take the same mutein and test it in four different assays in which there are different accessory cells and short-term or long-term responses, you have to worry about whether the mutein is less stable than the wild-type and is degraded in a long-term assay or in an assay in which macrophages are present. Care is needed in the interpretation of results from one mutein in different types of assays.

*Lotem:* Dr Sehgal's results (p 17) mean that such caution is also necessary with assay of non-mutated IL-6. The physical form of the IL-6, whether it's monomeric or in a large complex, can affect the range of cells which react to it. There are cells which recognize IL-6 as part of a high molecular mass complex but other cells do not recognize the complex as IL-6 and respond only to the low molecular mass form.

*Nicola:* You have to consider what else might be in that complex. We know much about synergistic effects of cytokines. If you see an effect only with a high molecular weight form you have to be sure that you are not looking at a synergy with another factor.

*Lotem:* You can block the response with the appropriate monoclonal antibody to IL-6.

*Nicola:* That means that IL-6 is necessary, but it may not be sufficient.

*Kishimoto:* I don't think that the functional pleiotropy of IL-6 can be explained by heterogeneity of the ligand molecule, because we have detected only one type of IL-6 receptor and one gp130. Functional pleiotropy might be generated by downstream molecules. Heterogeneity might affect the sensitivity, but I doubt it can explain pleiotropy.

*Heinrich:* Natural human IL-6 is $N$- and and $O$-glycosylated. This probably has a bearing on the circulatory half-life and might also be involved in a loose type of receptor recognition. We have compared the biological activity of glycosylated and unglycosylated IL-6 in the B9 cell proliferation assay as well as in the fibrinogen induction assay in HepG2 cells. The native glycosylated protein is 3–4-fold more active.

*Lotem:* That might be because it's more stable.

*Sehgal:* We have compared specific activities in the B9 assay of the human natural fibroblast form of IL-6, which is $N$-glycosylated, $O$-glycosylated and heavily phosphorylated on Ser-81, and the *E. coli* reference standard that is distributed by the NIH (preparation 88/514). I agree with Professor Heinrich. The glycosylated human version is 4–5-fold more active on a mass basis than the *E. coli* IL-6.

*Heinrich:* This is not a simple experiment because you have to compare the same amounts of glycosylated and non-glycosylated IL-6 in the bioassays. We started with [$^{35}$S]methionine-labelled $N$- and $O$-glycosylated human IL-6 and deglycosylated this material with neuraminidase, $O$-glycanase and peptide-$N^4$-($N$-acetyl-β-glucosaminyl)asparagine amidase (PNGase F). We tested equal amounts, in terms of radioactivity, of the glycosylated and deglycosylated IL-6 in the B9 cell proliferation assay and in the γ-fibrinogen induction assay in HepG2 cells.

*Metcalf:* What is the basis of this difference? Only a few things can be altered—the half-life of the molecule, the on–off kinetics of binding, or the rate of internalization of bound receptors, or there could be some qualitative differences in the ligand. Which of these do you think applies?

*Heinrich:* I cannot tell you whether the on–off kinetics or the rates of internalization are different for glycosylated and non-glycosylated human IL-6. We have studied the half-life in plasma of human IL-6 after intravenous injection into a rat; the half-life of the deglycosylated form was about 3 min, whereas that of the glycosylated form was about 5 min (Castell et al 1990). We do not know whether there is a difference in half-life between glycosylated and deglycosylated human IL-6 in cell culture. The specific biological activities differ; 1 µg glycosylated IL-6 has a 3–4-fold higher biological activity than 1 µg unglycosylated IL-6.

*Gauldie:* Rat IL-6 is very lightly glycosylated, if at all. We see little difference between the recombinant material produced in COS cells and IL-6 produced in *E. coli*. There appear to be no qualitative differences in their effects—the same sets of genes appear to be regulated—but there may be quantitative differences resulting from effects on receptor binding.

*Heath:* The glycosylated material might be more stable structurally, because the carbohydrate forms a sort of cloud around the protein. Some structural verification of the identity of glycosylated and non-glycosylated material is needed.

*Sehgal:* The purified glycosylated IL-6 exists as an 85–100 kDa trimeric complex, whereas the *E. coli* material is monomeric, or is artifactually dimeric. Your point is important. The different preparations do appear to have intrinsically different secondary and tertiary structures.

*Gearing:* Professor Gauldie has pointed out that IL-6 from different sources shows no qualitative differences. I think it's now clear that with LIF from different sources there are qualitative differences. In a retroviral infection assay using CFU-S, Fred Fletcher has demonstrated that LIF seems to promote retroviral infection, but only when produced in COS cells (Fletcher et al 1990, 1991). COS cell-conditioned media have been suggested to contain other factors; it is worth noting that some papers on LIF involved COS cell-conditioned media so synergy with other factors may have influenced the results.

*Aarden:* Professor Kishimoto has just said there is only one type of IL-6 receptor, yet hybridoma cells are 200 times more sensitive to one type of IL-6 than other cells. Does anyone have an explanation for that finding?

*Metcalf:* The first question to ask is how many receptors there are per cell.

*Aarden:* B9 cells have a few hundred IL-6 receptors. If you use 1 pg/ml of IL-6, at the point of half maximal proliferation there is less than one receptor occupied.

*Nicola:* You can make arguments about the efficiency of the coupling between the receptor and the ligand, and about how much substrate there is in different cells. However, you cannot easily relate receptor affinity to receptor occupancy under biological conditions. Most cells internalize and degrade ligand at 37 °C, but affinity is measured at 4 °C, on ice. Consequently, under steady-state conditions, there is higher receptor occupancy than you would predict. The situation is analogous to an enzyme–substrate reaction where the product of the reaction is taken away. Although you may think that there is only one receptor occupied at a certain ligand concentration, in fact under steady-state conditions there could be an average of 10 occupied per unit time. The other major contributor to differences in dose–response curves, even with identical receptors which are processed differently by different cells, is the extent to which the rate of utilization of the ligand is the limiting factor. In some cells there can be a rapid rate of degradation, so you have to provide a reservoir of IL-6 for a seven-day assay. I'm not saying that these are the answers. I am just saying it's not a simple argument to extrapolate from affinity to biological dose–response curves.

*Aarden:* Your last point, at least, doesn't apply to the B9 system because plasmacytomas as well as hybridomas have been shown not to consume IL-6.

*Nicola:* That is unusual. Most ligands are internalized.

*Gauldie:* Dr Sehgal's results with glucocorticoid suppression of IL-6 gene expression (Ray et al 1990) raise an important issue. We've studied some 5′ non-translated region constructs of the rat IL-6 gene with Wolfgang Northemann and Georg Fey, looking particularly at the inhibition of IL-6 gene expression by corticosteroid. The region that Dr Sehgal showed to be important for suppression in the human gene is the same as the region we have found in the rat. The important finding was that the upstream putative glucocorticoid response elements (GREs) appear to do nothing whatsoever; although there are GRE-like sequences they are non-functional. The only sequence that appears to be important is the Inr region adjacent to the transription start site that Dr Sehgal has described. Similarly, the AP1-like sites in the IL-6 gene seem not to be functional—if we remove them there is no change in the inducibility or the intrinsic expression of the IL-6 gene in L929 cells.

*Fey:* It depends on how you define a GRE site, whether you mean the hexanucleotide core or the real consensus sequence, as defined by Nordeen et al (1990), which is 14 bp long. If you search for the hexanucleotide core only, instead of the 14 bp binding sequence, you will find it in many places. You need to show functional binding to be certain that you have a GRE.

*Aarden:* I would suggest that the outcome of these experiments must be dependent on the cell type that you use. For example, endothelial cells are stimulated to produce IL-6 by 10 ng/ml TPA whereas production of IL-6 by monocytes is completely inhibited. However, if you use 0.1 ng/ml TPA monocytes are stimulated to produce IL-6.

*Sehgal:* I would like to reinforce this point about differences among cell types. Investigators tend to use their favourite cell lines for transfection assays and get different kinds of answers from others. Some report that the NF$\kappa$B site in the IL-6 promoter is the most important in induction, and others, like ourselves and Professor Kishimoto, think a different region, the MRE region, is important. However, all my results are restricted to HeLa cells. Someone else could unravel the same system in a different cell type and come up with quite different answers.

*Gauldie:* We did our transfection studies with the rat IL-6 gene constructs in L929 cells, so the glucocorticoid inhibitory element is similar and functional in at least two cell lines.

*Fey:* It's clear that there are many parallel features in the regulation of the LIF and IL-6 genes. As Dr Gough showed, the upstream sequences are highly conserved. There is strong conservation between the LIF and IL-6 genes in the arrangement of the upstream sequences, yet the sequences reveal nothing about potentially important regulatory sites in the LIF gene. It is not enough to do computer comparisons of sequences—you have to do functional comparisons. I predict that comparable functional studies on the LIF gene will come up with a picture similar to that for the IL-6 gene.

*Sehgal:* Professor Kishimoto's Fig. 1, as most people's introductory slides do these days, included the IL-6-induced neuronal differentiation of PC12 cells. I am not aware that that work (Satoh et al 1988) has been repeated. We have had mixed results. In a certain PC12 clone we saw the differentiation effect, with neurite outgrowth, but many other PC12 cell lines and subclones which differentiate with nerve growth factor (NGF) did not respond to IL-6. I know of other investigators who find it difficult to observe IL-6-induced neuronal differentiation. Professor Kishimoto, has your group done any more work on the effects of IL-6 on the nervous system?

*Kishimoto:* I think that the second signal transducer for the NGF receptor is the *trk* oncogene product, which has tyrosine kinase activity. We think that gp130 interacts with an intracellular tyrosine kinase, and if a PC12 cell line lacks this tyrosine kinase, Trk can transduce signals whereas no response to IL-6 is possible.

*Sehgal:* Is that your experience or theory?

*Kishimoto:* It's a hypothesis.

*Metcalf:* I suspect that, as with M1 cells, there are differences between sublines of this particular cell line.

*Sehgal:* Does human IL-6 induce neuronal differentiation in any human cells?

*Kishimoto:* We have no model systems in which to test that.

*Gough:* Is there any evidence that IL-6 has actions in normal neuronal differentiation?

*Patterson:* The most relevant work is ours on primary cultures of dissociated rat sympathetic neurons and adrenal chromaffin cells (M.-J. Fang & P. H. Patterson, unpublished results). We find that IL-6 does not appear to exert NGF-like effects, nor does it alter neuronal gene expression in the way that LIF does. There is evidence, however, suggesting that IL-6 has NGF-like effects on cultured

98                                                                    Discussion

rat septal neurons (Hama et al 1989). IL-6 has been shown to bind to hypothalamic membranes (Cornfield & Sills 1991) and to stimulate release of corticotropin-releasing factor from hypothalamic explants *in vitro* (Navarra et al 1991). Injection of recombinant IL-6 into the preoptic-anterior hypothalamus can induce fever *in vivo* (Blatteis et al 1990). Furthermore, IL-6 stimulates the release of anterior pituitary hormones *in vitro*, and cultured pituitary cells release this cytokine (Spangelo et al 1990).

**References**

Blatteis CM, Quan N, Xin L, Ungar AL 1990 Neuromodulation of acute-phase responses to interleukin-6 in guinea pigs. Brain Res Bull 25:895–901
Brakenhoff JPJ, Hart M, Aarden LA 1989 Analysis of human IL-6 mutants expressed in *Escherichia coli*. Biologic activities are not affected by deletion of amino acids 1–28. J Immunol 143:1175–1182
Castell J, Klapproth J, Gross V et al 1990 Fate of IL-6 in the rat: involvement of skin in its catabolism. Eur J Biochem 189:113–118
Cornfield LJ, Sills MA 1991 High affinity interleukin-6 binding sites in bovine hypothalamus. Eur J Pharmacol 202:113–115
Fletcher FA, Williams DE, Maliszewski C et al 1990 Murine leukemia inhibitory factor enhances retroviral-vector infection of hematopoietic progenitors. Blood 76:1098–1103
Fletcher FA, Moore KA, Ashkenazi M et al 1991 Leukemia inhibitory factor improves the survival of retroviral vector-infected hematopoietic stem cells in vitro, allowing efficient long-term expression of vector-encoded human adenosine deaminase in vivo. J Exp Med 174:837–845
Gootwine E, Webb CG, Sachs L 1982 Participation of myeloid leukaemic cells injected into embryos in haematopoietic differentiation in adult mice. Nature (Lond) 299:63–65
Hama T, Miyamoto M, Tsukui H, Nishio C, Hatanaka H 1989 Interleukin-6 as a neurotrophic factor for promoting the survival of cultured basal forebrain cholinergic neurons from postnatal rats. Neurosci Lett 104:340–344
Hoffman-Liebermann B, Liebermann D 1991a Interleukin-6-induced and leukemia inhibitory factor-induced terminal differentiation of myeloid leukemic cells is blocked at an intermediate stage by constitutive c-*myc*. Mol Cell Biol 11:2375–2381
Hoffman-Liebermann B, Liebermann DA 1991b Suppression of c-myc and c-myb is tightly linked to terminal differentiation induced by IL-6 or LIF and not growth inhibition in myeloid cells. Oncogene 6:903–909
Lotem J, Sachs L 1981 *In vivo* inhibition of the development of myeloid leukemia by injection of macrophage and granulocyte inducing protein. Int J Cancer 28:375–386
Lotem J, Sachs L 1983a Coupling of growth and differentiation in normal myeloid precursors and the breakdown of this coupling in leukemia. Int J Cancer 32:127–134
Lotem J, Sachs L 1983b Control of *in vivo* differentiation of myeloid leukemic cells. III. Regulation by T lymphocytes and inflammation. Int J Cancer 32:781–791
Lotem J, Shabo Y, Sachs L 1988 Role of different normal hematopoietic regulatory proteins in the differentiation of myeloid leukemic cells. Int J Cancer 41:101–107
Navarra P, Tsagarakis S, Faria MS, Rees LH, Besser GM, Grossman AB 1991 Interleukins-1 and interleukins-6 stimulate the release of corticotropin-releasing hormone-41 from rat hypothalamus in vitro via the eicosanoid cyclooxygenase pathway. Endocrinology 128:37–44

Nordeen SK, Suh BJ, Kühnel B, Hutchinson CA 1990 Structural determinants of a glucocorticoid recognition element. Mol Endocrinol 4:1866–1873

Ray A, LaForge KS, Sehgal PB 1990 On the mechanisms for efficient repression of the IL-6 promoter by glucocorticoids: enhancement of TATA-box and RNA start site (Inr motif) occlusion. Mol Cell Biol 10:5736–5746

Resnitzky D, Kimchi A 1991 Deregulated c-*myc* expression abrogates the interferon and interleukin-6-mediated $G_0/G_1$ cell cycle arrest but not other inhibitory responses in M1 myeloblastic cells. Cell Growth & Differ 2:33–41

Resnitzky D, Yarden A, Zipori D, Kimchi A 1986 Autocrine β-related interferon controls c-*myc* suppression and growth arrest during hematopoietic cell differentiation. Cell 46:31–40

Satoh T, Nakamura S, Taga T et al 1988 Induction of neuronal differentiation in PC12 cells by B-cell stimulatory factor 2/interleukin 6. Mol Cell Biol 8:3546–3549

Schneider E, Ploemacher RE, Navarro S, Van Beurden C, Dy M 1991 Characterization of murine hematopoietic progenitor subsets involved in interleukin-3 induced interleukin-6 production. Blood 78:329–338

Schooltink H, Stoyan T, Lenz D et al 1991 Structural and functional studies on the human hepatic interleukin-6 receptor. Biochem J 277:659–665

Spangelo BL, MacLeod RM, Isakson PC 1990 In: Oppenheim JJ, Powanda MC, Kluger MJ, Dinarello CA (eds) Molecular and cellular biology of cytokines. (Prog Leukocyte Biol Ser) Wiley-Liss, New York, p 433–438

Valtieri M, Tweardy DJ, Carraciolo D et al 1987 Cytokine-dependent granulocyte differentiation. Regulation of proliferative and differentiative responses in a murine progenitor cell line. J Immunol 138:3829–3835

Webb CG, Gootwine E, Sachs L 1984 Developmental potential of myeloid leukemia cells injected into mid-gestation embryos. Dev Biol 101:221–224

# The action of interleukin 6 and leukaemia inhibitory factor on liver cells

H. Baumann, S. Marinkovic-Pajovic, K.-A. Won, V. E. Jones, S. P. Campos, G. P. Jahreis and K. K. Morella

*Department of Molecular and Cellular Biology, Roswell Park Cancer Institute, Elm and Carlton Streets, Buffalo, New York 14263, USA*

*Abstract.* The hepatic action of cytokines has generally been analysed in terms of the acute-phase response of the liver. The qualitative and quantitative changes in the expression of plasma proteins serve as defining criteria for cytokine function. Interleukin 6 (IL-6) and leukaemia inhibitory factor (LIF) are representatives of a group of cytokines which display strikingly similar effects in both human and rodent liver cells. Hallmarks of the action of these cytokines are the stimulation of type 2 acute-phase plasma proteins and enhancement of the effect of interleukin 1 (IL-1) or tumour necrosis factor $\alpha$ (TNF-$\alpha$) on type 1 acute-phase plasma proteins. The transcriptional activation of the various acute-phase plasma protein genes involves common *cis*-acting regulatory elements whose sequences and location relative to the transcription start site vary from gene to gene. The activity of the IL-6- and LIF-responsive genes depends in part on transcription factors including several members of the C/EBP family, JunB and the glucocorticoid receptor. The expression of these transcription factors is in turn under cytokine-specific control. In a few cases, expression is temporally correlated with the activation of 'late' acute-phase protein genes. The finding that structurally distinct cytokines interact with separate receptors but elicit an almost identical liver cell response demands a reassessment of the contribution of each factor to the *in vivo* acute-phase response.

*1992 Polyfunctional cytokines: IL-6 and LIF. Wiley, Chichester (Ciba Foundation Symposium 167) p 100–124*

There are two general approaches taken in studies of the biological action of cytokines. One is purification and cloning of a cytokine based on one specific functional criterion, or just on sequence similarities. The resulting recombinant product is then used to search for and define its physiological target by treatment of a battery of cell types in tissue culture or by its injection into test animals. The second approach is to select a specific biological process and then to identify all the cytokines which are causally involved. Although the former approach is well suited for generating information on the pleiotropic action of cytokines, the latter can more readily indicate the biological relevance of the cytokine as well as the relationship between its actions and those of other factors.

The characterization of the liver-regulating cytokines, such as interleukin 6 (IL-6) and leukaemia inhibitory factor (LIF), has relied in almost all instances on the second type of approach. The hepatic action of the cytokines has been defined in the context of the transcriptional regulation of acute-phase plasma protein genes (Fey & Gauldie 1990). Without question, the profound changes in the expression of many plasma protein genes induced by inflammatory stimuli have provided an exquisite model for assessing immediate cytokine response (Koj 1974, Kushner 1982, Baumann 1989). Indeed, the striking magnitude of cytokine-regulated expression of plasma proteins has been marvelled at not only by scientists specializing in liver gene regulation, but also by researchers active in other disciplines such as immunology, developmental biology and molecular endocrinology.

Although the focus of research so far has been on acute-phase protein genes, the action of most, if not all, liver-controlling cytokines will probably prove to be far broader, perhaps even involving many of the primary metabolic pathways (Fey & Gauldie 1990). In view of the central role of the liver in systemic homeostasis, the significance of the potential modulating influence of cytokines on metabolism may be equal to, if not greater than, that of the change in plasma protein expression. Although studies of metabolic changes are more complex and less glamorous than those involving gene expression, they are necessary to establish the hepatic action of cytokines completely.

Here, we describe some of the regulatory properties of IL-6 and LIF on hepatic cells. The findings indicate that the cellular responses elicited by these two cytokines share remarkable similarities. The defined response pattern was subsequently recognized as being representative of an entire group of structurally unrelated factors. Although the actions of the cytokines appear similar, the broad spectrum of changes in gene expression that occurs in liver cells offers the opportunity to delineate cytokine-specific properties.

## Liver cell responses

The coordinate activation of a subset of acute-phase plasma protein genes by systemic tissue injury became the hallmark of the hepatic acute-phase response (Birch & Schreiber 1986). The qualitative and quantitative changes observed at the level of gene transcription, mRNA accumulation, protein synthesis and plasma concentration have defined the reaction in various species (Kushner & Mackiewicz 1987, Fey & Gauldie 1990). For identification of the factors communicating the inflammatory signal to the liver, experimental tissue culture systems have been established which have allowed near faithful reproduction of the *in vivo* regulation (Baumann 1989). We have selected a subline of Reuber H-35 rat hepatoma cells as a tissue culture model (Baumann et al 1989a). These cells have retained the ability to respond to cytokines by increasing the expression of all major acute-phase protein genes within 24 h to the same level as acute-phase rat liver cells. Using these cells, we determined that optimal expression of the entire set of acute-phase protein genes required, at a minimum, the combination of IL-1, IL-6 and glucocorticoids (Fig. 1). Treatments with individual factors and pairwise combinations indicated that the plasma protein

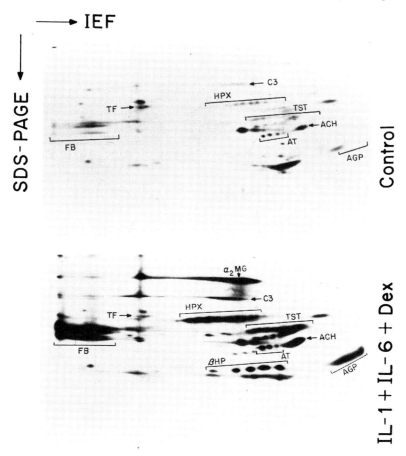

FIG. 1.   Regulation of plasma protein synthesis in H-35 cells. Duplicate cultures of confluent H-35 cells (clone T-7-18; Baumann et al 1989a) were treated for 18 h with serum-free medium alone (control) or containing a mixture of 100 U/ml IL-1β, 100 U/ml IL-6 and 1 μM dexamethasone (Dex). The media were replaced by fresh, methionine-low media containing 1 mCi [$^{35}$S]methionine per ml. After a further 6 h incubation, a 50 μl aliquot from each culture was subjected to two dimensional polyacrylamide gel electrophoresis (isoelectric focusing then SDS–PAGE) followed by fluorography. The fluorograms after 18 h exposure are shown. The positions of the following plasma proteins are indicated: ACH, $\alpha_1$-antichymotrypsin; AGP, $\alpha_1$-acid glycoprotein; AT, $\alpha_1$-antitrypsin; C3, complement component C3; FB, fibrinogen; βHP, β-haptoglobin; HPX, haemopexin; $\alpha_2$MG, $\alpha_2$-macroglobulin; TF, transferrin; TST, thiostatin.

genes regulated fall into two categories: maximal stimulation of type 1 acute-phase proteins, including $\alpha_1$-acid glycoprotein, haptoglobin, haemopexin, complement component 3 and serum amyloid A, depended on IL-1 and IL-6, whereas production of type 2 proteins, including fibrinogens, $\alpha_2$-macroglobulin, thiostatin, $\alpha_1$-antitrypsin and $\alpha_1$-antichymotrypsin, required only IL-6 (Baumann et al 1987, Gauldie et al 1987, Baumann et al 1989a, Baumann & Gauldie 1990). The response to cytokines in each group was generally enhanced by dexamethasone. A prominent permissive action of the steroid was noted for $\alpha_1$-acid glycoprotein and $\alpha_2$-macroglobulin. None of the treatments was found to cause appreciable morphological changes or major growth rate modulation.

By extending the analysis to other cytokines and factors, we found that TNF-$\alpha$ functioned similarly to IL-1 (Baumann et al 1987), whereas the actions of LIF (Baumann & Wong 1989) and IL-11, the recently discovered factor (Paul et al 1990) which has IL-6-like effects on liver cells (Baumann & Schendel 1991), were similar to those of IL-6. No significant changes in the pattern of proteins expressed were achieved by treatment with IL-2, IL-3, IL-4, IL-5, GM-CSF (granulocyte-macrophage colony-stimulating factor), G-CSF, M-CSF, insulin, glucagon, growth hormone, epidermal growth factor, triiodothyronine, $\gamma$-interferon, adrenaline, noradrenaline, 5'-bromocyclic adenosine monophosphate, ionomycin, arachidonic acid or phorbol ester. However, many of these seemingly 'inactive' components need to be tested in combinations with each other and with the known active factors before one can reach a conclusion about their effectiveness in acute-phase protein regulation.

**Responses of plasma proteins to LIF and IL-6**

Preliminary characterization of the hepatic action of human recombinant LIF, formerly also known as hepatocyte-stimulating factor III (Baumann et al 1989b), has shown a substantial overlap with IL-6 in the profile of plasma proteins regulated (Baumann & Wong 1989). Like IL-6, LIF stimulated the synthesis of all type 2 proteins. However, levels produced by each were not always identical (Baumann & Gauldie 1990), and differences between the two cytokines emerged when the factors were assayed in the presence of dexamethasone (Fig. 2A). Although the extent of stimulation of plasma proteins was comparable in the absence of the steroid, a much lower synergy with dexamethasone was achieved with LIF than with IL-6. The modulation of LIF's action by dexamethasone varied from gene to gene; for example, $\alpha_2$-macroglobulin production was stimulated to half the level achieved with IL-6 plus dexamethasone, whereas dexamethasone reduced LIF-stimulated thiostatin synthesis (Fig. 2A). As seen with IL-6 and IL-1, LIF and IL-1 had additive effects on type 1 acute-phase protein genes. The maximal stimulation attained with LIF was in all instances less than with IL-6.

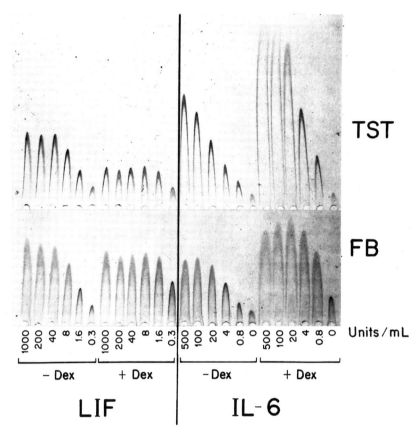

FIG. 2(A). Effect of LIF and IL-6 on plasma protein expression; dose–response relationship. H-35 cells in 24-well cluster plates were treated for 24 h with serum-free medium (300 µl) containing the indicated concentrations of cytokine with or without 1 µM dexamethasone (Dex). After 24 h duplicate 50 µl aliquots of the culture media were analysed by rocket immunoelectrophoresis for thiostatin (TST) and fibrinogen (FB).

Taken together, the functional properties established for LIF and IL-6 seemed to be qualitatively identical. A major distinction between the two cytokines was in their efficacy of stimulation. The analyses done in H-35 cells were extended to primary cultures of rat and mouse hepatocytes and to human hepatoma HepG2 cells, and comparable results were obtained (Baumann et al 1989b, V. E. Jones, S. Marinkovic-Pajovic & H. Baumann, unpublished work). Therefore, we concluded that the cytokine-mediated regulation of plasma proteins in H-35 cells reflected general properties of IL-6 and LIF's actions on hepatic cells.

IL-6 and LIF interacted with separate receptor systems in liver cells (Baumann et al 1989b). These receptors appeared to function independently of each other,

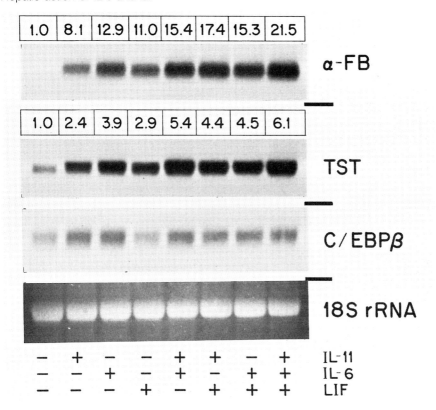

FIG. 2(B).   Additive effects of cytokines. Confluent monolayers of H-35 cells in 10 cm diameter culture dishes were treated for 6 h with serum-free medium containing the indicated cytokines at the following concentrations: 1000 U/ml IL-11, 100 U/ml IL-6 and 100 U/ml LIF. Total cellular RNA was extracted and 15 µg aliquots were analysed by Northern blot hybridization for mRNAs encoding α-fibrinogen (α-FB), thiostatin (TST) and C/EBPβ. The numbers above the autoradiograms indicate the change relative to the untreated control. The lowest panel shows the ethidium bromide staining pattern of the 18S ribosomal RNA, to document identical gel loading.

as indicated by the additive response of the cells to the cytokines (Fig. 2B). Moreover, the effect of both IL-6 and LIF was enhanced by IL-11 (Baumann & Schendel 1991). However, no combination of these three cytokines, either with or without dexamethasone, stimulated expression exceeding the maximal level achieved with IL-6 and dexamethasone for type 2 acute-phase genes or with the combination of IL-1, IL-6 and dexamethasone for type 1.

**LIF and IL-6 act at the level of transcription**

The cytokine-induced changes seen in plasma protein secretion (Figs 1 and 2A)

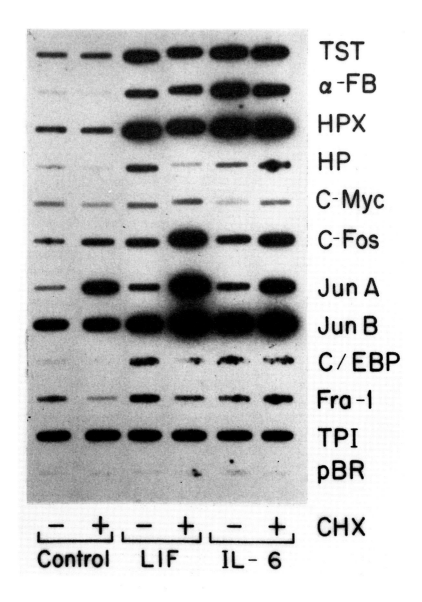

TST
α-FB
HPX
HP
C-Myc
C-Fos
Jun A
Jun B
C/EBP
Fra-1
TPI
pBR

−   +   −   +   −   +   CHX
Control    LIF    IL-6

were reflected in a proportional change in the corresponding mRNA (Fig. 2B; Baumann et al 1990a). By nuclear run-on reaction, we found that LIF and IL-6 increased the rates of transcription of plasma protein genes similarly. This effect on the 'early' acute-phase reactant genes, such as those encoding thiostatin, α-fibrinogen or haemopexin, was independent of protein *de novo* synthesis (Fig. 3). IL-6 and LIF had effects on transcription factor genes ranging from minor to non-detectable, with the exception of 'superinduction' of c-*fos*, *junA* and *junB* genes in the presence of cycloheximide.

**Differential effects of LIF and IL-6 on transcription factor mRNA**

Short term (30 min) treatment of H-35 cells with LIF or IL-6 resulted in a reproducible 3–5-fold increase of *junB* gene transcription, but no increase in transcription of CAAT/enhancer binding protein α (C/EBPα) (Fig. 3), C/EBPβ or C/EBPδ. These four transcription factors received our attention because their expression in rat liver is altered during the acute-phase response (Baumann et al 1991, H. Baumann, K. K. Morella, S. L. McKnight & G. P. Jahreis, unpublished work). Although the transcriptional changes in H-35 cells were minimal, the cytokines substantially increased mRNA for JunB, C/EBPα and C/EBPβ (C/EBPδ is not expressed in H-35 cells) (Fig. 4A). Because the extent of the effect on mRNA levels vastly exceeded that on transcription rates, we assumed that both IL-6 and LIF enhance the stability of cytoplasmic mRNA or processing of the nuclear transcript, or both.

IL-6 produced an immediate and transient increase in C/EBPα mRNA, with a peak value at 30 min treatment. The effect on JunB mRNA had similar kinetics, but the initial increase was 10–50-fold higher than the increase in transcription (compare Figs 4A and 3). The change of C/EBPβ was delayed by 30 min to 1 h and reached a maximum (five-fold above control) after 2 h. LIF had markedly different effects. There was essentially no change in C/EBPα and C/EBPβ mRNA. The rise in JunB mRNA, however, was initially greater than that in IL-6 treated cells, but this peak was followed by a return to basal level by 2 h. This transient peak of JunB mRNA was also seen in LIF-treated HepG2 cells (Fig. 4B).

---

FIG. 3.   Activation of transcription by LIF and IL-6. Duplicate confluent monolayers of H-35 cells in 15 cm diameter culture dishes were treated for 30 min with serum-free medium (Control) or medium containing 100 U/ml LIF or 100 U/ml IL-6. One culture of each set had been pretreated for 5 min with 10 μg/ml cycloheximide (CHX) and was maintained in presence of this inhibitor of translation throughout the subsequent 30 min hormone treatment. Nuclei were prepared and subjected to transcription run-on reaction (Baumann et al 1990a). The $^{32}$P-labelled transcripts ($5 \times 10^7$ c.p.m.) were hybridized to slot blot-immobilized cDNA or genomic sequences encoding the following proteins: TST, thiostatin; α-FB, α-fibrinogen; HPX, haemopexin; HP, haptoglobin; TPI, triose-phosphate isomerase (internal control) and the indicated transcription factors. pBR322 DNA (pBR) served as an indicator of non-specific hybridization.

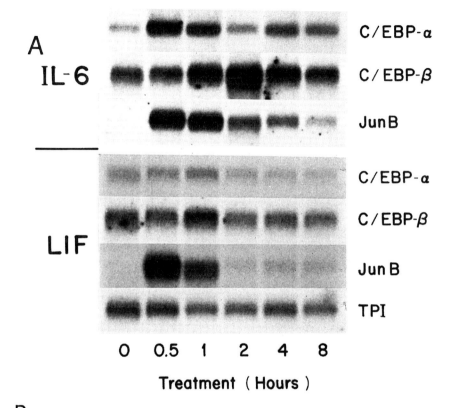

A
IL-6

C/EBP-α
C/EBP-β
JunB

LIF

C/EBP-α
C/EBP-β
Jun B
TPI

0    0.5   1    2    4    8

Treatment ( Hours )

B

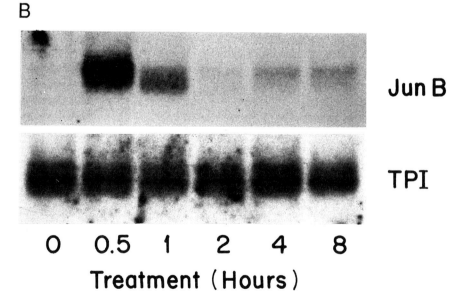

Jun B

TPI

0    0.5   1    2    4    8

Treatment ( Hours )

This modulation of JunB and C/EBP mRNA levels by LIF and IL-6 is considered to be relevant, because the proteins encoded by these genes have been implicated in the control of both basal and cytokine-regulated expression of acute-phase plasma protein genes (Poli et al 1990, Baumann et al 1991).

### LIF and IL-6 utilize common intracellular regulatory elements

Over the past few years we have identified the IL-6 responsive *cis*-acting regulatory elements of the rat genes encoding $\alpha_1$-acid glycoprotein (Won & Baumann 1990), haptoglobin (Marinkovic & Baumann 1990) and β-fibrinogen (Baumann et al 1990b). Each of these gene sequences had similar activity when tested in rat and human hepatoma cells. In transiently transfected hepatoma cells these elements mediated a several-fold increase in expression of a linked gene induced by IL-6 or LIF. This is shown in Fig. 5, using the example of the IL-6-responsive element (IL-6RE) of the β-fibrinogen gene linked to a heterologous promoter upstream of the bacterial chloramphenicol acetyl transferase (CAT) gene. The cytokine-independent expression of the co-transfected plasmid with the major urinary protein gene served as an internal reference marker (Fig. 5). The chimeric CAT gene constructs displayed a dose–response relationship equivalent to that of the endogenous fibrinogen gene. The additivity of IL-6 and LIF's effects was maintained in this system (Fig. 5B).

It has been proposed that members of the C/EBP family, specifically C/EBPβ (IL-6DBP; Poli et al 1990) or NF-IL6 (Akira et al 1992, this volume) are instrumental in signalling the IL-6 stimulus to the acute-phase protein genes. Indeed, the IL-6RE of β-fibrinogen, but not those of the rat $\alpha_1$-acid glycoprotein or haptoglobin genes (Baumann et al 1991), proved to be sensitive to *trans* activation by C/EBPs derived from co-transfected expression vectors (Fig. 5B). The remarkable regulatory element specificity of C/EBP action makes an involvement of this type of transcription factor in acute-phase protein gene expression seem likely. Less certain, however, was whether any of the C/EBP proteins has a direct role in communicating the IL-6 or LIF signal. The *trans* activation by C/EBPα, β or δ proved to be independent of IL-6 or LIF treatments (Fig. 5B, bottom panel). Moreover, the introduced transcription factors, at any concentration, failed to improve the magnitude of response to cytokine (Baumann et al 1991, H. Baumann, K. K. Morella, S. L. McKnight

---

FIG. 4. The effect of cytokines on transcription factor mRNA levels. (A) H-35 cells were treated for the indicated periods of time with either 100 U/ml IL-6 or 100 U/ml LIF. Polyadenylated RNA was isolated and 15 µg samples from each preparation were analysed by Northern blotting. The membrane-immobilized RNAs were sequentially probed for C/EBPα (Baumann et al 1991), C/EBPβ (Cao et al 1991), JunB and triose-phosphate isomerase (TPI). (B) HepG2 cells were treated with LIF and analysed for changes in JunB mRNA as described above for H-35 cells.

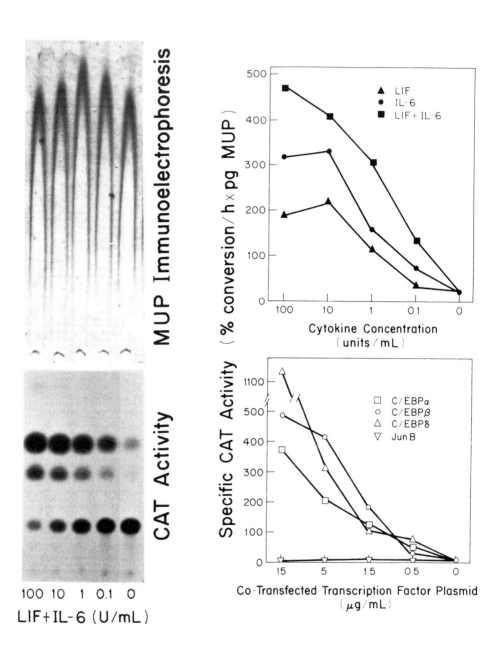

A

B

& G. P. Jahreis, unpublished work). With high levels of co-transfected C/EBPs, the transcriptional activity of the IL-6RE-containing reporter construct reached a maximum that was not increased further by IL-6 or LIF treatment.

From these results, we concluded that C/EBP-related factors are involved in the regulatory mechanism of acute-phase protein gene expression. The cytokine-enhanced expression of C/EBPα and β (Fig. 4A) in H-35 cells would account for activation of late acute-phase protein genes such as those encoding $\alpha_1$-acid glycoprotein and $\alpha_2$-macroglobulin (Won & Baumann 1991, Baumann et al 1991). However, because IL-6 and LIF stimulation of early genes occurs before new C/EBP has been synthesized (Fig. 3), a functional modification or activation of preexisting C/EBPs could be proposed (Poli et al 1990, Akira et al 1990). Arguing against this model is the fact that C/EBPs derived from co-transfected plasmids did not require such a cytokine-dependent activation (Fig. 5B).

An alternative model invokes a cytokine-dependent removal of a C/EBP-inhibiting factor, analogous to the mechanism of NF$\varkappa$B activation (Leonardo & Baltimore 1989). In cells transiently transfected with C/EBP expression vector, the amount of C/EBP protein synthesized is probably greatly in excess of the amount of inhibitor, so cytokine-independent *trans* activation would result. At present, we have no clues as to the biochemical identity of such a hypothetical C/EBP inhibitor in untreated hepatic cells. The change in the expression of mRNA for JunB (Fig. 4A) channelled our attention to the significance of this

---

FIG. 5.    Regulation of transcription in transiently transfected HepG2 cells. (A) HepG2 cells in 6-well cluster plates were transfected with calcium phosphate precipitates of 20 µg plasmid DNA per ml. The DNA mixture was composed of 18 µg pβFb (2 × IL-6RE)-CT (which contains two tandem copies of the 35 bp IL-6 response element of the rat β-fibrinogen gene in the polylinker region of the pCT, a reporter gene construct containing adenovirus major late promoter and chloramphenicol acetyl transferase gene; Baumann et al 1990b) and 2 µg pIE-MUP (internal transfection marker plasmid containing the immediate early promoter of human cytomegalovirus promoter/enhancer linked to the structural gene of mouse major urinary protein; Prowse & Baumann 1988). After 36 h the cells were treated for 24 h with IL-6 and LIF. The cell extracts were assayed for the activity of the CAT enzyme (*bottom*), and the media were assayed for the amounts of secreted MUP by rocket immunoelectrophoresis (*top*). (B) *Top panel*: For one experimental series of cultures that have been treated with the indicated cytokines CAT activity is expressed as % conversion of substrate to product per h, normalized to pg MUP produced by the same culture (Marinkovic & Baumann 1990). *Bottom panel*: *Trans* activation by transcription factors. HepG2 cells were transfected as in A, but the mixtures of plasmid DNAs were composed of (per ml) 5 µg pβFb(2 × IL-6RE)-CT, 2 µg pIE-MUP, and 15 µg expression vector pCDpoly (Pruitt 1988) alone or proportionally substituted with pCDpoly vector containing the cDNAs encoding mouse JunB and C/EBPα (Baumann et al 1991), C/EBPβ and C/EBPδ (Cao et al 1991). The activity of the CAT enzyme was determined in cells *not* treated with any hormones. The values were normalized to the amount of internal transfection marker MUP.

factor in the regulation of acute-phase protein genes. Co-transfection of JunB expression vector produced no *trans*-activating activity (Fig. 5B). However, with IL-6 or LIF treatment, as well as in the presence of co-transfected C/EBPs, JunB acted as an inhibitor of cytokine stimulation occurring through the IL-6RE sequence (Baumann et al 1991). Therefore, JunB may well function at a later stage in the response to cytokines as a modulator of the co-stimulated C/EBPs (Fig. 4A). However, the low level of JunB in unstimulated cells (Fig. 4A) seems to rule out a role for control of C/EBPs at the initial stage of cytokine action. The identification of the molecular event taking place within the first 30 min after IL-6 or LIF treatment will be the key to understanding of the intercellular action of these two liver-regulating cytokines.

## Conclusion

The changes in the expression of plasma protein genes in liver cells are sensitive and characteristic indicators for the molecular actions of multiple cytokines. IL-6 and LIF induce intracellular events with common properties, leading to an immediate increase in transcription of genes encoding acute-phase plasma proteins. Although transcription factors potentially involved in controlling the target genes have been identified, the molecular nature of the intracellular signals that are generated by IL-6 and LIF and affect the transcription components still remains to be defined.

*Acknowledgements*

We are greatly indebted to Dr S. L. McKnight for providing C/EBPβ and C/EBPδ expression vectors, Dr S. Dower, Immunex Corp. for recombinant human IL-1β, and Drs G. G. Wong and P. Schendel, Genetics Institute, for human recombinant LIF-expressing CHO cells and human recombinant IL-11, respectively. We thank M. Held and L. Scere for secretarial assistance. The studies in the authors' laboratory are supported by National Institute of Health grants CA26122 and DK33886.

## References

Akira S, Isshiki H, Sugita T et al 1990 A nuclear factor for IL-6 expression (NF-IL6) is a member of a C/EBP family. EMBO (Eur Mol Biol Organ) J 9:1897–1906
Akira S, Isshiki H, Nakajima T et al 1992 Regulation of expression of the IL-6 gene: structure and function of the transcription factor NF-IL6. In: Polyfunctional cytokines: IL-6 and LIF. Wiley Chichester (Ciba Found Symp 167) p 47–67
Baumann H 1989 Hepatic acute phase reaction *in vivo* and *in vitro*. In Vitro Cell & Dev Biol 25:115–126
Baumann H, Gauldie J 1990 Regulation of hepatic acute phase plasma protein genes by hepatocyte-stimulating factors and other mediators of inflammation. Mol Biol & Med 7:147–159
Baumann H, Schendel P 1991 Interleukin-11 regulates the hepatic expression of the same plasma protein genes as interleukin-6. J Biol Chem 266:20424–20427

Baumann H, Wong GG 1989 Hepatocyte-stimulating factor-III shares structural and functional identity with leukemia inhibitory factor. J Immunol 143:1163–1167

Baumann H, Onorato V, Gauldie J, Jahreis GP 1987 Distinct sets of acute phase plasma proteins are stimulated by separate human hepatocyte-stimulating factors and monokines in rat hepatoma cells. J Biol Chem 262:9756–9768

Baumann H, Prowse KR, Marinkovic S, Won KA, Jahreis GP 1989a Stimulation of hepatic acute phase response by cytokines and glucocorticoids. Ann NY Acad Sci 557:280–295

Baumann H, Won KA, Jahreis GP 1989b Human hepatocyte-stimulating factor-III and interleukin-6 are structurally and immunologically distinct but regulate the production of the same acute phase plasma proteins. J Biol Chem 264:8046–8051

Baumann H, Morella KK, Jahreis GP, Marinkovic S 1990a Distinct regulation of the interleukin-1 and interleukin-6 response elements of the rat haptoglobin gene in rat and human hepatoma cell. Mol Cell Biol 10:5967–5976

Baumann H, Jahreis GP, Morella KK 1990b Interaction of cytokine-and glucocorticoid-response elements of acute phase plasma protein genes. J Biol Chem 265: 22275–22281

Baumann H, Jahreis GP, Morella KK et al 1991 Transcriptional regulation through cytokine- and glucocorticoid-response elements of rat acute phase plasma protein genes by C/EBP and JunB. J Biol Chem 266:20390–20399

Birch H, Schreiber G 1986 Transcriptional regulation of plasma protein synthesis during inflammation. J Biol Chem 261:8077–8080

Cao Z, Umek RM, McKnight SL 1991 Regulated expression of three C/EBP isoforms during adipose conversion of 3T3-L1 cells. Genes & Dev 5:1538–1552

Fey G, Gauldie J 1990 The acute phase response of the liver in inflammation. Prog Liver Dis 9:89–116

Gauldie J, Richards C, Harnish D, Lansdorp P, Baumann H 1987 Interferon beta 2/B-cell stimulating factor type 2 shares identity with monocyte-derived hepatocyte-stimulating factor and regulates the major acute phase protein response in liver cells. Proc Natl Acad Sci USA 84:7251–7255

Koj A 1974 Acute phase reactants—their synthesis, turnover and biological significance. In: Allison AC (ed) Structure and function of plasma proteins. Plenum Press, London, vol 1:73–131

Kushner I 1982 The phenomenon of the acute phase response. Ann NY Acad Sci 389:39–48

Kushner I, Mackiewicz A 1987 Acute phase proteins as disease markers. Dis Markers 5:1–11

Leonardo MJ, Baltimore D 1989 NF$\varkappa$B: a pleiotropic mediator of inducible and tissue-specific gene control. Cell 58:227–229

Marinkovic S, Baumann H 1990 Structure, hormonal regulation, and identification of the interleukin-6 and dexamethasone-responsive elements of the rat haptoglobin gene. Mol Cell Biol 10:1573–1583

Paul SR, Bennett F, Calvetti JA et al 1990 Molecular cloning of a cDNA encoding interleukin-11, a stromal cell-derived lymphopoietic and hematopoietic cytokine. Proc Natl Acad Sci USA 87:7512–7516

Poli V, Mancini FP, Cortese R 1990 IL-6DBP, a nuclear protein involved in interleukin-6 signal transduction, defines a new family of leucine zipper proteins related to C/EBP. Cell 63:645–653

Prowse KR, Baumann H 1988 Hepatocyte-stimulating factors, beta 2 interferon, and interleukin-1 enhance expression of the rat alpha-1-acid glycoprotein gene via a distal upstream regulatory region. Mol Cell Biol 8:42–51

Pruitt SC 1988 Expression vectors permitting cDNA cloning and enrichment for specific
    sequences by hybridization/selection. Gene (Amst) 66:121–134
Won KA, Baumann H 1990 The cytokine response element of the rat alpha-1-acid
    glycoprotein gene is a complex of several interacting regulatory sequences. Mol Cell
    Biol 10:3965–3978
Won KA, Baumann H 1991 NF-AB, a liver specific and cytokine-inducible nuclear factor
    that interacts with the interleukin-1 response element of the rat alpha-1-acid
    glycoprotein gene. Mol Cell Biol 11:3001–3008

## DISCUSSION

*Patterson:* I believe you said that the effects of IL-6 and LIF and IL-11 are
additive at maximal doses. In Fig. 2B it appeared that you were adding sub-
maximal doses.

*Baumann:* Dose–response analysis showed that a LIF concentration of 100 U/ml
gave a maximal response for all acute-phase protein genes (Fig. 2A). At concentrations above 1000 U/ml LIF became inhibitory. 100 U/ml IL-6 produced a
maximal stimulation of some acute-phase protein genes, such as fibrinogen, but
expression of other genes, such as thiostatin, did not reach a maximum plateau.
The failure to achieve maximal expression seems to be a problem with rat
hepatoma cell lines—the more IL-6 you add, the higher is the expression of some
genes. This feature has also been noted for the less active IL-11. We have not
tested high concentrations of IL-11 because of scarcity of purified recombinant
protein. However, 1000 U/ml IL-11 gave an overall response that could be rated
as close to maximal. I have to stress that the additive action of the cytokines is
more prominent after a few hours treatment than after one or two days
treatment. It appears that long-term treatment leads to down-regulation of
receptor function.

*Sehgal:* Could you clarify what you mean when you say 1000 units; what is
that in terms of mass?

*Baumann:* In our studies we express cytokine activity in 'hepatocyte-
stimulating factor', or HSF, units. One unit is defined as the concentration
required for half-maximal stimulation of fibrinogen in HepG2 or H-35
cells. In general, one HSF unit is equal to one nanogram of IL-6 or
LIF per ml.

*Sehgal:* So 1000 units is 1 µg per/ml. That is a large amount.

*Baumann:* You are correct, but it should be realized that maximal stimulation
of primary hepatocyte cultures and most hepatoma cells is actually achieved
with about 50 ng/ml (or 50 HSF U/ml), so 1000 U/ml is a truly maximal
dose.

*Sehgal:* Is that load achievable in the circulation?

*Baumann:* As far as I know, yes.

*Heinrich:* We also find a weaker stimulation of $\alpha_1$-antichymotrypsin synthesis in HepG2 cells with LIF than with IL-6. The effect of LIF, however, is amplified strongly by the addition of soluble human IL-6 receptor.

*Martin:* Glucocorticoids caused a much greater enhancement of the response of the fibrinogen gene to IL-6 than that to LIF. Could you comment on the mechanism of that? When you put LIF and IL-6 together do you get additivity at maximum concentrations?

*Baumann:* As shown in Fig. 2A, dexamethasone does not significantly increase maximal LIF stimulation of fibrinogen synthesis, but greatly enhances that induced by IL-6. The combination of dexamethasone and IL-6 resulted in the maximum possible expression of fibrinogen. Addition of LIF, rather than increasing, actually reduced the stimulation to a level intermediate between that induced by IL-6 and that by LIF. Not much is known about the molecular mechanism of this. Analysis of the β-fibrinogen gene promoter has indicated the presence of a strong glucocorticoid response element 5′ to the IL-6/LIF response element. The latter element is recognized by C/EBP-related factors (Fig. 5) and its control might therefore be dependent on the regulatory mechanisms described by Dr Akira. Simultaneous recognition of the two regulatory elements by relevant transcription factors could explain synergism. We have no idea as yet what the reason is for the difference in regulation between IL-6 and LIF.

*Kishimoto:* C/EBPβ may be identical to NF-IL6; what is C/EBPδ?

*Baumann:* C/EBPβ is identical to AGP/EBP, and is the mouse equivalent of LAP, IL-6DBP and, according to you, NF-IL6. C/EBPδ is the third member of the C/EBP family of the mouse cloned by Cao et al (1991). It is probably the equivalent of your NF-IL6β.

*Kishimoto:* Which cDNA did you use?

*Baumann:* We used SV40 promoter-containing expression vectors, into which we inserted the cDNA encoding C/EBPα from *Mus caroli*, and cDNA encoding C/EBPβ and C/EBPδ provided by Dr S. L. McKnight.

*Kishimoto:* The effects of IL-6 and LIF on expression of C/EBPβ (NF-IL6) are different—IL-6 augments expression whereas LIF depresses it—so the signals may differ. Do the effects of IL-6 and LIF on C/EBPδ differ?

*Baumann:* C/EBPδ is not expressed in our hepatoma cells. It is expressed only in liver, and even there it is expressed at a low level. C/EBPα expression is stimulated in cytokine-treated H-35 cells, whereas it is reduced in the liver during the acute phase.

*Lotem:* To what extent is the acute-phase response influenced *in vivo* by the length of time IL-6 is present in the serum? Is the acute-phase response induced by injection of recombinant IL-6 different from that induced by injection of lipopolysaccharide, for example, which will alter the length of time IL-6 is present in the serum?

*Gauldie:* When you give a bolus of IL-6, either concomitantly with corticosteroid or with appropriate cortisteroid augmentation, depending on the animal model, you get a pronounced immediate acute-phase response. If you give LPS you get a good, sustained liver response, simply because it is present for a long time. If you give IL-6 progressively, over a period of days, you get a constantly increasing acute-phase response. I think it's the ability of IL-6 to continue to stimulate the liver that leads to the extent and duration of the acute-phase response. If IL-6 is present in raised amounts for a long period you will get a strong response; if it's applied repetitively, even though the half-life is short, you will get a similar strong response.

*Heinrich:* We have measured the plasma clearance of iodinated human IL-6 after intravenous injection into the rat. The kinetics of clearance are biphasic. There is a rapid initial disappearance of $^{125}$I-labelled recombinant human IL-6, with a half-life of about 3 min, and a second, slower decrease, with a half-life of about 55 min (Castell et al 1988).

*Kishimoto:* Dr Baumann, although the LIF receptor and gp130 are similar, the effects of LIF and IL-6 on transcription factors differ greatly. How can you explain that?

*Baumann:* One of the points I was trying to make was that LIF and IL-6 must have different internal signalling pathways. Certainly, there must be some convergence of the pathways, to account for the regulation of the same acute-phase plasma protein genes via identical gene elements. LIF and IL-6 must generate a whole set of different signals and I am fairly sure that IL-11 does that too. Jack Gauldie has an even more striking example, a newly discovered cytokine, Oncostatin M, that functions similarly, but not identically, to IL-6, IL-11 and LIF. In the case of Oncostatin M, there is the additional complication of species-specific differences in the hepatic response pattern.

*Gauldie:* Carl Richards and I have collaborated with Heinz Baumann to identify another factor like LIF and IL-11, Oncostatin M (Richards et al 1992). In rat liver and human and rat hepatoma cells Oncostatin M has similar effects to IL-6, stimulating the genes for type 2 acute-phase proteins. This class is interesting because it includes all the antiproteases, as well as fibrinogen and haptoglobin. In HepG2 cells Oncostatin M is as potent as, and perhaps more potent than, IL-6. Like the others (IL-6, IL-11 and LIF) it will act in synergy with IL-1 on type 1 acute-phase proteins and its effects on type 2 proteins are synergistically enhanced by dexamethasone. In primary rat hepatocytes we cannot differentiate between the activity of Oncostatin M and that of LIF, though there are some differences in human cells. Perhaps in rat hepatocytes Oncostatin M is working through the LIF receptor, or perhaps there are common receptor-mediated pathways of activation. Oncostatin M is thus another member of a family of cytokines that stimulate acute-phase protein genes in the liver, apparently acting at similar or identical sites of the promoter region of the genes. In rat hepatocytes, IL-6 in the presence of corticosteroid is the most potent stimulator for the entire set of type 2 acute-phase protein genes, and the maximum stimulation of type 1 acute-phase proteins is achieved with a combination of IL-6, IL-1 and corticosteroid.

*Gough:* You suggested that Oncostatin M and LIF might signal through the same receptor in the rat, but said there were slight differences between their effects in human cells. Does that imply that in human cells they are not operating through the same receptor? Also, is there competition for receptor binding between LIF and Oncostatin?

*Gauldie:* I don't yet know the answer to your second question. In human cells, Oncostatin M is a more potent stimulator than LIF. Perhaps it binds better to a LIF receptor, or perhaps it actually binds to an Oncostatin receptor. Alternatively, it could be acting through the LIF receptor *and* its own receptor.

*Nicola:* LIF and Oncostatin M show fairly significant sequence identity, about 30%, and there's some indication that the arrangement of disulphide bonds is the same in the two molecules. They have some overlapping actions, such as induction of M1 cell differentiation and what you have just described. However, Rose & Bruce (1991) have shown that they have some distinct actions; Oncostatin M is active on certain tumour cells on which LIF is not active, and Oncostatin M is not active on embryonic stem cells, whereas LIF is. If that is true, there cannot be just one common receptor.

*Lee:* Does Oncostatin M induce differentiation of M1 cells?

*Gauldie:* I am unaware of any results on M1 cells. It induces LDL receptors on HepG2 cells, but I don't know if LIF does.

*Heath:* How does IL-11 fit into this picture?

*Gauldie:* IL-11, IL-6, Oncostatin M and LIF are a class of regulators that regulate the expression of both the type 2 and the type 1 acute-phase proteins in the liver—there may be some quantitative differences between them, but qualitatively, their effects are the same. The factors regulate the same sets of genes, probably through the same set of nuclear proteins. There are some differences but these may be evident only when you analyse a particular cell. It is important, as I think Dr Baumann would agree, not to target only one gene in your analysis. When you look at the genes collectively, and also look across species, you see the common message, which is that there are at least four regulators. The next question to answer is whether or not they are all important in the circulation.

*Heath:* Does IL-11 have similar functions to those of IL-6 and LIF in other systems?

*Gauldie:* IL-11 stimulates the proliferation of IL-6-dependent plasmacytoma cells (Paul et al 1990).

*Kishimoto:* IL-11 was originally characterized as a plasmacytoma growth factor which was different from IL-6.

*Metcalf:* IL-11 also stimulates platelet formation.

*Baumann:* We have analysed all possible combinations of IL-6, IL-11, LIF and oncostatin for their effects on rat and human hepatic cells. Our results indicated the presence of only three types of receptor responses—to IL-6, LIF and IL-11. Oncostatin M always behaved like LIF.

*Fey:* We have studied the mechanism of target gene regulation by IL-6 and LIF using the example of the rat $\alpha_2$-macroglobulin gene (Hocke et al 1992; M. Z. Cui, G. Hocke, G. Baffet, A. Goel & G. H. Fey, unpublished work). This is one of the most strongly reacting rat liver acute-phase protein genes and responds to both IL-6 and LIF. The IL-6 response element (IL-6RE) of this gene has been mapped previously (Kunz et al 1989, Ito et al 1989, Hattori et al 1990) in the region between $-150$ and $-210$ bp upstream of the transcription start site. The element was found to consist of two subregions, called the core and core homology regions, located 20 bp apart. Each of these was approximately 18 to 20 bp long and contained one protein-binding site. Five of the six nucleotides in the central hexanucleotides of the two regions were identical and therefore these were thought to be binding sites for the same DNA-binding protein. The IL-6RE was both necessary and sufficient for the response of the $\alpha_2$-macroglobulin gene to IL-6; when the element was destroyed by mutagenesis, responsiveness to IL-6 was lost. When multiple copies of the element were linked to the minimal enhancerless promoter of the $\alpha_2$-macroglobulin gene or heterologous promoters, responsiveness to IL-6 was conferred on these normally unresponsive promoters. A series of contructs had been prepared to map the IL-6RE. We attempted to use these contructs to identify a LIF response element (LIF-RE) in the $\alpha_2$-macroglobulin gene (Fig. 1, Table 1).

The constructs were transfected into FAO rat and human hepatoma cells, which were then treated with LIF. Construct p4×T.$\alpha_2$MLuc (Fig. 1) carried the minimal promoter of the $\alpha_2$-macroglobulin gene and four tandem copies of the 18 bp core region of the IL-6RE directing the expression of a luciferase reporter. When this contruct was transiently transfected into HepG2 human hepatoma cells IL-6 treatment caused a 187-fold stimulation of reporter gene activity. Treatment with LIF caused a 37.9-fold activation (Table 1, Fig. 2A). No synergistic action between glucocorticoids and either IL-6 or LIF was observed in this cell line, probably because it had low levels of glucocorticoid receptors. When the level of glucocorticoid receptors was increased by transfection of an expression vector coding for the receptor, synergism between IL-6 and glucocorticoids was observed. Therefore, the 18 bp region containing the IL-6RE also contained a LIF-RE.

Are these response elements overlapping or identical? We addressed this question by mutagenesis. The wild-type construct carried two tandem copies of the 18 bp core region of the IL-6RE driving the $\alpha_2$-macroglobulin gene minimal promoter and a luciferase reporter (p2×T.$\alpha_2$MLuc, Fig. 1). The expression from this construct ws induced 44.6-fold by IL-6 and 7.7-fold by LIF (Table 1). When the hexanucleotide cores of both 18 bp elements were substituted by an unrelated sequence (construct p2×mT.$\alpha_2$MLuc; Fig. 1) the

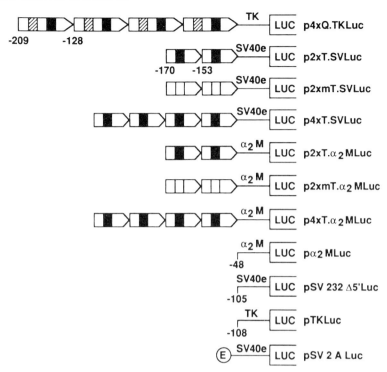

FIG. 1. (*Fey*) Constructs used to demonstrate that the 18 bp IL-6 response element (RE) of the $\alpha_2$-macroglobulin ($\alpha_2$M) gene functions as a LIF response element. All constructs contained the cDNA coding sequences of the firefly luciferase (Luc) gene as a reporter (de Wet et al 1987). In some contructs the minimal, enhancerless promoter of the herpes simplex virus thymidine kinase gene (TK) was used; in others the minimal promoter of the simian virus 40 early gene (SV40e) or that of the rat $\alpha_2$-macroglobulin itself (coordinates $+54$ to $-48$, relative to the transcription start site; Hocke et al 1992) was used. E, enhancer of the SV40e gene; black box, 18 bp IL-6RE core region; hatched box, 18 bp IL-6RE core homology region; open box, mutated 18 bp core region with central hexanucleotide replaced, rendering it functionally inactive. Construct p4×Q.TKLuc carried four tandem repeats of the 82 bp genomic DNA fragment, Q, that contained a complete IL-6RE including one core and one core homology region in its natural configuration.

response to both IL-6 and LIF was lost (Table 1). Thus, the same six base pairs in the centre of the 18 bp region were essential for responsiveness to both IL-6 and LIF. Protein–DNA binding studies showed that the same 6 bp sequence was required for specific binding of a nuclear protein.

These results suggested that the IL-6 and LIF response elements overlapped and were probably identical. It is still not known, however, whether the effects of IL-6 and LIF are mediated by the same or different proteins. It is conceivable

**TABLE 1  The 18 bp IL-6 response element core element confers LIF-responsiveness to three different minimal promoters**

| Construct transfected[a] | Luciferase activity after treatment[b] | | | Relative induction[d] | |
| --- | --- | --- | --- | --- | --- |
| | Medium alone | LIF[c] | IL-6 | LIF | IL-6 |
| p4×Q.TKLuc | 519 232 | 1 536 514 | 2 858 572 | 3.0 | 5.5 |
| p2×T.SVLuc | 79 100 | 218 564 | 523 602 | 2.8 | 6.6 |
| p2×mT.SVLuc | 92 477 | 90 064 | 114 684 | 1.0 | 1.2 |
| p4×T.SVLuc | 261 866 | 1 083 518 | 2 434 943 | 4.1 | 9.3 |
| p2×T.$\alpha_2$MLuc | 7234 | 55 477 | 322 938 | 7.7 | 44.6 |
| p2×mT.$\alpha_2$MLuc | 5064 | 5285 | 7616 | 1.0 | 1.5 |
| p4×T.$\alpha_2$MLuc | 7743 | 293 685 | 1 450 274 | 37.9 | 187.3 |
| p$\alpha_2$MLuc | 9207 | 9796 | 14 499 | 1.1 | 1.6 |
| pSV232Δ5'Luc | 103 247 | 111 534 | 122 361 | 1.1 | 1.2 |
| pTKLuc | 238 884 | 259 950 | 293 822 | 1.1 | 1.2 |
| pSV2ALuc | 19 777 802 | 18 964 833 | 23 483 519 | 1.0 | 1.2 |

[a]Constructs were transfected into HepG2 human hepatoma cells ($3.5 \times 10^6$ cells/10 cm diameter dish; 15 μg of plasmid DNA/dish).
[b]After incubation with the transfected constructs for a total of 36 h and subsequent treatment with cytokines (60 BSF-2 units/ml of IL-6 and 625 units/ml of LIF for 4 h), cell extracts were prepared and luciferase activities were measured in standard assays using 200 μg of protein extract/assay. All values are averages over two dishes. Each experiment was repeated three times. Luciferase assays were performed as described by de Wet et al 1987.
[c]The source of LIF was the supernatant of a CHO cell line secreting approximately 625 000 units/ml of recombinant human LIF (Moreau et al 1988), diluted 1:1000 (10 μl supernatant in 10 ml of HepG2 culture medium, corresponding to approximately 625 units/ml of LIF). Recombinant human IL-6 was purchased from Amgen.
[d]Relative inductions (fold-induction) are multiples of the values obtained after treatment with medium alone.

that two different proteins could bind specifically at the same DNA sequence. From our results, we favour the hypothesis that the two proteins involved are very similar or are identical. We did similar experiments by transfection of the same constructs into Hep3B human hepatoma cells (Fig. 2B). In this cell line a substantial response to IL-6 and a clear cut synergism of IL-6 with glucocorticoids were observed, as expected because this line is known to carry high levels of glucocorticiod receptor. There was no response to LIF in Hep3B cells. The experiments in HepG2 cells had suggested that the intracellular signal cascades for IL-6 and LIF converged to a common end point, binding of a common factor to the IL-6RE/LIF-RE. This common end point must also have been functional in Hep3B cells, because these cells responded to IL-6 (Table 1, Fig. 2B). The absence of a response to LIF therefore indicated that Hep3B cells must lack a component that acts early in the LIF response cascade, upstream

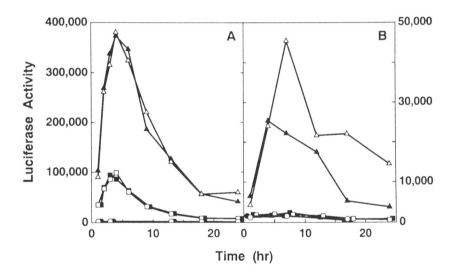

FIG. 2. (*Fey*) The 18 bp LIF-RE functions in HepG2 but not in Hep3B human hepatoma cells. Construct p4×T.$\alpha_2$MLuc was transfected into HepG2 (A) and Hep3B (B) cells. The cells were then treated with 60 BSF2 units/ml recombinant human IL-6 (filled triangles), IL-6 plus $10^{-6}$ M dexamethasone (open triangles), 625 units/ml recombinant human LIF (filled squares) or LIF plus $10^{-6}$ M dexamethasone (open squares). Cellular protein extracts were then prepared and analysed for luciferase activity. Activity is given in light units per 100 µg cellular protein extract in a standard luciferase assay. Filled circles give results for transfected cells treated with medium alone and open circles represent cells treated with medium plus $10^{-6}$ M dexamethasone. Results are from one experiment and represent averages from two dishes containing cells transfected separately. The experiment was repeated twice, with reproducible results.

of its point of convergence with the IL-6 response cascade. The convergence point is not known, but the structural similarity between the LIF receptor and gp130, the IL-6 signal transducer, suggests that the convergence point may be the receptor itself or an early post-receptor step. To investigate this, we did studies with radioiodinated LIF. We found that HepG2 cells carry significant numbers of high affinity LIF receptors whereas Hep3B cells had none. Surprisingly, LIF receptors were also found on HeLa cells. After transfection of the $p4 \times T.\alpha_2MLuc$ construct and appropriate controls into HeLa cells we found that HeLa cells showed a significant response to LIF but not to IL-6, consistent with the fact that HeLa cells express very few IL-6 receptors.

The nuclear proteins binding to the IL-6RE/LIF-RE, the IL-6RE-binding protein and the LIF-RE-binding protein, were studied by gel mobility shift experiments using a probe carrying two tandem copies of the 18 bp IL-6RE/LIF-RE core region and nuclear protein extracts from a variety of cells. With nuclear extracts from acute-phase rat livers, but not from normal rat livers, a specific complex, called complex II, was obtained. Gel shift analysis showed that it bound with sequence specificity to the hexanucleotide core sequence. A complex II of indistinguishable mobility was also formed between the same probe and nuclear extracts from IL-6-treated Hep3B, HepG2 and FAO rat hepatoma cells, but not with extracts from untreated HepG2 or Hep3B cells. A similar complex was also evident with extracts from LIF-treated HepG2 cells, but not from LIF-treated Hep3B cells, consistent with their lack of LIF receptors. A similar complex was also found with extracts from IL-6-treated U-266 and CESS human B lymphoid cell lines, which are responsive to IL-6 but not to LIF. The complex was not seen in LIF-treated U-266 and CESS cells, as expected. Our own experiments (G. Hocke, unpublished work) confirmed that U-266 and CESS cells lack high affinity LIF receptors. Complex II was also generated with extracts from LIF-treated HeLa cells but not from IL-6-treated HeLa cells, as would be anticipated. Finally, complex II was also obtained with nuclear extracts from LIF-treated ES1 mouse embryonic stem cells, but not with extracts from ES1 cells transferred to LIF-free medium. In this medium the cells began to differentiate and the ability to form complex II was rapidly lost.

From all these results we have concluded that: (a) the LIF response element-binding protein must have a molecular mass very similar to that of the IL-6 response element-binding protein from various cell types and species (rats, mice and humans); thus, these two proteins are likely to be identical; (b) formation of complex II with extracts from various cell types occurred whenever the cell carried a complete signal cascade for one of these cytokines and was activated by binding of the cytokine to its receptor; therefore, formation of complex II could be considered an indicator of the successful arrival and processing of an IL-6 or LIF signal in the nucleus of one of their target cells; (c) the IL-6/LIF response element-binding protein was not the same as NF$\varkappa$B or NF-IL6/IL-6DBP/LAP or DBP. The first of these participates in the regulation of a broad range of inflammation-controlled genes, the second in the control of a subgroup of IL-1- and IL-6-responsive genes, including the IL-6 gene itself, and the third is a general transcription factor for hepatic genes. The protein that binds to the IL-6 and LIF response elements is an apparently as yet undescribed transcription factor important for induction of IL-6 and LIF target genes in a variety of tissues. It has been reported here (p 117) that IL-11 and Oncostatin M also have effects similar to those of IL-6 and LIF on a similar spectrum of liver acute-phase protein genes. It is therefore to be expected that the intracellular signal cascades for IL-11 and Oncostatin M will converge with those for IL-6 and LIF, and may use a similar or common nuclear end point, including the same *cis* element and the nuclear factors that bind to this element. We shall continue to characterize and to try to clone and sequence this factor and to apply the knowledge gained from this study in liver cells to other cells that are targets of IL-6 and LIF, including embryonic stem cells and myeloid leukaemic cells.

## References

Cao ZD, Umek RM, McKnight SL 1991 Regulated expression of three C/EBP isoforms during adipose tissue conversion of 3T3-L1 cells. Genes & Dev 5:1538–1552

Castell JV, Geiger T, Gross V et al 1988 Plasma clearance, organ distribution and target cells of interleukin-6/hepatocyte stimulating factor in the rat. Eur J Biochem 177:357–361

de Wet JR, Wood KV, DeLuca M, Helinski DR, Subramani S 1987 Firefly luciferase gene: structure and expression in mammalian cells. Mol Cell Biol 7:725–737

Hattori M, Abraham LJ, Northemann W, Fey GH 1990 Acute-phase reaction induces a specific complex between hepatic nuclear proteins and the interleukin 6 response element of the rat $\alpha_2$-macroglobulin gene. Proc Natl Acad Sci USA 87:2364–2368

Hocke G, Barry D, Fey GH 1992 Synergistic action of interleukin 6 and glucocorticoids is mediated by the interleukin 6 response element of the rat $\alpha_2$-macroglobulin gene. Gene Mol Cell Biol, in press

Ito T, Tanahashi H, Misumi Y, Sakaki Y 1989 Nuclear factors interacting with an interleukin 6 responsive element of rat $\alpha_2$-macroglobulin gene. Nucleic Acids Res 17:9425–9435

Kunz DR, Zimmermann R, Heisig M, Heinrich PC 1989 Identification of the promoter sequences involved in the interleukin 6 dependent expression of the rat $\alpha_2$-macroglobulin gene. Nucleic Acids Res 17:1121–1138

Moreau J-F, Donaldson DD, Bennett F, Witek-Giannotti J, Clark SC, Wong GG 1988 Leukaemia inhibitory factor is identical to the myeloid growth factor human interleukin for DA cells. Nature (Lond) 336:690–692

Paul SR, Bennett F, Calvetti JA et al 1990 Molecular cloning of a cDNA encoding interleukin 11, a stromal cell-derived lymphpoietic and haemopoietic cytokine. Proc Natl Acad Sci USA 87:7512–7516

Richards CD, Brown TJ, Shoyab M, Baumann H, Gauldie J 1992 Recombinant
  oncostatin M stimulates the production of acute phase proteins in HepG2 cells and
  rat primary hepatocytes in vitro. J Immunol 148:1731–1736
Rose TM, Bruce G 1991 Oncostatin M is a member of a cytokine family that includes
  leukemia-inhibitory factor, granulocyte colony-stimulating factor and interleukin-6.
  Proc Natl Acad Sci USA 88:8641–8645

# Further studies of the distribution of CDF/LIF mRNA

Paul H. Patterson and Ming-Ji Fann

*Biology Division, California Institute of Technology, Pasadena, CA 91125, USA*

*Abstract.* Differentiation choices in the haemopoietic and nervous systems are controlled in part by instructive factors. The cholinergic differentiation factor (CDF, also known as leukaemia inhibitory factor, LIF) affects the development of cultured cells from both systems. To understand the role of CDF/LIF during normal development *in vivo*, we have begun to localize its mRNA in the late fetal and postnatal rat. Application of reverse transcriptase–polymerase chain reaction and RNase protection methods reveals that CDF/LIF mRNA levels are developmentally modulated in both haemopoietic and neural tissues. A target tissue of cholinergic sympathetic neurons, the footpads that contain the sweat glands, express high levels of this mRNA (relative to mRNA for actin and $\beta_2$-microglobulin). Levels in targets of noradrenergic neurons are lower, but do undergo significant changes during development. Signals are also detected in selective regions of the adult brain, and in embryonic skeletal muscle. This finding in muscle may be significant for motor neurons, because CDF/LIF is a trophic factor for these neurons in culture. Embryonic liver, neonatal thymus and postnatal spleen express CDF/LIF mRNA, and expression in gut is the highest of all tissues examined. The selective tissue distribution and developmental modulation of CDF/LIF mRNA expression support a role for this factor in the normal development of several organ systems.

*1992 Polyfunctional cytokines: IL-6 and LIF. Wiley, Chichester (Ciba Foundation Symposium 167) p 125–140*

There are a number of parallels between the haemopoietic and nervous systems in the generation of the diverse arrays of cellular phenotypes found in these tissues. In both systems, multipotential stem cells give rise to progenitor cells that are committed to distinct sublineages (Anderson 1989, Nawa et al 1990). Moreover, differentiation decisions made by these progenitors can be guided by intercellular signals that influence proliferation and expression of lineage markers (Nicola 1989, Nawa & Patterson 1990, Nawa et al 1990). In fact, at least one of these instructive factors can influence differentiation choices in cultured cells of both systems (Yamamori et al 1989). The cholinergic differentiation factor (CDF) can switch the phenotype of post-mitotic sympathetic neurons from noradrenergic to cholinergic, without affecting neuronal survival or growth (Patterson & Chun 1977, Fukada 1985). This protein also promotes survival and differentiation of sensory neurons (Murphy et al 1991). CDF is identical to leukaemia inhibitory factor (LIF), a protein that inhibits the proliferation of certain myeloid cell lines and promotes the development of macrophage characteristics (Gearing et al 1987). It is now clear that CDF/LIF, also known by several other names, is a cytokine that is capable of acting on many different tissues, at various times during development and in maturity (Gough & Williams 1989).

Little is known, however, about the role CDF/LIF plays in the normal development of the haemopoietic and nervous systems *in vivo*. An understanding of this problem will require determination of where the protein is produced, which cells express receptors for it, as well as perturbation studies *in situ*. We have therefore begun to examine the tissue distribution of CDF/LIF mRNA, focusing in particular on the nervous and haemopoietic systems. The initial findings using the reverse transcriptase–polymerase chain reaction (RT–PCR) method indicated that levels of this mRNA are very low in a number of postnatal rat tissues (Yamamori 1991). Expression is, however, much higher in a target tissue of cholinergic sympathetic neurons, the footpads that contain the sweat glands, than it is in target tissues of noradrenergic sympathetic neurons, such as various glands and cardiac muscle, at least at postnatal ages. This finding strengthened the possibility that CDF/LIF may, in fact, be the instructive factor responsible for the noradrenergic-to-cholinergic phenotypic conversion that sympathetic neurons undergo when they innervate sweat glands during normal development (Landis 1990). Using the RT–PCR and RNase protection methods, we have extended these studies to include the tissue distribution of CDF/LIF mRNA in the rat embryo and at several early postnatal ages. The results support a role for this cytokine in the development of the nervous and immune systems.

## Methods

### Isolation of mRNA

Tissues were dissected and quickly diced with scalpels, frozen in petri dishes on dry ice, and stored at −80 °C. Tissues were homogenized (approximately 1 g wet weight for most tissues) with glass homogenizers in buffer containing sodium dodecylsulphate (SDS) and proteinase K, as provided in the Invitrogen Fast Track kit (San Diego, CA). mRNA was isolated using oligo(dT) cellulose, as described in this kit, and yields were estimated by measuring absorbance at 260 nm.

### RT–PCR

2–10 µg of RNA from each tissue was used for RT reaction. For samples with very low RNA levels, 20 µg glycogen was added so that visible pellets were formed. Each 20 μl RT reaction contained 1 × RT buffer (BRL, Grand Island, NY), 0.5 mM dNTP (Boehringer Mannheim, Indianapolis, IN), 1 μg random primer (Promega, Madison, WI), 5 mM dithiothreitol (DTT) (BRL), 40 μl RNasin (Promega) and 200 U M-MLV-RT (BRL). Reactions were run at 37 °C for 90 min, and stopped by addition of 1 μl 50 mM EDTA.

Each PCR tube contained a final volume of 20 μl, consisting of 1 × PCR buffer (Promega), 0.5 units Taq DNA polymerase (Promega), 0.25 mM dNTP, and primers (see below) for β-actin (200 nM), $\beta_2$-microglobulin (200 nM) or CDF/LIF (200 nM). In some experiments multiple sets of primers were included in each tube. All tubes and reagents were kept on ice until the start of the PCR. Before beginning the PCR, 40 μl mineral oil was added to each tube. Reactions were run in Perkin-Elmer and Cetus machines under the following conditions: actin and $\beta_2$-microglobulin, 1 cycle of 94 °C for 1 min, 20 cycles of 95 °C for 1 min, 57 °C for 2 min, and 72 °C for 3 min, 1 cycle of 72 °C for 10 min; CDF/LIF, same conditions, except that 30 cycles were run instead of 20. Samples were stored at 4 °C. An 8 μl sample from each PCR run was analysed on a 2% agarose gel, which was visualized with ethidium bromide staining and UV irradiation.

*Primers*

The CDF/LIF primers used were
5'-CCGTGTCACGGCAACCTCATGAACCAGATC-3' (in the second exon; Yamamori et al 1989, Stahl et al 1990) and
5'-GGGGACACAGGGCACATCCACATGGCCCAC-3' (in the third exon); the expected size of the PCR product with these primers is 396 bp. The primers used in the β-actin assay were
5'-TCATGAAGTGTGACGTTGACATCCGTAAAG-3' and
5'-CCTAGAAGCATTTGCGGTGCACGATGGAGG-3' (Nudel et al 1983); the expected size of the product is 285 bp. The primers used to detect $\beta_2$-microglobulin were 5'-GTGATCTTTCTGGTGCTTGTC-3' and
5'-TTTGGGCTCCTTCAGAGTGAC-3' (Mauxion & Kress 1987); the expected size of the product is 318 bp.

*RNase protection assay*

CDF/LIF, γ-actin and $\beta_2$-microglobulin probes were labelled with [α-$^{32}$P]-GTP in reaction mixtures containing 1 μg plasmid [pMJL for CDF/LIF (Yamamori et al 1989); B2-7 for $\beta_2$-microglobulin (D. Stemple, unpublished); human γ-actin (Gunning et al 1983, Enoch et al 1986)], 7.5 μl of [$^{32}$P]GTP (10 mCi/ml, 850 Ci/mmol, 12.5 μM: New England Nuclear, Boston, MA), 3 μl 5 × transcription buffer (Promega), 2 μl nucleotides (100 mM DTT, 2.5 mM A, C, UTP; Promega), 0.5 μl RNasin (Promega), and 15 units SP6 RNA polymerase

(Promega). Incubation was at 37 °C for 60 min; reactions were stopped by addition of 1 unit DNase (RQ1, Promega) and incubation was continued at 37 °C for 15 min. After addition of 2 µl 0.5 M EDTA, 20 µl of a formamide dye mixture (10 mM EDTA, 1 mg/ml xylene cyanol FF, 1 mg/ml bromphenol blue, 80% formamide) was added and after incubation at 95 °C for 5 min, tubes were immediately placed on ice. The labelled probes were purified by electrophoresis on a 6% polyacrylamide gel; after autoradiography for 30 s, bands were cut out and eluted in 1 ml of buffer (for 5 ml: 100 µl 1 M Tris pH 7.5, 0.1 ml 0.5 M EDTA, 0.25 ml 10 M NH$_4$OAc, 5 µl 10 mg/ml tRNA, 4.65 ml H$_2$O). Elution was for 90 min at room temperature, on a rotator. Yields were 50 000–75 000 c.p.m./µl.

Hybridization mixtures contained 33 µl probe (20 µl 10 M NH$_4$OAc, 3 µl 10 mg/ml tRNA, $5 \times 10^5$ c.p.m. in 10 µl) and 67 µl of tissue mRNA plus TE buffer (10 mM Tris pH 7.5, 1 mM EDTA). After mixing, 100 µl isopropanol was added and the tubes were incubated at 4 °C for 10 min. After centrifugation in a microfuge for 15 min at 4 °C, the supernatants were removed and 100 µl 70% ethanol (−20 °C) was added. Solutions were mixed and after centrifugation for 15 min at 4 °C, the supernatants were removed and the samples dried in a Speedvac (Savant, Farmingdale, NY). Pellets were dissolved in 28 µl buffer (for 1 ml: 0.8 ml deionized formamide, 20 µl 0.5 M EDTA, 40 µl 1 M PIPES pH 6.5, 140 µl H$_2$O); the solutions were mixed and centrifuged briefly in a microfuge. After addition of 2.4 µl 5 M NaCl, mixing, and centrifugation, the tubes were incubated for 5 min at 95 °C, then overnight at 50 °C.

After brief centrifugation at 4 °C, 300 µl of RNase mix [for 10 ml: 100 µl 1 M Tris pH 7.5, 100 µl 0.5 M EDTA, 0.6 ml 5 M NaCl, 10 µl RNase A (10 µg/ml; Boehringer Mannheim), 1.67 µl RNase T1 (1800 U/µl; BRL), 9.2 ml H$_2$O] was added to each tube. After incubation at 42 °C for 20 min, 200 µl stop solution [for 7 ml: 0.95 ml 10% SDS, 0.42 ml 5 M NaCl, 5.52 ml H$_2$O, 105 µl proteinase K (20 mg/ml; Boehringer Mannheim)] was added. The mixture was incubated at 37 °C for 15 min. 600 µl phenol solution (for 25 ml: 0.5 ml isoamyl alcohol, 12 ml CHCl$_3$, 12.5 ml H$_2$O-saturated phenol, made in advance) was added to each tube in a fume hood. After thorough mixing, samples were centrifuged at 4 °C for 5 min. Most (470 µl) of each upper phase was carefully extracted into tubes containing 2 µl (10 µg) tRNA. After isopropanol addition (640 µl), the samples were incubated for 10 min at room temperature, centrifuged for 15 min at 4 °C, and the supernatants were removed. The pellets were washed with 100 µl 70% ethanol (−20 °C), and dried in the Speedvac.

Pellets were dissolved in 6 µl of the same formamide loading dye described above, incubated at 95 °C for 5 min and placed on ice immediately. Samples were analysed by 6% PAGE. After drying, gels were exposed to either (or both, sequentially) Kodak XOMAT film or Molecular Dynamics Phosphor screens. Exposure was for 1–2 weeks for CDF/LIF bands, or overnight for γ-actin and β$_2$-microglobulin bands. Radioactivity was quantitated in arbitrary units by cutting out and weighing peaks from scans [LKB Ultrascan XL (Piscataway, NJ)] of the films, or by scanning the Phosphor screens with a Phosphor-imager 400S (Molecular Dynamics) and computing with ImageQuant software.

In some experiments, tissue mRNAs were incubated with either CDF/LIF or $\beta_2$-microglobulin probes; in other experiments these two probes plus the $\gamma$-actin probe were incubated in the same tubes with a given tissue sample. The three probes and their protected products are clearly distinguishable in this gel system (CDF/LIF initial probe 213 bp, protected product 169 bp; $\beta_2$-microglobulin initial probe 158 bp, product 114 bp; $\gamma$-actin initial probe 112 bp, products 82 and 72 bp).

**Results**

*RT–PCR*

Our first study of the distribution of mRNA for CDF/LIF using the RT–PCR method and total RNA from various postnatal rat tissues found a detectable signal only in the footpad and the visual areas of the mature brain (Yamamori 1991). The ratio of the signal observed in footpad tissue to that potentially present (but not detected) in the other tissues was estimated to be greater than 20:1. We have re-examined many of these same tissues using oligo(dT)-purified mRNA and RT–PCR. This more sensitive procedure reveals a reproducible signal in several tissues. For example, a band corresponding to the size expected for CDF/LIF is found in extracts from postnatal day 1 and 11 (P1 and P11) heart and from adult brain (Fig. 1).

The levels of CDF/LIF mRNA in these and other tissues undergo significant changes during development. For example, there is more CDF/LIF mRNA in P1 heart than in adult heart, and the signal from adult brain is stronger than that from P1 and embryonic day 16 (E16) brain (Fig. 1). Although not quantified, these differences are observed under conditions where the levels of mRNA for $\beta$-actin and $\beta_2$-microglobulin from the same extracts are similar at all ages (Fig. 1) (at the levels of input RNA and number of amplification cycles used, the actin and $\beta_2$-microglobulin signals vary according to the amount of RNA in the tissue sample). Developmental modulations of CDF/LIF mRNA levels are also observed in liver, spleen and lung (results not shown). The fact that CDF/LIF mRNA expression differs among various tissues and changes with age indicates that the band is unlikely to arise from a contamination of reagents or procedures with a CDF/LIF plasmid in the laboratory or from other exogenous sources. The identity of the band has been confirmed by Southern blotting and by sequencing (T. Yamamori & X-Y. Xie, unpublished work).

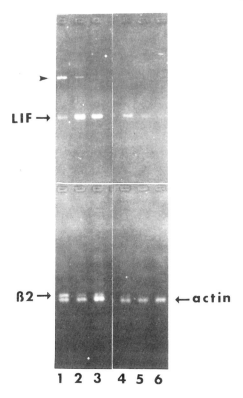

FIG. 1.   The reverse transcriptase–polymerase chain reaction (RT–PCR) method was
used to assay the relative levels of mRNA for CDF/LIF (upper gel), $\beta_2$-microglobulin
(upper arrow, lower gel) and $\beta$-actin (lower arrow, lower gel) in rat heart (lanes 1–3)
and whole brain (lanes 4–6). The ages tested were adult (lane 1), P11 (lane 2) and P1
(lane 3) for heart, and adult (lane 4), P1 (lane 5) and E16 (lane 6) for brain. The arrowhead
on the upper left indicates the position of a band resulting from amplification of genomic
CDF/LIF. Because primers for all these transcripts were included in all tubes, the lack
of $\beta_2$-microglobulin in the brain samples suggests that these contained less RNA than
the heart samples. A reproducible finding was that in the brain the CDF/LIF signal
increases with age, relative to $\beta_2$-microglobulin and actin, whereas in the heart the
CDF/LIF signal decreases with age.

*RNase protection*

To test further and quantify the results obtained by RT–PCR, we used the RNase
protection method, using $\beta_2$-microglobulin and $\gamma$-actin as controls for mRNA
recovery. The presence of CDF/LIF mRNA in adult rat brain is confirmed by
this technique, as are the relatively strong signals in neonatal gut, kidney, lung
and thymus, P11 spleen, and embryonic skeletal muscle and liver (Table 1).
In Table 1 the tissues are listed in order of their CDF/LIF: $\beta_2$-microglobulin

**TABLE 1  RNase protection assay of CDF/LIF mRNA levels in various tissues**[a]

| Tissue | LIF:$\beta_2$-micro | LIF:actin | LIF | $\beta_2$-micro | Actin |
|--------|---------|-----------|-----|----------|-------|
| P1 gut | 0.108 | 0.191 | 276 | 2546 | 1440 |
| P1 kidney | 0.055 | 0.004 | 131 | 2365 | 29190 |
| P1 lung | 0.034 | 0.009 | 151 | 4478 | 17010 |
| E17 skeletal muscle | 0.017 | 0.009 | 61 | 3564 | 6720 |
| P1 thymus | 0.015 | 0.011 | 69 | 4650 | 6543 |
| Adult brain | 0.013 | 0.006 | 91 | 6855 | 16160 |
| E17 liver | 0.013 | 0.004 | 70 | 5530 | 16821 |
| P11 spleen | 0.005 | 0.005 | 162 | 32370 | 31520 |
| E17 brain | <0.005 | <0.001 | 0 | 978 | 37940 |

[a]The levels of CDF/LIF, $\beta_2$-microglobulin ($\beta_2$-micro) and $\gamma$-actin mRNA were assayed by the RNase protection method, as described in the methods section, and the data are presented in arbitrary units. Because the specific activities of the three probes used are different, the numbers for the three mRNAs are not directly comparable. Although the recoveries of the mRNAs in each tissue are not quantified absolutely, the CDF/LIF/$\beta_2$-microglobulin and CDF/LIF:actin ratios are a measure of the relative abundance of CDF/LIF mRNA in each tissue. These data were obtained by Phosphoimager quantification; similar conclusions were derived from an independent experiment in which results were quantified by scanning an autoradiogram of the same type of gel.

ratios. This rank order is similar to that obtained using the CDF/LIF:actin ratios; the exception is kidney, which has a significantly higher actin:$\beta_2$-microglobulin ratio than the other tissues.

**Discussion**

Previous studies have established the presence of CDF/LIF transcripts in the early mouse embryo, in preimplantation blastocysts and extra-embryonic membranes (Conquet & Brûlet 1990), and at the egg cylinder stage (Rathjen et al 1990). These findings are consistent with the fact that this protein can control the proliferation of embryonic stem cells (Evans & Potten 1991). Given its effects on certain myeloid lines, a role for CDF/LIF in haemopoiesis has also been considered. The expression of CDF/LIF mRNA in human bone marrow stromal cells in culture (Wetzler et al 1991) supports this notion, but results from tissue extracts have not previously been reported. Moreover, in bone marrow the activity of CDF/LIF could be on bone deposition and resorption (cf. Gough & Williams 1989) rather than on haemopoiesis. Our results from the RT–PCR and RNase protection assays provide evidence that haemopoietic organs do, in fact, express this mRNA. Expression in liver is prominent during embryogenesis, when the number of colony-forming cells is at its highest and the tissue serves a haemopoietic function (Burgess & Nicola 1983). Our finding that CDF/LIF mRNA is expressed in spleen is further evidence in support of a haemopoietic role for this cytokine. The relatively high levels in neonatal kidney could also be physiologically relevant; CDF/LIF has recently been found to exert striking effects on the development of metanephric mesenchyme in organ culture (Bard & Ross 1991).

In the nervous system, culture studies have shown that CDF/LIF stimulates the generation of neurons from neural crest, and supports the generation or survival of sensory neurons from dorsal root ganglia (Murphy et al 1991). Moreover, this protein can regulate the expression of a number of neuropeptide and neurotransmitter genes in these neurons (Nawa et al 1990). We also found that this cytokine can control the phenotype of post-mitotic, sympathetic neurons in culture (Yamamori et al 1989). In particular, CDF/LIF acts as a cholinergic differentiation factor, up-regulating expression of the acetylcholine biosynthetic enzyme, choline acetyltransferase, as well as that of the neuropeptide often associated with acetylcholine, vasoactive intestinal polypeptide, while down-regulating expression of the noradrenergic biosynthetic enzyme, tyrosine hydroxylase, and the frequently co-localized neuropeptide, neuropeptide Y (Nawa & Patterson 1990, Nawa et al 1990). Such findings suggested that CDF/LIF may be the factor that converts noradrenergic sympathetic neurons to the cholinergic phenotype when they innervate the sweat glands in the footpad of the rat (Landis 1990). The other primary candidate for the footpad cholinergic factor is ciliary neurotrophic factor (CNTF), a protein that has identical effects to CDF/LIF on a number of neurotransmitter and neuropeptide genes in cultured sympathetic neurons (Saadat et al 1989; M. Rao, S. C. Landis & P. H. Patterson, unpublished work). In fact, we have detected both a CNTF-like protein and CDF/LIF mRNA in extracts of footpads (Rao & Landis 1990, Yamamori 1991). As mentioned previously, the fact that CDF/LIF transcript levels are higher in footpads than in noradrenergically innervated tissues of similar ages lends credence to a role for this cytokine in cholinergic differentiation *in vivo*.

CDF/LIF mRNA expression in embryonic skeletal muscle supports the possibility that this factor may be a target-derived trophic factor for motor neurons. CDF/LIF enhances choline acetyltransferase activity in, and the survival of, cultured embryonic rat motor neurons purified by fluorescence-activated cell sorting [J. C. Martinou, I. Martinou, P. H. Patterson, A. Kato 1991 Third IBRO World Congress of Neuroscience, 4.45 (abstr)]. The presence of higher levels of CDF/LIF message in adult brain than in embryonic and neonatal brain suggests another potential role for this cytokine in the central nervous system. As postulated for CNTF (Stöckli et al 1991), CDF/LIF could be produced for the purpose of wound repair. Alternatively, CDF/LIF could be involved in the ongoing functioning of the mature brain. Yet another possibility, related to both of these notions, is that CDF/LIF could be a sprouting factor involved in the production of new connections in response to incoming sensory stimuli. Physical connectivity has been shown to be altered in several learning paradigms in adult animals (e.g. Schacher & Montarolo 1991).

The overlap between the activities of CDF/LIF and CNTF in the control of a battery of phenotypic markers in sympathetic neurons has prompted a search for shared molecular themes in the sequences of these two differentiation factors. This analysis revealed a sequence motif, D2, shared between CDF/LIF and CNTF, as well as a putative tertiary structure shared with IL-6 (Bazan 1991). The molecular relationship between these three factors has now been extended to their receptors; the CNTF receptor is most closely related to the ligand-binding subunit of the IL-6 receptor (Davis et al 1991), and the CDF/LIF receptor is related to the IL-6 signal-transducing subunit gp130 (Gearing et al 1991). Further tightening this set of interrelationships is the considerable overlap in the effects of CDF/LIF and IL-6 in a number of non-neuronal tissues described in this volume. The redundancy in function and similarity in structural motifs among these and several other regulatory factors emphasize the importance of localizing these agents *in vivo* (Patterson 1992).

## Acknowledgements

We thank David Anderson for the use of his equipment and Derek Stemple and Susan Birren for advice and reagents. Tetsuo Yamamori initiated the studies of postnatal tissues and Xiao-Yi Xie participated in the beginning of the studies of embryonic tissues. This work was supported by a Jacob Javits Investigator Award from the National Institute of Neurological Disorders and Stroke to P.H.P., and by fellowships to M.-J.F. from the Helen G. and Arthur McCallum Foundation, the Ministry of Education, Taiwan, R.O.C., and a Li Ming Scholarship.

## References

Anderson DJ 1989 The neural crest cell lineage problem: neuropoiesis? Neuron 3:1–12

Bard JBL, Ross ASA 1991 LIF, the ES-cell inhibition factor, reversibly blocks nephrogenesis in cultured mouse kidney rudiments. Development 113:193–198

Bazan JF 1991 Neuropoietic cytokines in the hematopoietic fold. Neuron 7:197–208

Burgess A, Nicola N 1983 Growth factors and stem cells. Academic Press, Sydney

Conquet F, Brûlet P 1990 Developmental expression of myeloid leukemia inhibitory factor gene in preimplantation blastocysts and in extraembryonic tissue of mouse embryos. Mol Cell Biol 10:3801–3805

Davis S, Aldrich TH, Valenzuela DM et al 1991 The receptor for ciliary neurotrophic factor. Science (Wash DC) 253:59–63

Enoch T, Zinn K, Maniatis T 1986 Activation of the human β-interferon gene requires an interferon-inducible factor. Mol Cell Biol 6:801–810

Evans GS, Potten CS 1991 Stem cells and the elixir of life. BioEssays 13:135–138

Fukada K 1985 Purification and partial characterization of a cholinergic neuronal differentiation factor. Proc Natl Acad Sci USA 82:8795–8799

Gearing DP, Gough NM, King JA et al 1987 Molecular cloning and expression of cDNA encoding a murine myeloid leukaemia inhibitory factor (LIF). EMBO (Eur Mol Biol Organ) J 6:3995–4002

Gearing DP, Thut CJ, VandenBos T et al 1991 Leukemia inhibitory factor receptor is structurally related to the IL-6 signal transducer, gp130. EMBO (Eur Mol Biol Organ) J 10:2839–2848

Gough NM, Williams RL 1989 The pleiotropic actions of leukemia inhibitory factor. Cancer Cells (Cold Spring Harbor) 1:77–80

Gunning P, Ponte P, Okayama H, Engel J, Blau H, Kedes L 1983 Isolation and characterization of full-length cDNA clones for human $\alpha$-, $\beta$-, and $\gamma$-actin mRNAs: skeletal but not cytoplasmic actins have an amino-terminal cysteine that is subsequently removed. Mol Cell Biol 3:787–795

Landis SC 1990 Target regulation of neurotransmitter phenotype. Trends Neurosci 13:344–350

Mauxion F, Kress M 1987 Nucleotide sequence of rat beta2-microglobulin cDNA. Nucleic Acids Res 15:7638

Murphy M, Reid K, Hilton DJ, Bartlett PF 1991 Generation of sensory neurons is stimulated by leukemia inhibitory factor. Proc Natl Acad Sci USA 88:3498–3501

Nawa H, Patterson PH 1990 Separation and partial characterization of neuropeptide-inducing factors in heart cell conditioned medium. Neuron 4:269–277

Nawa H, Yamamori T, Le T, Patterson PH 1990 Generation of neuronal diversity: analogies and homologies with hematopoiesis. Cold Spring Harbor Symp Quant Biol 55:247–253

Nicola NA 1989 Hemopoietic cell growth factors and their receptors. Annu Rev Biochem 58:45–77

Nudel U, Zakut R, Shani M, Neuman S, Levy Z, Yaffe D 1983 The nucleotide sequence of the rat cytoplasmic $\beta$-actin gene. Nucleic Acids Res 11:1759–1771

Patterson PH 1992 The emerging neuropoietic cytokine family: first CDF/LIF, CNTF and IL-6, next ONC, MGF, GCSF? Curr Opin Neurobiol 2, in press

Patterson PH, Chun LLY 1977 The induction of acetylcholine synthesis in primary cultures of dissociated rat sympathetic neurons. I. Effects of conditioned medium. Dev Biol 56:263–280

Rao MS, Landis SC 1990 Characterization of a target-derived neuronal cholinergic differentiation factor. Neuron 5:899–910

Rathjen PD, Nichols J, Toth S, Edwards DR, Heath JK, Smith AG 1990 Developmentally programmed induction of differentiation inhibiting activity and the control of stem cell populations. Genes & Dev 4:2308–2318

Saadat S, Sendtner M, Rohrer H 1989 Ciliary neurotrophic factor induces cholinergic differentiation of rat sympathetic neurons in culture. J Cell Biol 108:1807–1816

Schacher S, Montarolo PG 1991 Target-dependent structural changes in sensory neurons of aplysia accompany long-term heterosynaptic inhibition. Neuron 6:679–690

Stahl J, Gearing DP, Willson TA, Brown MA, King JA, Gough NM 1990 Structural organization of the genes for murine and human leukemia inhibitory factor. Evolutionary conservation of coding and non-coding regions. J Biol Chem 265:8833–8841

Stöckli KA, Lillien LE, Näher-Noé M et al 1991 Regional distribution developmental changes and cellular localization of CNTF-mRNA and protein in the rat brain. J Cell Biol 115:447–459

Wetzler M, Talpaz M, Lowe DG, Baiocchi G, Gutterman JU, Kurzrock R 1991 Constitutive expression of leukemia inhibitory factor RNA by human bone marrow stromal cells and modulation by IL-1, TNF-$\alpha$, and TGF-$\beta$. Exp Hematol (NY) 19:347–351

Yamamori T 1991 Localization of CDF/LIF mRNA in the rat brain and peripheral tissues. Proc Natl Acad Sci USA 88:7298–7302
Yamamori T, Fukada K, Aebersold R, Korsching S, Fann M-J, Patterson PH 1989 The cholinergic neuronal differentiation factor from heart cells is identical to leukemia inhibitory factor. Science (Wash DC) 246:1412–1416


## DISCUSSION

*Metcalf:* You didn't say much about IL-6.

*Patterson:* We have tested the effects of IL-6 on the same set of neurotransmitter and neuropeptide genes that respond to CDF/LIF, and the results were negative. Ming-Ji Fann now has a method by which he can quickly screen factors for their effects on a dozen different neuronal phenotypic markers *in vitro*, and he has observed little response to the addition to sympathetic neurons of cytokines, including oncostatin and stem cell factor. We tested IL-6 several times on these neurons because they are related to PC12 cells and Satoh et al (1988) found that IL-6 exerts an NGF-like effect on PC12 cells.

*Dexter:* Are there behavioural changes in animals after administration of LIF, CNTF or IL-6?

*Patterson:* Don Metcalf and David Gearing found that injection of high doses of recombinant LIF, or cells secreting it, induced hypermobility and irritability in mice (Metcalf & Gearing 1989, Metcalf et al 1990). These effects could, of course, be indirect or the result of general toxicity. We have begun to study the effects on neurons *in vivo* of local injections of the protein as well as of antibodies that block its activity.

*Dexter:* In these differentiation pathways are any of the commitment steps reversible?

*Patterson:* The three major branches of the sympathoadrenal lineage— sympathetic neurons, adrenal chromaffin cells and small intensely fluorescent (SIF) cells—have a common progenitor. SIF cells and chromaffin cells, although they are terminal phenotypes in the adult animal, can be transformed into sympathetic neurons by appropriate manipulation of the instructive signals in their environment (Patterson 1990). This switch involves an intermediate stage in which the cells resemble the sympathoadrenal progenitor itself; therefore, the step to chromaffin and SIF cell differentiation can be regarded as reversible with the progenitor cell stage (Doupe et al 1985, Anderson 1989). This phenotypic reversibility can also be observed with cells taken from the adult animal, and can occur after transplantation of the adrenal medulla to the appropriate location *in vivo* (Olson 1970, Doupe et al 1985). In contrast, we have not been able to reverse neuronal differentiation back to the progenitor cell stage.

*Nicola:* Do you know anything about the frequency at which this occurs? Are you dealing with a population of stem cells that can differentiate into either a chromaffin cell or a neuron?

*Patterson:* When we switch SIF cells or chromaffin cells to conditions that favour neuronal differentiation, virtually all of the former, and a majority of the latter, become neurons (Doupe et al 1985). The reversibility experiments involve both following the fates of individual cells over time and statistical analysis of the fates of large populations of cells.

*Heath:* Has this been done clonally? I can't see how you can get three different lineages directly from a single cell. To define a cell as a multipotential progenitor, by analogy with the haemopoietic system, you need to tag one cell and demonstrate three different types of progeny.

*Patterson:* The conclusion that a single cell can give rise to all three derivatives comes from fluorescence-activated cell sorting (FACS) experiments. Sympathoadrenal progenitors from embryonic adrenal and sympathetic ganglia were identified by immunohistochemical staining *in vivo*, and purified with the identifying antibodies by FACS (Anderson & Axel 1986, Carnahan & Patterson 1991). Individual cells, as well as populations, were followed under neuron-promoting (fibroblast growth factor plus nerve growth factor) and chromaffin-promoting (corticosteroid) conditions. It was found that the fates of the progenitor cells could be controlled, so that nearly all cells could become either neurons or chromaffin cells, depending on the signals present. These experiments ruled out the alternative explanation, that the two populations arose by selective survival of populations of predetermined precursor cells for neurons and chromaffin cells. In addition, as I mentioned, fully differentiated chromaffin and SIF cells could be transformed into sympathetic neurons. I should also stress that many of these experiments were done with postmitotic cells, so that differential proliferation of predetermined, separate precursors for chromaffin cells or neurons could not have given rise to the cells seen.

*Lee:* It is difficult to detect LIF mRNA in total brain because it is localized in particular regions. Do you know where it is localized, and have you looked for the protein by testing for biological activity or by immunoassay?

*Patterson:* We have examined this in a crude way, by dissecting the brain and running the RT–PCR assay. In this way we localized CDF/LIF mRNA in two structures in the visual system, the superior colliculus and the visual cortex of the adult rat brain (Yamamori 1991). In addition, Minami et al (1991) have detected a signal in the cerebral cortex and hippocampus. It's interesting that this mRNA appears to be selectively expressed in certain parts of the brain, and that the message is more prominent in the adult than in younger animals. We have not yet looked for CDF/LIF activity or protein in these areas.

*Gauldie:* Jean Marshall, Ron Stead and John Bienenstock have been looking at mast cell–nerve interactions, and how these are involved in inflammation. They have found a very close physical association between mast cells and nerves in the tissue (Stead et al 1989). Even in a chronically parasite-infected animal in which there's a major proliferation of mast cells, there is associated proliferation of the nerves. Degranulating peritoneal mast cells express LIF as well as IL-6 (J. S. Marshall, J. Gauldie & L. Nielsen, unpublished work). Perhaps mast cell-derived LIF plays a role in the association between those two cell types.

*Williams:* Does LIF have any developmental effects in terms of directing the growth of the neuron down along developing arterioles?

*Patterson:* There is good evidence for *tropic* effects of diffusible proteins in the nervous system. NGF, for example, can directionally guide axon outgrowth both *in vivo* and in culture. I know of no evidence that CDF/LIF can function in that way. Also, it does not have a *trophic*, or nutritive, effect on postmitotic sympathetic neurons. There is good evidence though that it can enhance the survival of sensory neurons in culture (Murphy et al 1991), and possibly also motor neurons (J. C. Martinou, I. Martinou, P. H. Patterson & A. Kato, unpublished work). Our work has centred on the control of phenotypic choices by CDF/LIF—its effects as a differentiation factor.

*Stewart:* Have you looked for expression of LIF in the olfactory nerves? I believe that's the only central nervous cell population which can extensively regenerate in the adult.

*Patterson:* We haven't looked at the olfactory nerve, which is the only progenitor cell population in mammals that constantly generates new neurons. If you are asking whether CDF/LIF or CNTF function not only as cholinergic differentiation factors but also to attract the neurons, we don't know.

*Nicola:* We have shown that after injection of radioactive LIF into the mouse footpad you see the same pattern as with NGF, accumulation of radioactivity at the point of sciatic nerve ligation, and labelling of a percentage of neurons in the dorsal root ganglia (Hendry et al 1991). That suggests that LIF acts as a classical neurotrophic factor in adults.

*Gauldie:* Mike Blennerhasset has been studying neural outgrowth in culture using time-lapse photography. When he puts a rat basophilic leukaemia (RBL) cell, which is capable of making LIF, to one side of the developing sympathetic neuron, the neuron turns and grows towards the RBL cell (Blennerhasset & Bienenstock 1990, Blennerhasset et al 1992). That's no guarantee that factors such as LIF are involved, but it's an indication that they might be.

*Gough:* Does LIF have effects on motor neurons?

*Patterson:* CDF/LIF and CNTF both act as survival factors for purified motor neurons in culture, and they also enhance choline acetyltransferase activity. Moreover, application of CNTF *in vivo* can promote survival of motor neurons in neonatal rats and embryonic chicks (Arakawa et al 1990, Oppenheim et al 1991). However, sensitive assays have so far failed to detect mRNA for CNTF in skeletal muscle. Thus, the current theory is that CNTF is not a classical, target-derived neurotrophic factor. Schwann cells contain high levels of CNTF and it could be that this factor is released when nerves are damaged, to promote neuronal regeneration. Our finding that embryonic rat skeletal muscle contains CDF/LIF mRNA leaves open the possibility that this protein acts as a target-derived motor neuron survival factor.

*Gough:* We have demonstrated production of LIF and IL-6 in primary astrocyte cultures after stimulation with bacterial lipopolysaccharide (LPS) or cytomegalovirus infection (Wesselingh et al 1990). Also, *in vivo*, after LPS stimulation we see a significant increase in the expression of LIF mRNA in the brain. What do you think is the significance of those results?

*Patterson:* That finding makes me even more interested to know whether it's present at low levels in astrocytes in normal brain. In reactive astrocyte syndrome many genes are up-regulated after damage to the brain, including those encoding matrix factors, such as laminin, as well as a variety of growth factors. It's thought this helps to prevent damage to neurons and facilitates repair. If LIF is part of that response syndrome, it is a possible candidate for a growth factor in the brain. The fact that it seems to be selectively localized in particular brain regions supports the idea that it plays a role. When a signal cannot be detected in a Northern or protection assay using whole brain extracts it is often assumed that the protein doesn't have a role there. The incredible diversity of phenotypes in the nervous system makes it difficult to detect even those growth factors that we know are present and active in the brain, such as NGF or BDNF, in extracts of whole brain, but if you look in the right place, at the right time of development, you can see a strong signal. This could also be true for LIF, IL-6 and oncostatin.

*Sehgal:* Is there any information about the induction of LIF by neuropeptides such as substance P? IL-6 can be induced in certain cell types by neuropeptides (Lotz et al 1988).

*Patterson:* I am not aware of any results on that.

*Heath:* Have you looked at LIF in microglia?

*Patterson:* No.

*Baumann:* Are there brain tumours which have increased LIF mRNA?

*Metcalf:* There are continuous cultures of childhood neuroblastoma tissue that produce LIF, but the cultures contain stromal cells so the exact cellular origin of the LIF has not been established.

*Baumann:* I have assayed supernatants from human astroglial cells in culture. Some do make a lot of LIF but again that could be a tissue culture artifact.

*Metcalf:* This is a problem with almost all the cytokines that we have looked at. When held in culture, cells exhibit a marked induction of cytokine transcription within hours. This makes it difficult to extrapolate the data to the normal functioning of that cell *in vivo*. The rise in transcription and rate of production of a protein can be of three orders of magnitude. You really need to look at mRNA levels in primary specimens taken directly from patients to answer this question.

Dr Patterson, I can envisage target tissue-generated LIF driving innervation and the type of signalling that results in neonatal animals, but do you think in earlier fetal development qualitatively different events might happen? We have to bridge the gap between the actions of LIF on embryonic stem cells, and this type of evidence, which is developmental but it's late developmental. What might happen in between?

*Patterson:* CDF/LIF can affect proliferation, survival and differentiation of dividing neuronal progenitors from the neural crest (Murphy et al 1991). In addition, CNTF was found to inhibit proliferation and to induce neuronal differentiation of progenitors in embryonic chick sympathetic ganglia (Ernsberger et al 1989).

*Williams:* Could you speculate, or do you know, whether these cytokines affect the threshold for firing, or the frequency of firing, or any process which might be part of the normal functioning of neurons?

*Patterson:* NGF is known to up-regulate sodium channel expression, so it can affect the ability of a neuron to fire. The effects of CDF/LIF on sympathetic neuron gene expression can be inhibited by calcium flux into the neurons (Walicke et al 1977, Walicke & Patterson 1981). Therefore, it is conceivable that the action of CDF/LIF could involve calcium traffic.

*Williams:* Are LIF levels altered in mice with nerve disorders?

*Patterson:* We are just beginning experiments with CDF/LIF and motor neurons *in vivo*, and would like to look at mouse models of motor neuron disease.

## References

Anderson DJ 1989 Cellular neoteny: a possible developmental basis for chromaffin cell plasticity. Trends Genet 5:174–178

Anderson DJ, Axel R 1986 A bipotential neuroendocrine precursor whose choice of cell fate is determined by NGF and glucocorticoids. Cell 47:1079–1090

Arakawa Y, Sendtner M, Thoenen H 1990 Survival effect of ciliary neurotrophic factor (CNTF) on chick embryonic motoneurons in culture: comparison with other neurotrophic factors and cytokines. J Neurosci 10:3507–3515

Blennerhassett MG, Bienenstock J 1990 Apparent innervation of rat basophilic leukemia (RBL-2H3) cells by sympathetic neurons in vitro. Neurosci Lett 120:50–54

Blennerhassett MG, Tomioka M, Bienenstock J 1992 Formation of contacts between mast cells and sympathetic neurons in vitro. Cell Tissue Res 265:121–128

Carnahan JF, Patterson PH 1991 Isolation of the progenitor cells of the sympathoadrenal lineage from embryonic sympathetic ganglia with the SA monclonal antibodies. J Neurosci 11:3520–3530

Doupe AJ, Patterson PH, Landis SC 1985 Small intensely fluorescent cells in culture: role of glucocorticoids and growth factors in their development and interconversions with other neural crest derivatives. J Neurosci 5:2143–2160

Ernsberger U, Sendtner M, Rohrer H 1989 Proliferation and differentiation of embryonic chick sympathetic neurons: effects of ciliary neurotrophic factor. Neuron 2:1275–1284

Hendry I, Murphy M, Hilton DJ, Nicola NA, Bartlett PF 1992 Binding and retrograde transport of leukemia inhibitory factor by the sensory nervous system. J Neurosci, in press.

Lotz M, Vaughan JH, Carson DA 1988 Effect of neuropeptides on production of inflammatory cytokines by human monocytes. Science (Wash DC) 241:1218–1221

Metcalf D, Gearing DP 1989 A fatal syndrome in mice engrafted with cells producing high levels of leukemia inhibitory factor. Proc Natl Acad Sci USA 86:5948–5952

Metcalf D, Nicola NA, Gearing DP 1990 Effects of injected leukemia inhibitory factor on haemopoietic and other tissues in mice. Blood 76:50–56

Minami M, Kuraishi Y, Satoh M 1991 Effects of kainic acid on messenger RNA levels of IL-1-β, IL-6, TNF-α and LIF in the rat brain. Biochem Biophys Res Commun 176:593–598

Murphy M, Reid K, Hilton DJ, Bartlett PF 1991 Generation of sensory neurons is stimulated by leukemia inhibitory factor. Proc Natl Acad Sci USA 88:3498–3501

Olson L 1970 Fluorescence histochemical evidence for axonal growth and secretion from transplanted adrenal medullary tissue. Histochemie 22:1–7

Oppenheim RW, Prevette D, Qin-Wei Y, Collins F, MacDonald H 1991 Control of embryonic motoneuron survival in vivo by ciliary neurotrophic factor. Science (Wash DC) 251:1616–1618

Patterson PH 1990 Control of cell fate in a vertebrate neurogenic lineage. Cell 62:1035–1038

Satoh T, Nakamura S, Taga T et al 1988 Induction of neuronal differentiation in PC12 cells by B-cell stimulatory factor 2/interleukin 6. Mol Cell Biol 8:3546–3549

Stead RH, Dixon MF, Bramwell NH, Riddell RH, Bienenstock J 1989 Mast cells are closely apposed to nerves in the human intestinal mucosa. Gastroenterology 97:575–585

Walicke PA, Patterson PH 1981 On the role of $Ca^{2+}$ in the transmitter choice made by cultured sympathetic neurons. J Neurosci 1:343–350

Walicke PA, Campenot RB, Patterson PH 1977 Determination of transmitter function by neuronal activity. Proc Natl Acad Sci USA 74:5767–5771

Wesselingh SL, Gough NM, Finlay-Jones JJ, McDonald PJ 1990 Detection of cytokine mRNA in astrocyte cultures using the polymerase chain reaction. Lymphokine Res 9:177–185

# Leukaemia inhibitory factor and bone cell function

T. J. Martin*, E. H. Allan*, R. S. Evely* and I. R. Reid†

University of Melbourne Department of Medicine and St. Vincent's Institute of Medical Research*, St. Vincent's Hospital, Melbourne, Australia, and University of Auckland Department of Medicine†, Auckland, New Zealand

*Abstract.* A bone-resorbing product of mouse spleen cells found to have differentiation-inducing activity was most probably leukaemia inhibitory factor (LIF). This revealed that LIF is a cytokine active on bone, in addition to its several other sites of action. In organ culture of newborn mouse bone, recombinant LIF promoted bone resorption by a prostaglandin-dependent process. Resorption by isolated rat osteoclasts was also promoted by LIF through an initial action on osteoblasts which was receptor-mediated. Incorporation of [$^3$H]thymidine into DNA was increased by LIF in cells (most probably osteoblasts) of the newborn mouse bones. Osteoblasts have been shown to produce LIF, and the amount is increased by treatment with retinoic acid or TNF-$\alpha$. LIF also acts directly on osteoblasts to inhibit plasminogen activator activity, by stimulating the synthesis of plasminogen activator inhibitor 1 mRNA and protein. The latter actions are very similar to those of TGF-$\beta$. Again like TGF-$\beta$, LIF was ineffective in promoting bone resorption *in vitro* in fetal rat long bones. These results, together with the *in vivo* data showing that high circulating levels of LIF in the mouse are accompanied by a substantial increase in trabecular bone mass, indicate that LIF is another cytokine with potent actions on bone and potentially important interactions with other osteotrophic factors.

*1992 Polyfunctional cytokines: IL-6 and LIF. Wiley, Chichester (Ciba Foundation Symposium 167) p 141–155*

The processes of bone formation and resorption are continuous throughout life, and need to be balanced if bone mass is to be maintained. This control is achieved through the actions of a number of circulating hormones, but especially through the local production and action within bone of several cytokines and growth factors. The very important role played by cytokines has been realized over the last few years. The term 'osteoclast-activating factor' (OAF) was applied to bone-resorbing activity detected in supernatants from lectin-transformed lymphocytes (Horton et al 1972); this activity was considered to explain the excessive bone resorption accompanying certain haematological malignancies. Since it has been discovered that certain cytokines are highly potent promoters of bone resorption *in vitro*, it has been accepted that the term 'OAF' describes those cytokines, notably interleukin 1 and tumour necrosis factor (TNF) $\alpha$ and $\beta$ (Martin et al 1989).

In the course of studying the bone-resorbing properties of products of mouse spleen cells Abe et al (1986) discovered a further bone-resorbing factor, which they denoted differentiation-inducing factor (DIF or D-factor). The biological assay used in the purification of DIF was the promotion of differentiation of M1 leukaemia cells. This same assay was used in the study and purification of leukaemia inhibitory factor (LIF) (Hilton et al 1988). In view of these earlier observations it was clearly of interest to study the effects of LIF on bone cell functioning when this cytokine had been cloned and expressed (Gearing et al 1987, 1989a,b, Gough et al 1988).

## Effects of LIF on bone resorption *in vitro*

Both recombinant mouse and human LIF stimulated bone resorption in mouse calvaria *in vitro* (Reid et al 1990), and, as with other bone-resorbing hormones and cytokines, this was associated with an increase in the number of active osteoclasts at the end of the experiment (e.g. control $360 \pm 50$, LIF-treated $710 \pm 50$ osteoclasts). Because the promotion of bone resorption by some cytokines in this system is dependent upon intra-osseous synthesis of prostaglandins, experiments with LIF were carried out in the presence of indomethacin (an inhibitor of prostaglandin synthesis), which was found to block the increase in resorption (Table 1). Thus, the effect of LIF on bone resorption in mouse calvaria is prostaglandin dependent, as are the effects of IL-1 and transforming growth factors (TGF) $\alpha$ and $\beta$ (Raisz 1988, Tashjian et al 1985).

However, in a fetal rat long bone organ culture system, in which resorption is considerably influenced by the generation of osteoclasts from precursors, LIF actually inhibited resorption (Lorenzo et al 1990). This pattern of responses is

**TABLE 1    Effect of indomethacin on LIF-induced bone resorption in mouse calvaria**

| Treatment[a] | $^{45}Ca^{2+}$ release (%) |
|---|---|
| Control | $18.6 \pm 0.3$ |
| Human recombinant LIF (5000 U/ml) | $22.5 \pm 0.5$* |
| Indomethacin ($10^{-7}$ M) | $16.8 \pm 0.1$ |
| Indomethacin + LIF | $18.7 \pm 0.4$ |

[a]Two-day-old mice were injected subcutaneously with 5 µCi $^{45}$Ca. Calvariae were removed four days later, incubated in medium for 24h then incubated in fresh medium containing indomethacin. LIF was added 2 h later. After a further 48 h the $^{45}Ca^{2+}$ content of the medium and of trichloroacetic acid extracts of calvariae was measured by liquid scintillation counting.
Data from Reid et al 1990.
*Significantly different from control, $P < 0.02$.

identical to that of TGF-β, which also inhibits bone resorption in fetal rat long bones in culture (Pfeilschifter et al 1988). Similar observations have been made by van Beek et al (1992), who used cultured fetal mouse metacarpal bones, in which resorption depends on the development of progenitors and precursors into mature osteoclasts. Mouse LIF inhibited resorption in this system, and it was noted also that LIF suppressed mineralization of the matrix during culture.

The stimulation of bone resorption by LIF in mouse calvarial cultures was achieved with concentrations (10–100 units/ml) similar to those which affect proliferation of M1 cells (Gearing et al 1989a) and the differentiation of embryonic stem cells (Williams et al 1988). Effects at such low concentrations indicate that LIF might have a role in normal bone cell functioning. The prostaglandin-dependent stimulation of resorption in mouse calvaria is in accord with earlier observations made by Abe et al (1986), who reported similar effects of partially purified D-factor on resorption. The same group has also shown prostaglandin-dependent actions of recombinant LIF on mouse calvarial resorption (T. Suda et al unpublished work), and, most interestingly, significant cooperative effects of LIF together with IL-1 and IL-6.

So far, no bone-resorbing agent has been shown to act directly on osteoclasts to promote their activity. All appear to act first on other cells, most probably those of osteoblast lineage, and osteoclast activation follows (Rodan & Martin 1981, Chambers 1985, Martin et al 1989). It seems likely also that the promotion of osteoclast formation requires the participation of stromal or osteoblastic cells (Takahashi et al 1988, Martin et al 1989), although this is more controversial.

We therefore carried out experiments to investigate further the mechanism of LIF stimulation of bone resorption. Receptor autoradiography using [125]I-labelled LIF clearly demonstrated the existence of receptors on osteoblasts (Allan et al 1990), but no evidence was obtained for the binding of LIF to cells with the phenotypic features of osteoclasts, either mono- or multi-nucleated. Furthermore, when we studied resorption by isolated osteoclasts on bone slices, using methods developed by Chambers (1985), we found that LIF promoted resorption only in cultures which were significantly contaminated with osteoblasts. Highly purified osteoclasts failed to respond (Table 2). In all these experiments the extent of the resorption response to LIF was consistently less than that to parathyroid hormone. The same applies to the magnitude of resorption observed in organ cultures of bone (Reid et al 1990).

We concluded, therefore, that the mechanism by which LIF stimulates osteoclasts to resorb bone resembles in general that of other bone-resorbing factors in that its action is not directly on osteoclasts, but mediated by osteoblasts. Furthermore, the prostaglandin dependence of this effect and LIF's inhibition of resorption when osteoclast generation is required are properties that are shared with TGF-β (Pfeilschifter et al 1988). For TGF-β, the latter effect may be explained by its ability to inhibit the proliferation of haemopoietic

**TABLE 2   Effects of LIF and parathyroid hormone (PTH) on resorption by isolated osteoclasts**

| Type of culture | Treatment | Area resorbed per bone slice $(\mu m^2 \times 10^{-3})$ |
|---|---|---|
| Highly purified osteoclasts | Control | $15.6 \pm 4.3$ |
| | LIF (1000 U/ml) | $11.8 \pm 2.1$ |
| | PTH (200 ng/ml) | $14.3 \pm 2.8$ |
| Osteoclasts and osteoblasts | Control | $9.8 \pm 4.1$ |
| | LIF | $21.2 \pm 3.0*$ |
| | PTH | $40.8 \pm 8.4**$ |

*Significantly different from control, $P < 0.05$; **$P < 0.001$.

precursors. One could predict that LIF will inhibit osteoclast formation, either by actions on specific precursors or through a general effect on the haemopoietic lineage, as TGF-β does. This is being investigated.

**Actions of LIF on bone *in vivo***

Overexpression of LIF was achieved in mouse haemopoietic FD cells transfected with a LIF cDNA. Injection of these cells into syngeneic DBA/2 mice resulted in very high circulating levels of LIF (Metcalf & Gearing 1989). The striking changes in these animals included a large increase in new bone formation and increased numbers of osteoblasts. There was also histological evidence for areas of increased osteoclastic resorption.

   Thus, the result of high LIF production in these animals was a substantial increase in bone turnover with a significant increase in bone mass. The levels of LIF reached were so high that there may have been effects on production of other cytokines or growth factors, which could have contributed to the impressive skeletal changes. The effect of LIF *in vivo* clearly requires further study. The method of delivery of the cytokine and the concentrations used will be important, especially because it seems likely that any effect LIF has on bone will, as with the other cytokines, result from paracrine actions.

**Production of LIF by osteoblasts**

A focal point in LIF's effects is the osteoblast, and it became clear early on that LIF receptors are present on cells of the osteoblast lineage (Allan et al 1990). Because several other cytokines are produced by osteoblasts, we sought evidence for production of LIF by osteoblast-like cells; we showed, by a biological assay (M1 differentiation inhibition) and a receptor binding assay, that LIF was indeed synthesized by osteoblasts, and that the amount of LIF mRNA and activity was increased by treatment with TNF-α (Allan et al 1990). We detected LIF mRNA

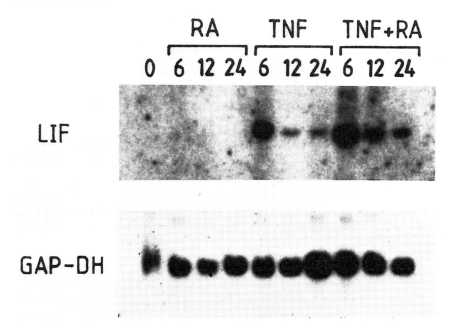

FIG. 1. Northern analysis showing the induction of LIF mRNA in pre-osteoblastic cells (UMR201) treated with TNF-α and retinoic acid, alone or in combination. RA, retinoic acid; TNF, tumour necrosis factor α; GAP-DH, glyceraldehyde-3-phosphate dehydrogenase. Numbers indicate hours after treatment. Reproduced from Allan et al 1990 with permission of Wiley-Liss, a division of John Wiley & Sons, Inc. Copyright ©1990 Wiley-Liss.

in calvarial osteoblasts, UMR106 osteogenic sarcoma cells and pre-osteoblastic UMR201 cells, with both the polymerase chain reaction and Northern analysis (Allan et al 1990). The example in Fig. 1 shows the stimulation by TNF-α of LIF mRNA in pre-osteoblastic UMR201 cells, and the augmentation of this by retinoic acid. It is interesting to note also that before LIF had been cloned and expressed, evidence was obtained for the synthesis of this biological activity by the mouse osteoblastic cell line MC3T3-E1 (Shiina-Ishimi et al 1986).

Thus, the production of LIF by osteoblastic cells in culture is established. It will be important to determine how LIF production in bone occurs *in vivo*— whether it is predominantly in precursor or more mature cells, or whether the marrow is the major source. One particularly interesting possibility is that LIF made by osteoblasts is located in the matrix; this prospect arises from the finding that LIF could be purified from the matrix laid down by cells in culture (Rathjen et al 1990). This should certainly be explored in studies of the role of LIF in bone, because there is an obvious parallel with the shortage in matrix of important bone growth factors (including TGF-β, insulin-like growth factors 1 and 2, fibroblast growth factor and granulocyte-macrophage colony-stimulating factor).

## Actions of LIF on osteoblasts

### Cell proliferation

In the course of investigating the effects of LIF on resorption in mouse calvarial tissue it was found that LIF treatment increased DNA synthesis, as assessed by incorporation of [3H]thymidine, an effect which was not prostaglandin dependent (Reid et al 1990; Table 3). Furthermore, incorporation of [3H]phenylalanine into protein synthesized by calvariae was also stimulated. Effects on cell replication in calvarial organ cultures are difficult to interpret because of their cellular heterogeneity. These studies need to be extended, to identify by [3H]thymidine autoradiography which cells exhibit the mitogenic response. The fact that receptor autoradiography has shown binding of LIF to osteoblasts or precursors but not to osteoclasts (Allan et al 1990) suggests that it is probably cells of the osteoblast lineage which respond in this manner in the calvarial organ cultures.

We therefore investigated the effects of LIF on DNA synthesis and replication of osteoblasts, using primary cultures of rat calvarial osteoblasts and the UMR106 osteogenic sarcoma cell line. LIF caused a dose-dependent increase in DNA synthesis and cell number in both growth-arrested and actively growing osteoblasts (Lowe et al 1991a). The response to LIF was most evident when cells were treated for up to 24 hours. By the time cells approached confluence, LIF slightly inhibited DNA synthesis. These observations were consistent with those from calvarial organ cultures. In both cases the stimulatory effects of LIF on DNA synthesis and cell replication were unaffected by blockade of

**TABLE 3   Effect of indomethacin on LIF-stimulated [3H]thymidine incorporation in neonatal mouse calvaria**

| Treatment | [3H] Thymidine incorporated[a] (c.p.m./mg × 10^{-3}) |
|---|---|
| Control | 44.1 ± 2.2 |
| Indomethacin ($10^{-7}$ M) | 44.2 ± 2.8 |
| LIF (5000 U/ml) | 63.8 ± 3.8* |
| Indomethacin + LIF | 62.4 ± 2.7* |

[a][3H]Thymidine incorporation was measured between 44 and 48 hours of incubation.
*Significantly different from control and from indomethacin-treated, $P < 0.001$.
Data from Reid et al 1990.

prostaglandin synthesis (Tables 3 and 4). The effects of LIF on calvarial DNA synthesis probably reflect proliferation of osteoblasts or their precursors, because in both osteoblast cultures and organ cultures low concentrations of LIF gave rise to a significant increase in DNA synthesis within 24 hours. The findings are in keeping with the results of *in vivo* studies (Metcalf & Gearing 1989), in which the increase in bone formation was accompanied by a clear increase in osteoblast numbers.

LIF inhibited DNA synthesis and cell replication in UMR106 osteogenic sarcoma cells (Lowe et al 1991a). A similar inhibitory effect was found in the clonal osteoblastic line MC3T3-E1 (Noda et al 1990), in which LIF also increased levels of mRNA for osteopontin, an acidic glycoprotein which is found in the extracellular matrix and contains the cell attachment sequence GRGD. TGF-$\beta$ is another cytokine which promotes DNA synthesis in normal osteoblasts (Centrella et al 1987), but inhibits growth in osteogenic sarcoma cells (Noda & Rodan 1987).

It is possible that the predominant effect of low concentrations of LIF in neonatal mouse calvaria is on DNA synthesis. This response is somewhat more sensitive to dose than the resorption response (Lowe et al 1991b). LIF causes a transient stimulation of DNA synthesis; at high concentrations this is an early response, seen at 24 hours, but after 72 hours of treatment LIF actually has inhibitory actions (Lowe et al 1991a). This biphasic response may explain why Lorenzo et al (1990) found that LIF did not influence DNA synthesis in prolonged culture of fetal rat long bones. Similar biphasic mitogenic responses have been noted with other cytokines, including TNF-$\alpha$ in fetal rat calvarial organ cultures (Canalis 1987) and TGF-$\beta$ in rat osteoblast cultures (Centrella et al 1987). Our interpretation of the results so far is that an early action of LIF in bone may be stimulation of cell replication and differentiation in a subpopulation of cells.

**TABLE 4  Effect of LIF and indomethacin on [$^3$H]thymidine incorporation in cultured rat osteoblasts**

| Treatment | [$^3$H]thymidine incorporation in 24 h (c.p.m./well $\times 10^{-3}$) |
|---|---|
| Control | $8.1 \pm 0.4$ |
| Indomethacin ($10^{-6}$ M) | $6.3 \pm 0.3$* |
| LIF (100 U/ml) | $9.9 \pm 0.1$* |
| Indomethacin + LIF | $10.0 \pm 0.5$** |

*Significantly different from control, $P<0.01$.
**Significantly different from indomethacin alone, $P<0.001$.
Data from Lowe et al 1991a.

*Other actions of LIF on osteoblasts*

In studies of the differentiation of pre-osteoblasts into more mature forms, it has been noted that LIF potentiates the increase in alkaline phosphatase mRNA and activity induced by retinoic acid in pre-osteoblastic cells (Allan et al 1990, Noda et al 1990). This effect, which resembles that of TNF-$\alpha$ (Ng et al 1989), may be the result of a stabilizing influence of LIF on alkaline phosphatase mRNA. Such a mechanism seems particularly suited to the action of a paracrine effector which would enhance stability of an mRNA whose transcription has been increased by a humoral agent. The stimulatory effect LIF has on osteopontin mRNA synthesis in MC3T3-E1 cells (Noda et al 1990) has not been studied in other osteoblast populations, but could reflect an anabolic or differentiation-enhancing effect on the osteoblast lineage.

It has been proposed that the plasminogen activator (PA) neutral protease system has a central role in local events in bone remodelling (Martin et al 1989). Plasmin generated as a result of PA activity can activate latent collagenase and thereby contribute to bone resorption, or can activate latent TGF-$\beta$ and contribute to bone formation (Allan et al 1990, 1991). LIF treatment of osteoblasts inhibited PA activity, over the same range of concentrations that were effective in promoting resorption and influencing osteoblast replication (Allan et al 1990). This inhibition was due to increased synthesis of PA inhibitor 1 (PAI-1) mRNA and protein under the influence of LIF. This action of LIF is virtually identical to that of TGF-$\beta$ in osteoblasts (Allan et al 1991), but it was clearly shown in these experiments, with the use of blocking antibodies against TGF-$\beta$, that the effect was not mediated through TGF-$\beta$ (Allan et al 1990).

The production of LIF by osteoblasts, and its specific actions on components of the tightly regulated PA–plasmin–PAI system of osteoblasts, draw attention to its likely role in influencing events around osteoblasts. Inhibition of plasmin formation would contribute to reduced activity of proteases activated by plasmin. Thus, the net effect of LIF, through its actions on osteoblasts, would be in in favour of bone formation.

**Summary**

The combined results identify LIF, a product of osteoblasts, as a further factor that may be involved in paracrine/autocrine regulation in bone. Its marked effects on osteoblast proliferation *in vitro* and *in vivo* may reflect important actions on the renewal of osteoblast populations. Its influence on the PA–plasmin neutral protease system, through its specific receptors on osteoblasts, indicates its potential involvement in local events. Finally, as with TGF-$\beta$, it may be a matrix-associated activity, and may be capable of bidirectional effects, influencing resorption also through prostaglandin-dependent mechanisms.

LIF has several actions which overlap with those of other cytokines but, as is the case with these other molecules, it has its own unique spectrum of activities, many of which identify it as important in bone, in addition to its roles in other organs.

## Acknowledgements

Work from the authors' laboratories was supported by the National Health and Medical Research Council of Australia and the Health Research Council of New Zealand. We acknowledge valuable collaboration with Drs D. Metcalf, N. Nicola, N. Gough, D. Hilton and M. Brown. We thank Dr T. Suda for providing data prior to publication.

## References

Abe E, Tanaka H, Ishimi Y et al 1986 Differentiation-inducing factor purified from conditioned medium of mitogen-treated spleen cell cultures stimulates bone resorption. Proc Natl Acad Sci USA 83:5958–5962

Allan EH, Hilton DJ, Brown MA et al 1990 Osteoblasts display receptors for and responses to leukemia inhibitor factor. J Cell Physiol 145:110–119

Allan EH, Zeheb R, Gelehrter TD et al 1991 Transforming growth factor beta inhibits plasminogen activator (PA) activity and stimulates production of urokinase-type PA, PA inhibitor-1 mRNA and protein in rat osteoblast-like cells. J Cell Physiol 149:34–45

Canalis E 1987 Effects of tumor necrosis factor on bone formation *in vitro*. Endocrinology 121:1596–1603

Centrella M, McCarthy TL, Canalis E 1987 Transforming growth factor β is a bifunctional regulator of replication and collagen synthesis in osteoblast-enriched cell cultures from fetal rat bone. J Biol Chem 262:2869–2874

Chambers TJ 1985 The pathobiology of the osteoclast. J Clin Pathol 38:217–229

Gearing DP, Gough NM, King JA et al 1987 Molecular cloning and expression of cDNA encoding a murine myeloid leukaemia inhibitory factor (LIF). EMBO (Eur Mol Biol Organ) J 13:3995–4002

Gearing DP, Nicola NA, Metcalf D et al 1989a Production of leukemia inhibitory factor (LIF) in *Escherichia coli* by a novel procedure and its use in maintaining embryonic stem (ES) cells in culture. Biotechnology 7:1157–1161

Gearing DP, King JA, Gough NM, Nicola NA 1989b Expression cloning of a receptor for human granulocyte–macrophage colony stimulating factor. EMBO (Eur Mol Biol Organ) J 8:3667–3676

Gough NM, Gearing DP, King JA et al 1988 Molecular cloning and expression of the human homologue of the murine gene encoding myeloid leukemia-inhibitory factor. Proc Natl Acad Sci USA 85:2623–2627

Hilton DJ, Metcalf D, Nicola NA 1988 Purification of a murine leukemia inhibitory factor from Krebs ascites cells. Anal Biochem 173:359–367

Horton JE, Raisz LG, Simmons HA, Oppenheim JJ, Meyerhager SE 1972 Bone resorbing activity in supernatant fluid from cultured human peripheral blood leukocytes. Science (Wash DC) 177:793–795

Lorenzo JA, Sousa SL, Leahy CL 1990 Leukemia inhibitory factor (LIF) inhibits basal bone resorption in fetal rat long bone cultures. Cytokines 2:1–5

Lowe C, Cornish J, Callon K, Martin TJ, Reid IR 1991a Regulation of osteoblast proliferation by leukemia inhibitory factor. J Bone Miner Res 6:1277–1283

Lowe C, Cornish J, Martin TJ, Reid IR 1991b Effects of leukemia inhibitory factor on bone resorption and DNA synthesis in neonatal mouse calvaria. Calcif Tissue Int 49:394–397

Martin TJ, Ng KW, Suda T 1989 Bone cell physiology. Endocr Metab Clin North Am 18:833–858

Metcalf D, Gearing DP 1989 A fatal syndrome in mice engrafted with cells producing high levels of leukemia inhibitory factor (LIF). Proc Natl Acad Sci USA 86:5948–5952

Ng KW, Hudson PJ, Power BE, Manji SS, Gummer PR, Martin TJ 1989 Retinoic acid and tumour necrosis factor-$\alpha$ act in concert to control the level of alkaline phosphatase mRNA. J Mol Endocrinol 3:57–64

Noda M, Rodan GA 1987 Type beta transforming growth factor (TGF beta) regulation of alkaline phosphatase expression and other phenotype-related mRNAs in osteoblastic rat osteosarcoma cells. J Cell Physiol 133:426–437

Noda T, Hasson DM, Vogel RL, Nicola NA, Rodan GA 1990 Leukemia inhibitory factor suppresses proliferation, alkaline phosphatase activity and type I collagen mRNA level and enhances osteopontin mRNA levels in murine osteoblast-like (MC3T3-E1) cells. Endocrinology 127:185–190

Pfeilschifter J, Seyedin SM, Mundy GR 1988 Transforming growth factor beta inhibits bone resorption in fetal rat long bone cultures. J Clin Invest 82:680–686

Raisz LG 1988 Local and systemic factors in the pathogenesis of osteoporosis. N Engl J Med 318:818–821

Rathjen PD, Toth S, Willis A, Heath JK, Smith AG 1990 Differentiation inhibiting activity is produced in matrix-associated and diffusible forms that are generated by alternate promoter usage. Cell 62:1105–1114

Reid IR, Lowe C, Cornish J et al 1990 Leukemia inhibitory factor: a novel bone-active cytokine. Endocrinology 126:1416–1420

Rodan GA, Martin TJ 1981 Role of osteoblasts in hormonal control of bone resorption—a hypothesis. Calcif Tissue Int 33:349–351

Shiina-Ishimi Y, Abe E, Tanaka H, Suda T 1986 Synthesis of colony-stimulating factor (CSF) and differentiation-inducing factor (D-factor) by osteoblastic cells, clone MC3T3-E1. Biochem Biophys Res Commun 134:400–406

Takahashi N, Akatsu T, Mdagawa N, Martin TJ, Suda T 1988 Osteoblastic cells are involved in osteoclast formation. Endocrinology 123:1504–1506

Tashjian AH Jr, Volken EF, Lazzaro M et al 1985 Alpha and beta transforming growth factors stimulate prostaglandin production and bone resorption in cultured mouse calvaria. Proc Natl Acad Sci USA 82:4535–4539

van Beek E, van der Wee-Pals L, Ruit M, Papapoulos SE, Nicola N, Lowik C 1992 Leukemia inhibitory factor (mLIF) inhibits bone resorption by blocking osteoclast formation and inhibits mineralisation in bone cultures. J Bone Miner Res, in press

Williams RL, Hilton DJ, Pease S et al 1988 Myeloid leukaemia inhibitory factor maintains the developmental potential of embryonic stem cells. Nature (Lond) 336:684–687

## DISCUSSION

*Heath:* I had thought Dr Nicola's work showed that in mice injected with radiolabelled LIF, the label was detected in osteoclasts.

*Nicola:* That's not correct. After injection you see clear labelling of cells that at least look like osteoblasts, which line the bone cavity. We have never seen labelling of osteoclasts *in vivo*.

*Lee:* How do osteoblasts activate osteoclasts? Is there a requirement for cell contact?

*Martin:* There is probably not a requirement for cell contact in the activation of existing osteoclasts. There is good evidence for a requirement for cell contact in the generation of new osteoclasts from precursors; when osteoclasts are formed *in vitro* from haemopoietic precursors (for example, from spleen), co-cultivations with osteoblasts or stromal cells have to be carried out on the same culture surface (Takahashi et al 1988, Udagawa et al 1989). With activation of existing osteoclasts, cell contact may not be necessary, but whatever the nature of the osteoclast activator produced by osteoblasts, it is probably labile. We and others have tried unsuccessfully to isolate from conditioned medium of stimulated osteoblasts a factor which is able to activate pure osteoclasts. This activity can be detected in conditioned medium irregularly, so irregularly that it is not feasible to purify it.

*Lotem:* In IL-6 transgenic mice there is no excess bone formation. Have you looked at the effects of IL-6 on bone formation in your system? Also, do osteoblasts have IL-6 receptors?

*Martin:* I am not aware of anyone having looked to see whether there are IL-6 receptors on osteoblasts. We have not looked at effects of IL-6 on bone formation. We have looked at its effects on plasminogen activator inhibitor; it has no effect on plasminogen activator inhibitor in the osteoblast, nor does it act in synergy with LIF or with IL-1 in this response. We have also looked at the effects of IL-6 on osteopontin expression in osteoblasts, which is stimulated by LIF; IL-6 has no effect on expression of mRNA for osteopontin.

There is some dispute about the effects of LIF on bone resorption in organ culture. Ishimi et al (1990) showed stimulation by IL-6 of resorption in fetal mouse calvariae, whereas Al-Humidan et al (1991) found no such effect in neonatal mouse calvariae. I suspect that differences in the amounts of endogenous IL-1 or IL-6 in the bones are responsible for the discrepancies. IL-6 has been shown to promote osteoclast formation from precursors (Lowik et al 1989a,b) and this could be the basis of any effect it has on resorption.

*Gauldie:* Carl Richards in our lab has recently investigated IL-6 and LIF's actions on synoviocyte cultures and fibroblast cultures. Martin Lotz had previously shown that IL-6 was capable of regulating protein synthesis of TIMP (tissue inhibitor of metalloproteinase) in similar cells (Lotz & Guerne 1991). Richards has found that both IL-6 and LIF are potent stimulators of the expression of TIMP at the RNA level and that they do not similarly stimulate collagenase, stromelysin, or any of the lytic activities, only the protective activity (C. D. Richards, R. Sebalt, M. Jordana & J. Gauldie, unpublished work). This is in contrast to IL-1 and TNF, which up-regulate those lytic enzymes in addition to TIMP. He has also shown that $PGE_2$ is capable of up-regulating the proteinase inhibitor in these same cells. I am not sure whether that will also be true in the osteoblast.

*Martin:* That is interesting. PGE actually increases plasminogen activator activity in osteoblasts by decreasing the synthesis of PAI-1, the opposite effect to that of LIF.

*Dexter:* Is the increased bone deposition due to increased numbers of osteoblasts or simply to increased activity?

*Martin:* We cannot tell from our experiments. We suspect that there is an increased number because of the effects LIF has on replication, particularly in pre-osteoblasts. Lowik et al (1989a) found that in the embryonic mouse limb bud cell metacarpal system, if they stripped away the periosteum, to prevent reabsorption from the generated osteoclasts, LIF caused a substantial increase in the number of osteoblasts at both ends of the bones, and inhibited mineralization in that organ culture.

*Dexter:* The evidence for LIF having a growth effect on pre-osteoblast cells appears to be controversial. Professor Alexander Friedenstein, who is currently on sabbatical in my department, has told me that LIF is not a proliferative stimulus for what he calls the osteoblast precursor cells, as measured by his *in vitro* CFU-F assay.

*Heath:* Maureen Owen has seen no overt effects of LIF in stromal cell cultures (personal communication).

*Dexter:* So, where do the extra osteoblasts come from? Is there really a real increase in these cells *in vivo*?

*Metcalf:* These osteoblasts are a mitotically active population. There were greatly increased numbers of osteoblasts in the bones of mice with excess LIF. However, I remain uncertain about the origin of these cells. Sections of the bone show streams of osteoblasts, suggestive of migration of cells through the foramina into the bone marrow, but there was no clear evidence for this process. There was mitotic activity in the osteoblasts, but not as much as I expected. Thus, there was a vast increase in cell number with local mitotic activity, which you don't normally see in osteoblasts in mice of that age, but there could have been some migration.

*Martin:* I was not aware of Professor Friedenstein's results to which Professor Dexter just referred. In primary cultures of osteoblasts from rat or mouse calvariae, the cells were clearly stimulated to replicate by LIF. In Lowik's organ cultures there was also clear stimulation (Lowik et al 1989a). However, in the MC-3T3 cell, which is a mouse osteoblastic cell line, there is clearly inhibition. Also, LIF inhibits replication of osteogenic sarcoma cells. There may be a sub-population of pre-osteoblasts that respond to LIF by increasing their growth rates. I don't know of any results from stromal cells.

*Nicola:* How many of the effects of LIF are prostaglandin dependent? Is the inhibition of osteoclast formation prostaglandin dependent?

*Martin:* I am not sure that question has been addressed. The only thing that is clearly prostaglandin dependent in our experiments is the stimulation of resorption. The effect on cell replication is not, nor is the effect on plasminogen activator inhibitor synthesis, nor that on osteopontin synthesis.

*Nicola:* Presumably there are antagonistic effects. You implied that LIF induces production of prostaglandins by osteoblasts, and that there are antagonistic effects of these two molecules.

*Martin:* I don't know how much prostaglandin production there is under these experimental conditions. There could be antagonistic effects.

*Romero:* Is there direct evidence that LIF and IL-6 stimulate prostaglandin production?

*Martin:* With LIF, there is direct evidence from organ cultures of mouse calvariae, in which there is an increase in prostaglandin E and its major metabolite in the medium. This seems to be true also with IL-6 (Ishimi et al 1990).

*Gauldie:* In human fibroblasts IL-1 increases PGE production but IL-6 and LIF do not.

*Heinrich:* In HepG2 cells transfected with the IL-6 receptor 80 kDa subunit we have good evidence that IL-6 is internalized after its binding (D. Zohlnhöfer, L. Graeve, S. Rose-John, H. Schooltink, E. Dittrich & P.C. Heinrich, unpublished work 1991). Is LIF also internalized?

*Nicola:* LIF is internalized by osteoblast cell lines, and in every other cell type we have looked at there is internalization and degradation. Experiments with primary cells have not been done.

*Heinrich:* Is the receptor recycled?

*Nicola:* We are not in a position to address that question.

*Lotem:* Our experience with M1 cells is that the IL-6 receptor is internalized and it is about eight hours before the receptors reappear at the cell surface, unlike insulin receptors, which are immediately recycled to the cell surface after internalization.

*Stewart:* Does LIF have any effects on cartilage formation or on chondrocytes?

*Nicola:* LIF, as does IL-1, causes increased release of glycosaminoglygans from chondrocyte cultures.

*Heinrich:* Does it cause degradation of the matrix?

*Nicola:* Yes.

*Gough:* From which haemopoietic cell is the osteoclast thought to be generated?

*Martin:* We do not know precisely what the haemopoietic precursor of osteoclasts is. Scheven et al (1984) generated osteoclasts from a single isolated precursor.

*Gough:* So osteoclasts can be generated in cultures containing enriched haemopoietic stem cell populations.

*Martin:* Osteoclasts can be generated if the appropriate accessory cells are provided. This can be achieved *in vitro* by using stromal cells (Udagawa et al 1989) or osteoblasts (Takahashi et al 1988).

*Gough:* Do osteoclasts display any cell surface markers of the monocyte/macrophage lineage? Do they never express LIF receptors, or is the expression switched off before they differentiate to morphologically identifiable osteoclasts?

*Martin:* Osteoclasts share a few antigens with mature macrophages but do not, on the whole, express the antigenic determinants of mature macrophages. We do not know whether their expression is switched off or is never switched on.

*Gough:* Do osteoclasts express c-*fms*?

*Martin: In situ* hybridization has shown that c-*fms* is expressed in mature osteoclasts (H. Fleisch, personal communication).

*Metcalf:* Professor Kishimoto, do transgenic IL-6 mice show evidence of excess bone formation?

*Kishimoto:* I don't think so.

*Gearing:* Were there any changes in $Ca^{2+}$ metabolism in IL-6 transgenic mice?

*Kishimoto:* We haven't studied that.

*Gearing:* We used two model systems to study the effects of excess LIF. One system involved the overproduction of LIF by engrafted FDC-P1 cells that had been engineered to secrete LIF, and the other was injection of recombinant LIF (Metcalf & Gearing 1989, Metcalf et al 1990). The engraftment models produced severe bone and tissue calcium disturbances and the injected LIF increased serum calcium levels.

## References

Al-Humidan A, Ralston SH, Hughes DE et al 1991 Interleukin-6 does not stimulate bone resorption in neonatal mouse calvaria. J Bone Miner Res 6:3–9

Ishimi Y, Miyaura C, Jin CH et al 1990 IL-6 is produced by osteoblasts and induces bone resorption. J Immunol 145:3297–3303

Lotz M, Guerne PA 1991 Interleukin-6 induces the synthesis of tissue inhibitor metalloproteinases/erythroid potentiating activity (TIMP-1/EPA). J Biol Chem 266:2017–2020

Lowik C, Van Der Pluijm G, Hoekman K, Aarden L, Bijvoet O, Papapoulos S 1989a IL-6 produced by PTH-stimulated osteogenic cells can be a mediator in osteoclast recruitment. J Bone Miner Res 4 (suppl 1):S257

Lowik CWGM, Van Der Pluijm G, Hoekman K, Bijvoet OLM, Aarden LA, Papapoulos SE 1989b Parathyroid hormone (PTH) and PTH-like protein (PLP) stimulate interleukin 6 production by osteogenic cells: a possible role for interleukin 6 in osteoclastogenesis. Biochem Biophys Res Commun 162:1546–1552

Metcalf D, Gearing DP 1989 A fatal syndrome in mice engrafted with cells producing high levels of the leukemia inhibitory factor. Proc Natl Acad Sci USA 86:5948–5952

Metcalf D, Nicola NA, Gearing DP 1990 Effects of injected leukemia inhibitory factor on hemopoietic and other tissues in mice. Blood 76:50–56

Scheven BAAA, van der Meer JWM, Nijweide PJ 1984 Osteoclast formation from mononuclear phagocytes: role of bone-forming cells. J Cell Biol 99: 1901–1906

Takahashi N, Akatsu T, Udagawa N et al 1988 Osteoblastic cells are involved in osteoclast
    formation. Endocrinology 123:2600–2602
Udagawa N, Takahashi N, Akatsu T et al 1989 The bone marrow-derived stromal cell
    lines MC3T3-G2/PA6 and ST2 support osteoclast-like cell differentiation in cocultures
    with mouse spleen cells. Endocrinology 125:1805–1813

# General discussion I

**Acute-phase responses**

*Metcalf:* I have never really understood what an acute-phase response is supposed to achieve. Everyone says that the antiproteases probably protect the body in the early stages of severe infection. Is that really true? An extended acute-phase response can go on for days and days. Is it clear from any studies whether such a response is 'good' for you? What do humans feel like in the middle of an acute-phase response? Do they feel awful, as in the early stages of an infection, or are there situations in which an acute-phase response can be demonstrated but the person feels well? Secondly, what are the long-term consequences for a liver cell of being forced to exhibit a sustained acute-phase response? Are there ultimately toxic consequences? In mice engrafted with LIF-producing cells, where sustained rises in LIF levels result, the acute-phase response is also sustained. Such mice, after several weeks, develop liver damage and even cirrhosis. Can anyone tell me what an acute-phase response feels like? Is it really a protective, not a destructive, response, and what are the consequences for the liver cells if the response is maintained for a long period?

*Baumann:* This question about the significance of the acute-phase response arises often. There are two situations to be considered—an acute phase with lethal consequences, such as in cases of severe endotoxaemia, and one that follows sublethal tissue damage. In the first situation, one could ask why people die from sepsis. Is the acute phase the cause of death, or is death a result of a failure to mount a hepatic acute-phase response? I guess that the lethal outcome is brought about by excess activation of all the suspected inflammatory mediators which in turn cause a failure in maintenance of systemic metabolic homeostasis. I understand that hypoglycaemia is common in lethal cases. Although the activation of hepatic acute-phase plasma protein production requires hours and therefore lags behind the metabolic changes, it is not clear whether the acute-phase plasma proteins would be of any benefit if present at raised levels at the onset of or during the initial course of a potentially lethal sepsis. In the second situation, non-lethal tissue injury, however, the acute-phase plasma proteins probably play the protective roles inferred from their biochemical functions. Indeed, this assumption seems justified in view of the fact that the acute-phase response has been strictly conserved through evolution. Many, if not all, vertebrate species display a hepatic acute-phase response that includes the regulation of similar sets of plasma proteins. The proteins always fulfil the same four basic functions—antiproteolysis, immune modulation, blood clotting and clearance of harmful compounds from the circulation. I imagine that failure

156

to execute one of these functions is lethal to the organism. That might be the reason why one does not find naturally occurring genetic variants that have lost the regulated expression of any of the major acute-phase proteins.

*Metcalf:* What's the advantage of, for example, a fall in the albumin level?

*Heinrich:* The fall in albumin compensates for the increase in acute-phase proteins and is probably important for the regulation of osmotic pressure.

*Gauldie:* The answer to Professor Metcalf's question about how you feel during an acute-phase response is that you feel rather terrible.

*Metcalf:* How do you know that it is actually the products of the liver cells which make you feel like that?

*Gauldie:* You could inject a pure cytokine, which would act on genes in the liver, and compare the results to those of an injection of endotoxin. I don't know if humans have been injected with IL-6. Primates, rats, mice and rabbits have been injected with huge amounts of IL-6 but it doesn't appear to be toxic. Rats don't get ruffled hair. They may get some fever, but it's not as dramatic as that induced with endotoxin. There are changes in levels of plasma proteins.

*Metcalf:* There are a number of potentially toxic agents present during an acute infection that could be making you feel unwell. I just wanted to know whether some of the symptoms are ascribable to agents like IL-6 or LIF or whether they result from the acute-phase response induced by these agents.

*Gauldie:* Over a period of 14 days, which is not long enough to be considered chronic, we did not see effects that were toxic, and with continued stimulation with homologous IL-6 there is a continued high level acute-phase response. I haven't gone on for longer to see whether there are detrimental effects.

*Fey:* There is a rat model of rheumatoid arthritis called adjuvant-induced chronic polyarthritis, which results from injection of complete Freund's adjuvant. A transient, initial acute-phase response is generated and then there is a secondary response which lasts for weeks. These rats have high plasma levels of $\alpha_2$-macroglobulin but to my knowledge there have been no reports of cirrhosis or long-term damage to the liver. I don't think that cirrhosis is a necessary consequence of a protracted liver acute-phase response. There may be another cause of cirrhosis in mice with raised LIF levels.

*Lotem:* We heard earlier that dexamethasone dramatically increases the effect of IL-6 on $\alpha$-fetoprotein and other acute-phase proteins. According to Professor Metcalf's ideas, bacterial lipopolysaccharide (LPS) plus dexamethasone should make you feel worse. In fact, if you inject LPS after mice have been treated with dexamethasone the mice are protected from the toxicity of LPS.

*Sehgal:* I think you are confusing the effect of dexamethasone on the induction of a cytokine such as IL-6, which is why mice injected with LPS plus dexamethasone 'feel better', with dexamethasone's synergy with a cytokine in inducing acute-phase plasma proteins. These are two totally different issues.

*Lotem: In vivo,* after injection of LPS into mice which have been receiving dexamethasone, IL-6, G-CSF and M-CSF are not completely eliminated. Their appearance in the circulation may be low, but they are present. The point I am making is that these mice injected with dexamethasone plus IL-6 'feel better' than dexamethasone-injected ones, yet they probably have higher levels of acute-phase proteins.

*Heinrich:* We have transfected HepG2 cells with cDNA for human IL-6. As a result, the cells synthesize and secrete IL-6 continuously; this can in turn act in an autocrine manner. However, the cells protect themselves from overstimulation by down-regulating their IL-6 receptors. The desensitized cells, which are exposed to high concentrations of IL-6, may be viewed as a model for a chronically inflamed liver cell. Interestingly, these cells behave like untransfected HepG2 cells in the absence of IL-6.

*Metcalf:* Are there any chronic clinical states in which there is no liver disease, but are sustained high levels of acute-phase proteins?

*Sehgal:* Rheumatoid arthritis would be a classic example. There is a chronic increase in C-reactive protein. Also, in psoriasis there is usually a chronic increase in C-reactive protein and in circulating IL-6.

Dr Baumann talked about lethal and sub-lethal infections; thus far, the focus has been on acute infections. I would like to know if circadian rhythms in acute-phase plasma protein levels have been investigated. We find that if we use a particular ELISA to measure IL-6 we observe a circadian rhythm of IL-6 levels in the blood of volunteers at the New York Hospital. At night the level is low and stable, but day-time levels show rapid spikes and increase towards afternoon and evening (S. Lowry, L. Moldawer & P. B. Sehgal, unpublished results).

*Metcalf:* That's not a new concept. We measured circadian fluctuations in the levels of colony-stimulating factors in urine. The levels did fluctuate, though not as dramatically as you have suggested for IL-6, but our experiments weren't done in New York, so perhaps the pressures were not as great!

*Baumann:* Short-term fluctuations in the specific concentration of acute-phase plasma proteins do not seem likely. Experiments over the last 20 years have shown that the half-life of a plasma protein typically exceeds days. An observed change in concentration might be due to dehydration. The acute-phase response itself has not been noted to affect the half-life of plasma proteins.

*Gough:* Dr Sehgal, is sunlight needed for these circadian rhythms? I remember you reporting that exposure to one second of u.v. light induces IL-6.

*Sehgal:* That has been published (Urbanski et al 1990). In this study the volunteers were indoors throughout, exposed to a suntan lamp.

*Heath:* It would logically follow from that that if you sit in bright sunlight you'll feel terrible!

*Gough:* Does LPS induce an acute-phase response directly, or only indirectly after induction of IL-6?

*Gauldie:* LPS has no direct activity on hepatocytes.

*Metcalf:* If you inject endotoxin in a C3H/HeJ mouse is there an acute-phase response?

*Baumann:* There is a response but only with certain types of LPS. TCA-extracted LPS is an excellent stimulator.

*Gauldie:* You can add LPS to primary hepatocyte cultures and it neither interferes with nor augments the response to cytokines.

Could I ask the chairman if he thinks we have adequately defended the concept of the acute-phase response?

*Metcalf:* I accept that the models with excess cytokine levels are complex. It is difficult to implicate an agent such as LIF, which has the potential to significantly alter the behaviour of many different systems, as the direct cause of cirrhosis. Nevertheless, the mice with excess LIF do develop quite severe cirrhosis. It was this which made me wonder whether a prolonged acute-phase pattern is known to be ultimately associated with hepatocyte damage. I suspect now that it's probably not. However, in mice with raised LIF levels the hepatocyte is forced to exhibit aberrant behaviour, and, because the origin of cirrhosis of any type is not clear, it remains possible that an induced, long-term alteration in the functional activity of liver cells does damage the cells and lead to cirrhosis.

## Reference

Urbanski A, Schwarz T, Neuner P et al 1990 Ultraviolet light induces increased circulating interleukin 6 in humans. J Invest Dermatol 94:808–811

# The role of interleukin 6 in megakaryocyte formation, megakaryocyte development and platelet production

N. Williams*, I. Bertoncello†, H. Jackson*, J. Arnold* and H. Kavnoudias*

*Department of Physiology, University of Melbourne, Parkville, Victoria 3051 and †Cell Biology Group, Peter MacCallum Cancer Institute, Little Lonsdale Street, Melbourne, Victoria 3000, Australia

*Abstract.* Megakaryocytopoiesis is the cellular amplification and differentiation of precursors into immature megakaryocytes, and the cytoplasmic maturation of these megakaryocytes, a process terminating in the release of platelets into the circulation. Interleukin 6 (IL-6) stimulates megakaryocytopoiesis in the bone marrow, increasing platelet numbers in the circulation. IL-6 alone is poorly active on the growth of stem cell populations, but acts in synergy with stem cell factor (c-*kit* ligand) to expand the committed myeloid progenitor compartments but not the megakaryocyte progenitors. IL-6 has a direct action on megakaryocyte progenitors but only in synergy with low doses of interleukin 3 (IL-3), increasing the number of immature megakaryocytes and enhancing the processes of development into mature megakaryocytes. IL-6 is about 10 times more active on megakaryocytes than on megakaryocyte progenitors in cell culture. It is active alone and will stimulate increases in cell size and DNA content. IL-6 does not appear to stimulate the process of platelet release. IL-6 is found in bone marrow, in both macrophage subsets and megakaryocytes, indicating that it may be an important physiological regulator of both paracrinal (microenvironmental) and autocrinal mechanisms controlling megakaryocyte development in bone marrow.

*Polyfunctional cytokines: IL-6 and LIF. Wiley, Chichester (Ciba Foundation Symposium 167) p 160–173*

## Humoral feedback control of megakaryocytopoiesis

Circulating platelet mass is known to profoundly influence megakaryocytopoiesis and platelet production, presumably by an endocrine pathway (reviewed by Ebbe 1976). An acute reduction in the number of platelets leads to a rebound thrombopoiesis, which is thought to result from the effect of increased circulating thrombopoietin levels on some unknown sensing mechanism (Odell et al 1976).

The action of plasma thrombopoietin, related factors and cytokines on megakaryocytopoiesis and platelet production has been recently reviewed (Hoffman 1989, Williams 1990). Haemopoiesis can be conceptually thought of in terms of three developmentally related cell pools: (1) a primitive or 'stem cell' pool with the capacity to sustain long-term haemopoiesis, (2) a committed progenitor pool of cells restricted in differentiation capacity, but retaining the potential to generate large numbers of mature blood cells, and (3) the morphologically recognizable cells which have restricted amplification capacity but are able to respond to immediate physiological demands.

Thrombopoietin is thought to act on the earliest stages of cells in pool (3), the morphologically recognizable megakaryocytes, stimulating an increase in cell size and DNA content of developing immature megakaryoblasts, and subsequently increasing platelet production (Penington & Olsen 1970, Odell et al 1976). The committed marrow progenitor cell populations are not involved in feedback mechanisms in response to platelet demand, because committed megakaryocyte progenitors (colony-forming cells) are not directly influenced by the onset of an acute thrombocytopenic state or administration of plasma fractions containing thrombopoietin (Burstein et al 1981, Levin et al 1982). From this and extensive findings *in vitro* the concept of multiple level regulation of megakaryocytopoiesis and platelet production was developed (Williams 1990).

## Bone marrow control of megakaryocytopoiesis

There is strong evidence to suggest that in addition to humoral feedback mechanisms mediated by thrombopoietin the marrow itself makes a substantial contribution to the overall control of megakaryocyte development and platelet production. A significant thrombocytosis is observed in animals which have altered bone marrow function but no detectable decrease in platelet numbers (Levin 1990, Ebbe 1990), implying that the marrow itself contributes to platelet production and release by processes not involving the classical thrombopoietin feedback mechanisms. 5-Fluorouracil (5-FU) severely reduces marrow cellularity, and before the rebound thrombocytosis induced by 5-FU treatment an increase in megakaryocyte number and size in the bone marrow is observed (Radley et al 1980). The additional megakaryocytes arise from the amplification of committed progenitors from primitive or stem cell populations (Bradley et al 1989, Ebbe et al 1989). Therefore, the major contribution to the thrombocytosis observed after 5-FU treatment results from an alteration in the types of responsive marrow cells and in the marrow microenvironment rather than from a feedback response to thrombocytopenia.

## Cytokines involved in megakaryocytopoiesis and thrombopoiesis

Although there has been intense study into the nature of plasma thrombopoietin (Hill & Levin 1989), it has still not been purified and cloned. A wide array of factors are known to stimulate megakaryocytopoiesis *in vitro* (Williams 1990) and several are active *in vivo* (interleukin 6 [IL-6], leukaemia inhibitory factor [LIF], IL-11 and IL-3; Ishibashi et al 1989a, Metcalf et al 1990, Ganser et al 1990, Carrington et al 1991). However, others, such as erythropoietin and granulocyte-macrophage colony-stimulating factor (GM-CSF), are active *in vitro* but not *in vivo* (Semenza et al 1989; Ishibashi et al 1990), indicating that we still know little of the biology of platelet production. It is clear that paracrine and endocrine factors, marrow–stromal cell interactions and inhibitory feedback mechanisms all contribute to the overall production and maintenance of constant platelet levels in steady-state conditions.

In 1989 IL-6 was reported to increase platelet numbers (Ishibashi et al 1989a, Asano et al 1990) and was recently reported to be a thrombopoietic stimulator (Hill et al 1990). There is now a substantial amount of information to suggest that IL-6 is a major stimulus of megakaryocytopoiesis and platelet production both *in vivo* and *in vitro*. The *in vitro* studies indicate that IL-6 alone has potent stimulator of megakaryocyte development but not of progenitor cell development (Ishibashi et al 1989b, Williams et al 1990). Thus, knowledge of the actions of IL-6 *in vivo* is central to the basic understanding of megakaryocytopoiesis and platelet formation, and the mechanisms by which these processes are regulated.

Recent results indicate that IL-6 has potent activity in the paracrine/autocrine stimulatory mechanisms of megakaryocyte maturation. A bone marrow-derived megakaryocyte growth factor (termed megakaryocyte potentiator) which is produced by a specific class of bone marrow accessory cells shares immunological properties with IL-6 (Banu et al 1990, Jackson et al 1992). Also, mRNA for IL-6 has been found in developing megakaryocytes (Navarro et al 1991).

The central theme of this meeting is the polyfunctional nature of this cytokine. IL-6 is one of the most potent known stimulators of platelet production and release. The results reviewed here indicate that IL-6 is likely to have a physiological role and that it can directly influence all stages of megakaryocytopoiesis and platelet production.

**Materials and methods**

*Assays for 'stem cells' or primitive myeloid progenitors (high proliferative potential colony-forming cells, HPP-CFCs) and committed myeloid progenitors (low proliferative potential colony-forming cells, LPP-CFCs)*

Mouse HPP-CFCs are amongst the most primitive progenitors yet detected in clonal agar cultures. For detection of these cells, cultures must contain colony-stimulating factor 1 (CSF-1), IL-1 and IL-3. HPP-CFCs were assayed in cultures containing purified CSF-1, recombinant mouse (rm) IL-3 and recombinant human (rh) IL-1 at pre-determined optimal doses (Bartelmez et al 1989).

Lineage-specific myeloid progenitors with low proliferative potential were assayed with purified CSF-1 as the sole stimulus (Bertoncello et al 1991a).

## Assays for megakaryocytic cells

Committed progenitor cells were assessed as megakaryocyte colonies grown from defined and enriched populations of bone marrow cells (Kavnoudias et al 1992). Megakaryocyte growth was assessed by counting detectable megakaryocytes grown from separated immature megakaryocyte populations (Williams et al 1990). In certain experiments bone marrow cell fractions were depleted of potential accessory cells using immuno-specific beads (Kavnoudias et al 1992). Megakaryocyte numbers and size *in vivo* were assessed after perfusion and fixation of the bone marrow (Arnold et al 1991).

## Cell separation

Enriched populations of bone marrow cells depleted of accessory cells were obtained by immuno-magnetic selection (Bertoncello et al 1991b) or by a combination of immuno-magnetic selection and sorting for cells bearing high levels of stem cell antigen (SCA-1) (Williams et al 1992).

## Growth factors

Recombinant human IL-6 (rhIL-6) was kindly provided by Ajinomoto Co. (Tokyo, Japan) and by Amgen Inc. (CA, USA). Units of rhIL-6 were standardized against Amgen IL-6. Recombinant rat stem cell factor (rrSCF) was provided by Amgen Inc. Recombinant mouse IL-3 (rmIL-3) was obtained from a genetically constructed cell line expressing high titres of the growth factor (Karasuyama & Melchers 1988).

## The actions of IL-6 on mouse megakaryocytopoiesis *in vitro*

### Effects of IL-6 and other growth factors on primitive haemopoietic cells (or at the 'stem cell' level)

IL-6 and stem cell factor (SCF) have a synergistic effect on enriched populations of primitive mouse bone marrow cells and committed progenitor cells (Bertoncello et al 1991a, Williams et al 1992), sustaining primitive myeloid progenitors while stimulating an expansion of the committed myeloid progenitor cell pools (Table 1). Because accessory cells were removed in the separation procedures, it appears that IL-6 was not acting indirectly via known accessory cell types. IL-6 was not active on primitive haemopoietic progenitors in suspension culture at the single cell level (Williams et al 1992) and sustained only 8% of input HPP-CFCs at higher cell densities (Table 1).

**TABLE 1   Effects of IL-6 and stem cell factor (SCF) on defined primitive haemopoietic cells**

| Treatment | Primitive myeloid progenitors[a] |
| --- | --- |
| | (% Input) |
| IL-6 (1000 U/ml) | 8 |
| SCF (100 U/ml) | 48 |
| IL-6 + SCF | 799 |

[a]Sorted mouse bone marrow cells (primitive myeloid progenitors, high proliferative potential colony-forming cells) (5000/ml) were incubated for six days in suspension culture with growth factor before assay for progenitor cell content.

## Effects of IL-6 and IL-3 on committed megakaryocyte progenitors

Bone marrow cells enriched in committed megakaryocyte progenitors and depleted of red cells, neutrophils, macrophages, CSF-1-responsive cells and both T and B cells were assayed for their ability to respond to rhIL-6. Megakaryocyte colonies were not observed in cultures containing rhIL-6 alone, but were detected when low doses of rmIL-3 and rhIL-6 were added together to the cultures (Table 2). This results agrees with that from liquid culture, where megakaryocyte progenitor cells were not sustained in the presence of rhIL-6 (Williams et al 1992).

## IL-6 acts directly on megakaryocyte development

The number and size of identifiable single megakaryocytes in culture are optimally increased by 30 U/ml rhIL-6 (Williams et al 1990). Its action seems to be a direct one on the developing megakaryocytes, because rhIL-6 is active on cultures of bone marrow cells containing immature megakaryocytes but depleted of mature megakaryocytes, macrophages, T cells and mature CSF-1 responsive cells (Table 3).

**TABLE 2   Effects of IL-6 and IL-3 on committed progenitor cells**

| Treatment | Committed megakaryocyte progenitors[a] |
| --- | --- |
| | (Number of colonies) |
| Medium | 0 |
| IL-6 (1000 U) | 0 |
| IL-3 (3 U) | $19.0 \pm 4.4$ |
| IL-3 + IL-6 | $47.7 \pm 6.4$ |

[a]$2 \times 10^4$ bone marrow cells were cultured after depletion of mature myeloid and lymphoid cells including macrophages, CSF-1-responsive cells and T cells. The results are the mean number of colonies $\pm$ SD from triplicate cultures.

**TABLE 3    Effect of rhIL-6 on megakaryocyte growth after accessory cell depletion**

| Cell type depleted | Number of megakaryocytes[a] | |
|---|---|---|
| | Control | rIL-6 (1000 U) |
| Polymorphonuclear leukocytes and monocytoid cells | $31.0 \pm 4.6$ | $64.5 \pm 7.0$[b] |
| T helpers | $13.9 \pm 1.5$ | $32.2 \pm 1.2$ |
| T suppressors | $16.7 \pm 1.3$ | $33.7 \pm 3.9$ |

[a]The results are mean $\pm$ SEM from three replicate cultures.
[b]All data from rhIL-6-stimulated cultures were significantly different from medium/FCS control values.

## The actions of rhIL-6 on megakaryocytopoiesis *in vivo*

*rhIL-6 stimulates both cellular amplification and megakaryocyte development* in vivo

Experiments were designed to determine whether rhIL-6 stimulates megakaryo-cytopoiesis and platelet production by preferentially enhancing the amplification of committed progenitors from stem cells, or by accelerating megakaryocyte development. A classical animal model was used, in which 5-FU (150 mg/kg) selectively removes proliferating cells. Rebound haemopoiesis (and thrombo-poiesis) has been shown to be generated from a biologically enriched select cohort of primitive cells. After three days there is a staged feedout from these cells into committed progenitors and the differentiated, morphologically recognizable cell classes (Fig. 1). In the megakaryocyte-platelet lineage, about 4% of committed progenitors remain after 5-FU treatment and there is a strong rebound in progenitor numbers between 4 and 10 days after treatment (Bradley et al 1989). At Day 5 a strong wave of megakaryocyte development commences, which peaks at Day 7. This is followed by a wave of platelet production and a marked thrombocytosis between Days 11 and 14 (Radley et al 1980).

Mice injected with 5-FU (Day 0) were injected with rhIL-6 either at Day 1, when the marrow contained enriched populations of primitive cells, or at Day 5, a time point immediately before the stage at which increased numbers of mature megakaryocytes are detected. The data in Table 4 show the rebound thrombocytosis (see Fig. 1) observed at Day 11 in 5-FU-treated animals. A single dose of 50 000 units IL-6 given one day after 5-FU administration was able to enhance the rate of platelet recovery, with a significant thrombocytosis being observed at Day 8. There was no significant change in white blood cell counts or in haematocrit readings, but a large increase in bone marrow cellularity was observed, indicating that the increased platelet production was probably due to accelerated marrow activity.

The effects of a single injection of rhIL-6 into 5-FU-treated mice at Day 1 or Day 5 are summarized in Table 5. Low doses of rhIL-6 (5000 units) injected

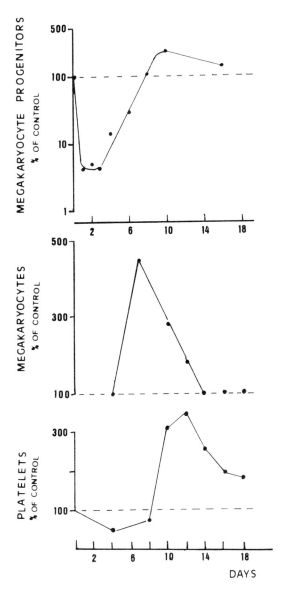

FIG. 1. Megakaryocytopoiesis and thrombopoiesis in mice treated with 5-fluorouracil showing the timed developmental feedout of mature megakaryocytes and platelets from a select cohort of primitive cells. *Upper panel*: the change in the number of committed megakaryocyte progenitors with time after treatment (megakaryocyte progenitor levels are plotted on a logarithmic scale). *Middle panel*: the relative increase in large mature megakaryocytes with time. *Lower panel*: the relative change in platelet count with time. The data were adapted from Bradley et al 1989 and Radley et al 1980.

**TABLE 4    IL-6 induces accelerated thrombopoiesis in mice treated with 5-fluorouracil**

| | IL-6 (50 000 U, Day 1) | Time (days) after 5-fluorouracil injection | | | |
|---|---|---|---|---|---|
| | | Untreated | Day 5 | Day 8 | Day 11 |
| Haematocrit | − | $0.51 \pm 0.02$ | $0.41 \pm 0.03$ | $0.34 \pm 0.02$ | $0.36 \pm 0.02$ |
| ± S.E.M. | + | | $0.39 \pm 0.02$ | $0.38 \pm 0.06$ | $0.38 \pm 0.05$ |
| Platelets $\times 10^{-9}$/l | − | $1.09 \pm 0.6$ | $0.61 \pm 0.7$ | $0.58 \pm 0.12$ | $1.84 \pm 0.54$ |
| | + | | $0.80 \pm 1.4$ | $1.42 \pm 0.25$ | $1.55 \pm 0.42$ |
| White Cells | − | $7.7 \pm 1.1$ | not done | $3.1 \pm 0.8$ | $3.3 \pm 0.3$ |
| $\times 10^{-6}$/l | + | | not done | $4.4 \pm 0.8$ | $4.5 \pm 0.7$ |
| Marrow Cellularity | − | 23.0 | 1.35 | 3.7 | 11.6 |
| $\times 10^{-6}$ | + | | 1.38 | 11.4 | 14.3 |

at Day 1 caused a large increase in the committed megakaryocyte progenitor compartment. This effect operated only within a narrow dose range, and no effect was observed when 20 000 units rhIL-6 were injected (N. Williams, unpublished work). When 5000 units rhIL-6 were injected into mice at Day 5, significant increases in megakaryocyte size were observed by Day 7. No increase in megakaryocyte number was evident, indicating that at that dose, rhIL-6 was not able to stimulate the immature megakaryocytes to undergo cell division.

*IL-6 stimulates platelet production but not release*

There is no culture system in which the cytoplasmic processes culminating in platelet release can be quantitated; all data have been obtained from *in vivo* experiments. There are convincing results showing that injection of IL-6 does not induce immediate platelet release. The most sensitive index to use when investigating factors that influence platelet release is the mean platelet volume, which is rapidly increased after induction of acute thrombocytopenia. No such increase in mean platelet volume is uniformly observed after IL-6 treatment (Hill et al 1991). Moreover, it appears that platelet production is not seen until

**TABLE 5    The effect of time of injection of 5000 units IL-6 on megakaryocytopoiesis in mice with rebound megakaryocytopoiesis and thrombopoiesis induced by 5-fluorouracil**

| IL-6 injection | Committed progenitors | Megakaryocytes | | Platelet number |
|---|---|---|---|---|
| | | Number | Size | |
| Day 1[a] | Increased[c] | Increased | No change | Increased |
| Day 5[b] | | No change | Increased | Increased |

[a]5-FU injections were on Day 0. Assays for mice injected with IL-6 on Day 1 were done on Day 6.
[b]Mice injected with IL-6 on Day 5 were assessed for megakaryocyte content and size on Day 7.
[c]Changes are in comparison with 5-FU-treated mice not given IL-6.

2–3 days after the injection regime has begun. The results indicate, therefore, that IL-6 does not act on the most mature megakaryocytes, but on maturing megakaryocytes, enhancing their maturation and accelerating platelet production. IL-6 can induce morphological and cytoplasmic changes in megakaryocytes that are consistent with those seen in the marrow of thrombocytopenic animals (Hill et al 1991). It is likely, therefore, that IL-6 acts on developing megakaryocytes and is not a potent stimulator of platelet release.

In summary, the results suggest that IL-6 can influence megakaryocytes at all stages of their formation, development and maturation. IL-6 does not appear to effect platelet release processes. It remains to be established whether IL-6 has dose-dependent effects on distinct stages of haemopoiesis, megakaryocytopoiesis and platelet production.

*Acknowledgement*

This work was supported by the National Health and Medical Research Council of Australia.

**References**

Arnold J, Ellis S, Radley JM, Williams N 1991 Compensatory mechanisms in platelet production: the response of Sl/Sl$^d$ mice to 5-fluorouracil. Exp Hematol 19:24–28

Asano S, Okano A, Ozawa K et al 1990 In vivo effects of recombinant human interleukin-6 in primates: stimulated production of platelets. Blood 75:1602–1605

Banu N, Fawcett J, Williams N, De Giorgio T, Withy R 1990 Tissue sources of murine megakaryocyte potentiator: biochemical and immunological studies. Br J Haematol 75:313–318

Bartelmez SH, Bradley TR, Bertoncello I et al 1989 Interleukin 1 plus interleukin 3 plus colony-stimulating factor 1 are essential for clonal proliferation of primitive myeloid bone marrow cells. Exp Hematol 17:240–245

Bertoncello I, Bradley TR, Hodgson GS, Dunlop JM 1991a The resolution, enrichment, and organization of normal bone marrow high proliferative potential colony-forming cell subsets on the basis of rhodamine-123 fluorescence. Exp Hematol 19:174–178

Bertoncello I, Bradley TR, Watt SM 1991b An improved negative immunomagnetic selection strategy for the purification of primitive hemopoietic cells from normal bone marrow. Exp Hematol 19:95–100

Bradley TR, Williams N, Kriegler AB, Fawcett J, Hodgson GS 1989 In vivo effects of interleukin-1 on regenerating bone marrow myeloid colony-forming cells after treatment with 5 fluorouracil. Leukemia (Baltimore) 3:893–896

Burstein SA, Adamson JW, Erb SK, Harker LA 1981 Megakaryocytopoiesis in the mouse: response to varying platelet demand. J Cell Physiol 109:333–341

Carrington PA, Hill RJ, Stenberg PE et al 1991 Multiple in vivo effects of interleukin-3 and interleukin-6 on murine megakaryocytopoiesis. Blood 77:34–41

Ebbe S 1976 Biology of megakaryocytes. Prog Thromb Haemostasis 3:211–229

Ebbe S 1990 Megakaryocyte size and ploidy in thrombocytopenic or megakaryocytopenic mice. In: Breton-Gorius J, Nurden A, Levin J, Williams N (eds) Molecular biology and differentiation of megakaryocytes. Alan R Liss, New York (Prog Clin Biol Res) p 133–144

Ebbe S, Yee T, Phalen E 1989 5-Fluorouracil-induced thrombocytosis in mice is independent of the spleen and can be partially reproduced by repeated doses of cytosine arabinoside. Exp Hematol 17:822–826

Ganser A, Lindemann A, Seipelt G et al 1990 Effects of recombinant human interleukin-3 in patients with normal hematopoiesis and in patients with bone marrow failure. Blood 76:666–676

Hill RJ, Levin J 1989 Regulators of thrombopoiesis: their biochemistry and physiology. Blood Cells (NY) 15:141–166

Hill RJ, Warren MK, Levin J 1990 Stimulation of thrombopoiesis in mice by human recombinant interleukin-6. J Clin Invest 85:1242–1247

Hill RJ, Warren MK, Stenberg P et al 1991 Stimulation of megakaryocytopoiesis in mice by human recombinant interleukin-6. Blood 77:42–48

Hoffman R 1989 Regulation of megakaryocytopoiesis. Blood 74:1196–1212

Ishibashi T, Kimura H, Shikama Y et al 1989a Interleukin-6 is a potent thrombopoietic factor *in vivo* in mice. Blood 74:1241–1244

Ishibashi T, Kimura H, Uchida T, Kariyone S, Friese P, Burstein SA 1989b Human interleukin 6 is a direct promoter of maturation of megakaryocytes *in vitro*. Proc Natl Acad Sci USA 86:5953–5957

Ishibashi T, Kimura H, Shikama Y, Uchida T, Kariyone S, Maruyama Y 1990 Effect of recombinant granulocyte-macrophage colony-stimulating factor on murine thrombocytopoiesis *in vitro* and *in vivo*. Blood 75:1433–1438

Jackson H, Williams N, Banu N 1992 The nature of the accessory cell in bone marrow stimulating murine megakaryocytopoiesis. Exp Hematol 20:241–244

Karasuyama H, Melchers F 1988 Establishment of mouse cell lines which constitutively secrete large quantities of interleukins 2,3,4, and 5 using modified cDNA vectors. Eur J Immunol 18:97–104

Kavnoudias H, Jackson H, Ettlinger K, Bertoncello I, McNiece I, Williams N 1992 Interleukin-3 directly stimulates both megakaryocyte progenitor cells and immature megakaryocytes. Exp Hematol 20:43–46

Levin J 1990 An overview of megakaryocytopoiesis. In: Breton-Gorius J, Nurden A, Levin J, Williams N (eds) Molecular biology and differentiation of megakaryocytes. Alan R Liss, New York (Prog Clin Biol Res) p 1–10

Levin J, Levin FC, Hull III DF, Penington DG 1982 The effects of thrombopoietin on megakaryocyte-CFC megakaryocytes and thrombopoiesis: with studies of ploidy and platelet size. Blood 60:989–998

Metcalf D, Nicola NA, Gearing DP 1990 Effects of injected leukemia inhibitory factor on hematopoietic and other tissues in mice. Blood 76:50–56

Navarro S, Debili N, Le Couedic J-P et al 1991 Interleukin-6 and its receptor are expressed by human megakaryocytes: in vitro effects on proliferation and endoreplication. Blood 77:461–471

Odell TT, Murphy JR, Jackson CW 1976 Stimulation of megakaryocytopoiesis by acute thrombocytopenia in rats. Blood 48:765–775

Penington DG, Olsen TE 1970 Megakaryocytes in states of altered platelet production: cell numbers, size and DNA content. Br J Haematol 18:447–463

Radley JM, Hodgson GS, Levin J 1980 Increased megakaryocytes in the spleen during rebound thrombocytosis following 5-fluorouracil. Exp Hematol 8:1129–1138

Semenza GL, Traystman MD, Gearhart JD, Antonarakis SE 1989 Polycythemia in transgenic mice expressing the human erythropoietin gene. Proc Natl Acad Sci USA 86:2301–2305

Williams N 1990 Stimulators of megakaryocyte development and platelet production. Prog Growth Factor Res 2:81–95

Williams N, De Giorgio T, Banu N, Withy R, Hirano T, Kishimoto T 1990 Recombinant interleukin 6 stimulates immature murine megakaryocytes. Exp Hematol 18:69–72

Williams N, Bertoncello I, Kavnoudias H, Zsebo K, McNiece I 1992 Recombinant rat stem cell factor stimulates the amplification and differentiation of fractionated mouse stem cell populations. Blood 79:58–64

## DISCUSSION

*Sehgal:* The dose of IL-6 that you injected into mice was 5000 units. What is that in terms of mass?

*Williams:* 10 units is 1 ng. So my 5000 units is equivalent to 500 of the units described by Dr Baumann (p 115).

*Gough:* In your *in vitro* experiments with and without IL-3, IL-6 did not cause direct colony stimulation, but in the other experiments you demonstrated an effect of IL-6 on megakaryocyte progenitors. If you add IL-6 initially, then add IL-3 24 hours later, will the IL-6 have maintained megakaryocyte colony-forming cells, even though it has not stimulated their proliferation and colony formation?

*Williams:* No, it won't.

*Lotem:* It has been reported that megakaryocytes produce IL-6 (Navarro et al 1991). In the colony formation assays, though you added IL-3 only you might actually have a combination of IL-3 and IL-6 in the plate; those cells which develop into mature megakaryocytes may begin production of IL-6 earlier, before the mature stage.

*Williams:* I don't think it is known if immature megakaryocytes can make IL-6.

*Lotem:* We showed that a monoclonal antibody to mouse IL-6 inhibits megakaryocyte colony formation induced by IL-3 (Lotem et al 1989). However, you work with a low number of a specific population of cells whereas we worked with whole bone marrow, which may make a difference. Our results with the antibody suggest that some of the megakaryocyte colony-inducing activity of IL-3 is mediated by IL-6.

*Williams:* I agree with that, but the IL-6 is not megakaryocyte-derived—it comes from a subset of macrophage progenitor cells. That's why we did our experiments on separated cells.

*Lotem:* In the bone marrow itself there is no IL-3. Therefore, there must be an active megakaryocyte-inducing factor in the bone marrow in addition to IL-6, unless IL-6 can act by itself. Our results indicate that if you seed about half a million bone marrow cells without anything else, only IL-6, you get megakaryocyte colonies. You don't get as many as with IL-3, but there isn't a large difference. When you add IL-3 plus IL-6 there is a synergistic effect.

*Gauldie:* Are there any differences between species in these responses? I have given 10 µg of IL-6 to rats each day for 10 days. This generates a good acute-phase response, but we saw no change in platelet numbers at that time.

*Williams:* I would advise you to give less IL-6. High doses of IL-6 injected into mice can suppress the system such that there is no increase in platelets.

*Lotem:* There is a molecule called 'megakaryocyte colony-stimulating activity'. Do you know anything about this? What is its relationship to IL-3 and IL-6? I have seen results with human bone plasma cells cultured in plasma clots and the megakaryocyte colonies contained 50–100 megakaryocytes—our IL-6- or IL-3-induced colonies contain 3–15 megakaryocytes per colony.

*Metcalf:* Are you referring to the molecule purified from urine by Ogata et al (1990) or to the factor in serum of patients with aplastic anaemia (Hoffman et al 1985)?

*Lotem:* I was thinking of the one purified from serum of humans with aplastic anaemia.

*Metcalf:* This factor has been reported to have physical properties different from those of known active molecules. Cloning of this factor is being attempted.

*Nicola:* Martin Murphy has said that sequencing suggests that megakaryocyte CSF is a novel molecule. He commented that after purification it stimulates megakaryocyte colonies of more modest size, suggesting that there are other potentiators in the crude material (M. Murphy, unpublished work, symposium on Blood Cell Growth Factors, Beijing, 21–24 August 1991).

*Sehgal:* Dr Williams, in your 5-FU model have you blocked the activity of endogenous IL-6 with an antibody?

*Williams:* No.

*Sehgal:* So, in the response to 5-FU, you don't know what the interplay of the different cytokines is.

*Williams:* That's correct.

*Nicola:* You indicated that the synergy between IL-6 and IL-3 occurred only at low doses. Am I correct in assuming that with a maximum dose of IL-3 you don't get additional colonies when you give IL-6 as well?

*Williams:* Yes.

*Baumann:* What are the effects of LIF in this system?

*Metcalf:* It has little action, actually. The changes with IL-6 are more dramatic than those with LIF. In particular, individual megakaryocytes stimulated by IL-6 reach a very large size.

*Baumann:* So that would lead you to position IL-11 closer to IL-6 than to LIF.

*Metcalf:* I think it's in between IL-6 and IL-3.

*Gauldie:* Do corticosteroids have any effects?

*Williams:* We haven't looked.

*Dexter:* We obviously have a major interest in the potential of IL-6 and LIF to enhance platelet regeneration after chemotherapy or bone marrow transplant. It's intriguing that in these patients there can be full regeneration of leukocytes in a relatively short period, yet platelets may take several months to return to normal levels. Does anyone have a mechanistic explanation for this delayed regeneration of platelets after the other progenitor cells have apparently returned to normal?

*Williams:* The growth factor levels are certainly not limiting in such patients. It's probably the responsiveness of the cell which is affected.

*Dexter:* Do you mean that megakaryocyte progenitor cells are produced in those patients but that they do not develop?

*Metcalf:* In patients injected with G-CSF there is a 100-fold rise in progenitor cells of all lineages in the peripheral blood. This is being exploited clinically; peripheral blood is harvested from patients scheduled to have chemotherapy followed by an autologous marrow transplant. Normally after such transplants the platelets are very slow to regenerate. By combining the harvested peripheral blood with the bone marrow you can greatly reduce subsequent requirements for platelets because you accelerate their regeneration.

*Williams:* That doesn't answer Professor Dexter's question.

*Metcalf:* It doesn't tell you what the limitation was. It is curious that in patients who received G-CSF also, and therefore must have developed increased numbers of progenitor cells in the blood, platelet regeneration is not accelerated as in those receiving the combination of peripheral blood and marrow cells.

*Sehgal:* With Frank Symington in Seattle we've been following about 20 bone marrow transplant patients. In these patients IL-6 appears in two bursts—there is an early burst about 2–3 weeks after transplant and a second burst which probably correlates with the onset of graft-verus-host disease or infection. In several individuals we saw a striking positive correlation between IL-6 levels and platelet numbers. The impression we got was that the recovery of platelet numbers was best in those patients who had, for whatever reason, the highest IL-6 levels.

*Lotem:* How do platelet numbers in mice injected with IL-6 compare to those in mice injected with IL-3?

*Williams:* IL-6 is a much stronger stimulus than IL-3.

*Lotem:* Doesn't that contradict the results on the induction of megakaryocyte colonies, where IL-3 was reported to be active but IL-6 was not active by itself? Do all the additional platelets in IL-6-treated mice come from already existing megakaryocytes stimulated by IL-6?

*Williams:* That doesn't have to be so; IL-3 will stimulate megakaryocytes.

*Lotem:* I agree, but then you would expect that with IL-3 treatment the number of platelets in the bloodstream would be higher than with IL-6 alone, because precursor cells would be stimulated to become platelet-forming megakaryocytes.

*Williams:* That assumes that both molecules would stimulate the cells equally.

*Lotem:* That's what I was asking.

*Dexter:* One of the problems with IL-3 is that *in vitro* it is a powerful stimulus for many myeloid cell lineages, but *in vivo* the effects can be modest. Clearly, the effects of IL-3 *in vitro* cannot be extrapolated to the *in vivo* situation.

*Lotem:* We have observed changes in myelopoiesis *in vivo* after injections of IL-3 in mice.

*Metcalf:* There are some cell populations that seem to respond to injected IL-3. Platelet responses, in our hands, have been almost undetectable in mice. This is curious, because we expected to get a response at least as good as that of white cells.

*Williams:* As far as I know, you have never worked with splenectomized mice. You don't know what spleen pooling is in those mice. There actually could be an effect on platelets that you cannot observe.

*Metcalf:* That's possible.

*Williams:* One issue which I didn't raise is the role of IL-6 in acute thrombocytopenia, and whether IL-6 is primarily an endocrine or a paracrine stimulus. Professor Gauldie's results with Jack Levin suggest that IL-6 levels do not correlate with the degree of thrombocytopenia. This suggests to me that whatever endocrine role IL-6 has in stimulating megakaryocytopoiesis, feedback mechanisms operate to stimulate platelet production; these mechanisms are probably independent of the IL-6 mechanisms, which are presumably restricted paracrine processes.

*Lotem:* Professor Kishimoto's group's IL-6 transgenic mice had bone marrow with an increased number of megakaryocytes. Did these mice have increased numbers of platelets?

*Kishimoto:* We didn't measure platelet numbers, so I don't know.

## References

Hoffman R, Yang HH, Bruno E, Straneva E 1985 Purification and partial characterization of a megakaryocyte colony-stimulating factor purified from human plasma. J Clin Invest 175:1174–1182

Lotem H, Shabo Y, Scahs L 1989 Regulation of megakaryocyte development by interleukin 6. Blood 74:1545–1551

Navarro S, Debili N, Le Couedic JP et al 1991 Interleukin-6 and its receptor are expressed by human megakaryocytes: *in vitro* effects on proliferation and endoreplication. Blood 77:461–471

Ogata K, Zhang Z-G, Abek K, Murphy MJ 1990 Partial purification and characterization of human megakaryocyte colony-stimulating factor. Int J Cell Cloning (suppl) 8:103–120

# Actions of leukaemia inhibitory factor on megakaryocyte and platelet formation

D. Metcalf, P. Waring and N. A. Nicola

*The Walter and Eliza Hall Institute of Medical Research, P.O. Royal Melbourne Hospital, 3050 Victoria, Australia*

*Abstract.* Leukaemia inhibitory factor (LIF) is able to potentiate megakaryocyte colony formation in cultures of mouse bone marrow cells in the presence of multi-CSF (interleukin 3). Membrane receptors for LIF are present on mouse megakaryocytes and receptor numbers increase with increasing maturation of the cells. When injected into normal mice at doses of 0.2–2 µg two to three times daily, LIF induced a rise in platelet numbers, which reached up to twice normal values during the second week of injections. This rise was preceded by a rise first in megakaryocyte progenitor numbers, then in mature megakaryocytes in the bone marrow and spleen. Injections of LIF also marginally accelerated platelet regeneration in mice pre-injected with 5-fluorouracil or subjected to whole-body irradiation and transplantation of marrow cells. In view of similar responses to LIF in parallel studies in primates, clinical trial of LIF in patients with thrombocytopenia is warranted.

*1992 Polyfunctional cytokines: IL-6 and LIF. Wiley, Chichester (Ciba Foundation Symposium 167) p174–187*

Because the haemopoietic colony-stimulating factors (CSFs) are known to have both proliferative and differentiation-inducing actions, it was initially expected that the leukaemia inhibitory factor (LIF), with its striking differentiation-inducing effects on the M1 myeloid leukaemic cell line (Metcalf et al 1988), would also have proliferative effects on at least some haemopoietic populations.

However, in semi-solid cultures of normal mouse fetal or adult haemopoietic populations LIF failed to stimulate cell proliferation, despite the presence of LIF receptors on monocyte/macrophages and a small subset of lymphocyte-like cells (Hilton et al 1991). Subsequent studies have shown that LIF does have a proliferative action on the mouse haemopoietic cell line DA-1 (Moreau et al 1988), and can potentiate the proliferation of certain *myc*-transformed mouse erythroid cell lines in cultures containing multi-CSF (interleukin 3) (Cory et al 1991).

Mice with chronically raised levels of LIF, produced by engrafting the mice with a LIF-producing continuous haemopoietic cell line, developed a complex disease state with osteosclerosis and occlusion of the bone marrow leading to a displacement of haemopoiesis to the spleen and liver (Metcalf & Gearing 1989).

Megakaryocytes became numerous in the spleen in such animals but apparently no more so than erythroid and other haemopoietic populations, in what appeared to be a simple compensatory response to the loss of marrow haemopoiesis (Metcalf & Gearing 1990).

It was not until mice were injected with recombinant LIF that the development of an obvious increase in the number of megakaryocytes in the spleen indicated the need to re-examine the effects of LIF on megakaryocyte and platelet formation.

### Actions of LIF *in vitro*

An analysis of the binding of [125]I-labelled LIF to mouse haemopoietic cells *in vitro* showed that megakaryocytes bind LIF, with the level of LIF binding rising progressively with megakaryocyte maturation. Although not completely blocked by an excess of unlabelled LIF, the binding of LIF appeared to indicate the presence of LIF receptors, at least on morphologically identifiable megakaryocytes (Metcalf et al 1991).

Three haemopoietic growth factors had previously been found to stimulate megakaryocyte colony formation in clonal cultures—multi-CSF (IL-3) (Metcalf et al 1987), IL-6 (Warren et al 1989) and, if used in high concentrations, granulocyte-macrophage CSF (GM-CSF) (Metcalf et al 1986). A re-analysis of the action of purified recombinant mouse LIF in concentrations up to $10^4$ units/ml in cultures of mouse bone marrow cells indicated not only a failure of LIF to stimulate megakaryocyte colony formation but also its inability to support the survival in such cultures of the occasional megakaryocyte present in the initial cell population being cultured.

However, as shown in the example in Table 1, stimulation of bone marrow cells by a mixture of LIF and multi-CSF did result in the formation of larger numbers

**TABLE 1  Potentiation of megakaryocyte colony formation *in vitro* by LIF or IL-6**

|  |  | Number of megakaryocyte colonies[b] | |
| --- | --- | --- | --- |
| *Cell source* | *Stimulus* | *Mature* | *Immature* |
| C3H/HeJ[a] | LIF $10^3$ units | 0 | 0 |
|  | Multi-CSF $10^3$ units | $4.2 \pm 3.0$ | $5.4 \pm 3.0$ |
|  | Multi-CSF $10^3$ units + LIF $10^3$ units | $6.9 \pm 2.9$ | $8.4 \pm 3.6$ |
| C57BL | Multi-CSF $10^3$ units | $16.0 \pm 6.0$ | $0.5 \pm 0.1$ |
|  | LIF $10^3$ units | 0 | 0 |
|  | IL-6 500 ng | $5.5 \pm 3.5$ | 0 |
|  | LIF $10^3$ units + IL-6 $10^3$ units | $8.0 \pm 3.7$ | 0 |
|  | Multi-CSF $10^3$ + IL-6 500 ng | $28.8 \pm 8.3$ | $1.8 \pm 1.0$ |

[a]Cultures contained 50 000 mouse bone marrow cells in 1 ml of medium.
[b]Acetylcholinesterase-positive colonies were scored after seven days of incubation. The table shows mean colony numbers $\pm$ standard deviations from quadruplicate cultures.

of megakaryocyte colonies than in cultures stimulated by multi-CSF alone. The range of megakaryocyte cell numbers in such colonies was similar to that in colonies stimulated by multi-CSF alone, implying that LIF was not merely acting on the more mature megakaryocyte precursors (forming small clones of large mature cells) but was also acting on megakaryocyte precursors at earlier stages of differentiation, including the multipotential precursors that can differentiate into cells of other lineages in addition to megakaryocytes (Metcalf et al 1991). Stimulation by a combination of IL-6 and multi-CSF resulted in a more obvious enhancement of megakaryocyte colony formation but with an evident bias towards the formation of megakaryocytes from more mature precursors.

Combination of LIF and IL-6 in the cultures used, which contained only 10 000 to 50 000 marrow cells per ml, did not result in consistently enhanced mega-karyocyte colony formation. Similarly, combination of LIF ($10^3$ units/ml) with stem cell factor (100 ng/ml), G-CSF ($10^3$ units/ml), GM-CSF ($10^3$ units/ml) or M-CSF ($10^3$units/ml) failed to result in the formation of either pure megakaryocyte colonies or mixed colonies containing megakaryocytes.

These observations supported the possibility that LIF, at least when associated with multi-CSF, could enhance the proliferation of megakaryocyte precursors, leading to the formation of additional megakaryocytes.

## *In vivo* responses to injected LIF

The intraperitoneal injection of purified recombinant mouse LIF in doses of up to 2 µg three times daily into adult DBA/2 or C3H/HeJ mice increased platelet counts up to two-fold compared with control mice (Metcalf et al 1990) (Fig. 1). This response had several characteristic features. Platelet levels were not increased acutely at six or 24 hours after injection, implying that LIF might not be able to induce platelet shedding from megakaryocytes or the release of sequestered platelets to the circulation. The increase in platelets did not begin until after 6–7 days of injections and levels reached plateau values at 10–14 days. One feature of this response, seen also in recent studies in primates (P. Mayer, personal communication), was a continued elevation of platelet counts for up to two weeks after cessation of LIF injections.

The platelets appearing in increased numbers in LIF-injected mice were of normal volume and appeared functionally intact, as assessed by aggregometry. Indeed, platelets from most LIF-injected mice were moderately hyperresponsive compared with platelets from control mice, an effect probably due to the development of fibrinogenaemia in response to the LIF injections.

Although a rise of up to two-fold in platelet counts may not seem remarkable in comparison with the 10–100-fold rises of white cell levels that can be achieved by injecting the CSFs, the rise is as high as that induced by IL-6, the only other agent with a significant capacity to augment platelet levels in normal mice

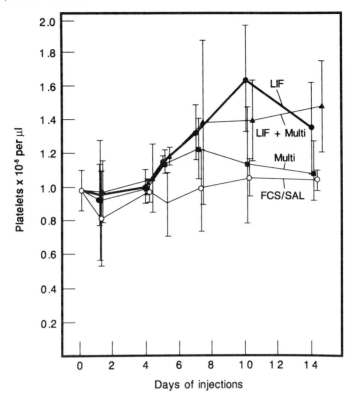

FIG. 1. Changes in platelet numbers in adult mice injected three times daily with 200 ng recombinant mouse LIF and/or 100 ng recombinant mouse multi-CSF (Multi). Control mice were injected with 0.2 ml of 5% fetal calf serum (FCS) in normal saline (SAL), the carrier solution for the LIF and multi-CSF. The graph shows mean values ± standard deviations from groups of eight mice per time point.

(Ishibashi et al 1989a). There are reasons why it might not be desirable for platelets to reach very high levels and it is possible that an effective negative feedback control system might become activated to restrict the magnitude of such responses or even to induce a rebound thrombocytopenia, as has been reported in mice injected with IL-6 (Carrington et al 1991). In mice with continuous excess levels of LIF resulting from engraftment with LIF-producing FDC-P1 cells platelet numbers are also increased, but by only 50%, to a level typical of that achieved in mice receiving non-toxic (200 ng) doses of LIF, three times daily. This observation could be interpreted as further evidence for the existence of some physiological barrier against higher responses.

Analysis of the marrow and spleen of mice at intervals during a two-week course of LIF injections revealed the development of a series of changes

(Metcalf et al 1990). The earliest change observed was a rise, evident after five days of injections, in the frequency of megakaryocyte precursors, particularly in the spleen and also in the marrow; this was followed by a progressive rise in the number of megakaryocytes in the marrow and spleen.

This sequence of changes suggests that the increased platelet numbers in LIF-injected mice result from raised progenitor cell numbers, followed by the formation of additional megakaryocytes, leading finally to a rise in platelet levels. The lag in development of these various responses suggests there was a near-normal temporal sequence of cell generation and maturation.

In preliminary studies the injection of multi-CSF (100 ng doses) combined with LIF did not further increase the number of platelets (Fig. 1), and multi-CSF alone did not elevate platelet counts. Similarly, combination of LIF with IL-6 (2.5 μg, twice daily) or with recombinant rat stem cell factor (SCF) (2 μg daily) failed to increase platelet levels above those induced by LIF alone.

Mice with chronically raised LIF levels develop a broad series of pathological changes including weight loss, osteosclerosis, elevated serum calcium levels, calcification of skeletal and cardiac muscle, cirrhosis, pancreatitis, thymus atrophy and failure of spermatogenesis or corpora lutea formation (Metcalf & Gearing 1989). This spectrum of pathology was not reproduced in mice injected with high doses of LIF (2 μg, three times daily) and the only changes detected were a decrease in body weight resulting from loss of subcutaneous and intraperitoneal fat and a reduction in weight of the thymus. Although the weight loss in the thymus suggests that the injection of LIF might induce some stress, moderate doses of LIF (200 ng, three times daily) induced no overt toxic effects in the mice and only a 5% loss of body weight. LIF therefore remains a potential candidate for therapeutic use for stimulation of platelet production.

## Actions of LIF in thrombocytopenic mice

Three model systems have been used in this laboratory to determine whether injection of LIF can accelerate the regeneration of platelets in mice with induced thrombocytopenia.

In the first system, adult C57BL mice were pre-injected with a single 150 mg/kg dose of 5-fluorouracil, then injected with 200 ng of LIF three times daily for up to two weeks. 5-Fluorouracil induces a moderate thrombocytopenia as a consequence of its lethal effects on dividing cells in the bone marrow. As shown in Table 2, the subsequent regenerative response is characterized by a marked overshoot in platelet numbers in the 14-day period following 5-fluorouracil injection. In mice treated with LIF the initial fall in platelet counts was not prevented and the rate of recovery was only marginally more rapid than in control mice. During the overshoot period, LIF-injected mice had higher numbers of platelets than control mice (though the differences were not statistically significant). In mice pretreated with 5-fluorouracil, treatment with

**TABLE 2   Influence of the injection of LIF on platelet and megakaryocyte numbers in mice preinjected with 5-fluorouacil**

| Days after 5-fluoro-uracil (150 mg/kg) | Bone marrow megakaryocytes | | Spleen megakaryocytes | | Blood platelets $\times 10^{-6}/\mu l$ | |
|---|---|---|---|---|---|---|
| | LIF[a] | NS[b] | LIF | NS | LIF | NS |
| 0 | $60 \pm 17$[c] | $51 \pm 17$ | $4 \pm 1$ | $5 \pm 2$ | $0.9 \pm 0.1$ | $0.8 \pm 0.1$ |
| 2 | $25 \pm 5$ | $29 \pm 9$ | $0.7 \pm 0.7$ | $0.3 \pm 0.2$ | $0.7 \pm 0.1$ | $0.6 \pm 0.2$ |
| 4 | $29 \pm 13$ | $15 \pm 4$ | $3 \pm 2$ | $0.7 \pm 0.7$ | $0.4 \pm 0.1$ | $0.5 \pm 0.1$ |
| 7 | $327 \pm 44$ | $185 \pm 68$ | $11 \pm 12$ | $7 \pm 4$ | $1.0 \pm 0.3$ | $0.7 \pm 0.2$ |
| 10 | $240 \pm 63$ | $149 \pm 64$ | $31 \pm 14$ | $61 \pm 82$ | $3.1 \pm 0.4$ | $2.3 \pm 0.5$ |
| 14 | $27 \pm 8$ | $20 \pm 9$ | $1027 \pm 218$ | $734 \pm 416$ | $2.7 \pm 0.4$ | $2.2 \pm 0.5$ |

[a]LIF-injected mice received 200 ng recombinant mouse LIF three times daily by intraperitoneal injection.
[b]Control mice (NS) were injected with the carrier solution (0.2 ml 5% fetal calf serum in normal saline).
[c]Five adult C57BL mice were examined at each time point. Mean values ± SD are shown.

LIF slightly increased the number of megakaryocytes in both the bone marrow and spleen.

In another model system severe acute thrombocytopenia was induced by a single injection of 100 μl of rabbit anti-mouse platelet serum. This procedure results in a dramatic fall in platelet counts to below 5% of normal levels within one hour. This is followed by a relatively rapid rebound of platelet levels within five days. In such mice injection of LIF did not significantly increase the rate of rise in platelet numbers. However, the number of mature megakaryocytes in the bone marrow at Days 1 and 2 was greater than in control mice. This raises the possibility that LIF may be able to accelerate megakaryocyte maturation, and this question is under further investigation.

In the third model, mice were subjected to whole-body irradiation ($2 \times 5.5$ Gy), then injected with either $10^6$ or $10^7$ syngeneic marrow cells. Recovery in this model is slow and normal levels of platelets are not achieved until three to eight weeks after irradiation, depending on the number of marrow cells transplanted. Injection of 200 ng LIF three times daily produced a marginal increase in platelet numbers between two and four weeks after irradiation, similar to that described in a study of responses to the injection of IL-6 (Patchen et al 1991).

## Discussion

Although LIF has no apparent ability, when acting alone, to stimulate mega-karyocyte formation in cultures of normal mouse bone marrow, it does have some capacity to enhance the formation of megakaryocyte colonies in cultures stimulated by multi-CSF. This *in vitro* action is relatively weak and it was surprising therefore to observe that injections of LIF into mice consistently increased platelet numbers in a response that was preceded by an obvious rise in mega-karyocytes in the marrow and spleen. The observed sequence of changes and their timing suggest that the basis of the platelet increase resulting from LIF stimulation *in vivo* is increased megakaryocyte formation, although there is preliminary evidence that LIF may also be able to accelerate the maturation of immature megakaryocytes, at least in mice recovering from thrombocytopenia induced by anti-platelet serum.

Preliminary studies in primates using recombinant LIF (P. Mayer, personal communication) and in rabbits (J. Campbell, personal communication) have indicated that LIF can increase platelet levels two- to three-fold in these species, again with a delay of about one week before the rise begins.

The reason for the discrepancy between these reproducible *in vivo* effects and the weak actions *in vitro* remains unresolved. Megakaryocytes have LIF receptors, so the *in vivo* effects could be due to a direct action of LIF on megakaryocyte precursors. However, the possibility remains either that the *in vivo* response depends on co-stimulation by some other factor, or that LIF acts indirectly by inducing the production of a direct-acting stimulus. Multi-CSF is not a good candidate for such a cofactor, because it has yet to be demonstrated *in vivo* in normal healthy animals. Other candidate cofactors, such as IL-6, stem cell factor or IL-1, do not enhance megakaryocyte colony formation *in vitro* when combined with LIF, and in preliminary *in vivo* experiments have not enhanced LIF-induced responses.

In the limited number of model systems so far tested in which thrombocytopenia has been induced, LIF has accelerated the recovery of platelet levels only moderately, but other experimental models are possible and these may show a more substantial difference.

Recently, a radioreceptor assay has been developed for human LIF. Preliminary assays on sera and plasmas from patients with aplastic anaemia, thrombocytopenia or thrombocythaemia have not revealed increased levels of LIF, so there is at present no supporting evidence for the involvement of LIF in clinical disorders of platelet formation.

The biology of LIF is obviously highly complex and one could argue that LIF is designed for local production and action in various tissues, rather than for action as a classical hormone. If this is so, future work will need to determine the capacity of stromal and other cells in the marrow to produce LIF in both normal and abnormal states, to establish whether any connection exists between LIF levels and megakaryocyte or platelet formation.

There is an urgent clinical need for haemopoietic growth factors which can increase platelet numbers. The two growth factors previously known to have some influence on platelet levels are multi-CSF (IL-3) and IL-6. Clinical responses to multi-CSF have included rises in platelet counts but these have not been rapid in onset or large (Ganser et al 1990). IL-6 has been shown to induce significant rises in platelet levels in normal mice (Ishibashi et al 1989a) and primates (Asano et al 1990), but produced only a marginal acceleration of platelet regeneration in mice treated with 5-fluorouracil (Takatsuki et al 1990) or subjected to irradiation (Patchen et al 1991).

The ability of LIF to increase platelet numbers is intriguing and is apparently achievable without major toxic consequences. However, the responses are slow in onset and of limited magnitude; it is uncertain whether responses with such characteristics would be of value in the usual clinical situations involving thrombocytopenia.

In principle, combination of two or more growth factors should result in enhanced stimulation of haemopoiesis. This may also prove true for factors that increase platelet levels, particularly since IL-6 appears to act mainly on relatively mature megakaryocyte precursors whereas LIF is more slowly acting but seems to act on a broader range of megakaryocyte precursors. Studies of the effects of combinations of growth factors on platelet numbers are not extensive, and some antagonistic effects have been reported (Ishibashi et al 1989b, Zeidler et al 1989, Carrington et al 1991).

Limited clinical trials of LIF seem warranted because responses in experimental animals, particularly mice, have not always proved reliable in indicating the magnitude of rises that can be elicited by haemopoietic growth factors in humans. However, in view of the actions of LIF on certain organs, such clinical trials would need to include very careful checking for changes that might occur in an unusually wide range of organs.

## Acknowledgements

This work was supported by the Carden Fellowship Fund of the Anti-Cancer Council of Victoria, the National Health and Medical Research Council, Canberra and the National Institutes of Health, Bethesda, grant no. CA-22556.

## References

Asano S, Okano A, Ozawa K et al 1990 *In vivo* effects of recombinant human interleukin 6 in primates: stimulated production of platelets. Blood 75:1602–1605

Carrington PA, Hill RJ, Stenberg PE et al 1991 Multiple *in vivo* effects of interleukin-3 and interleukin-6 on murine megakaryocytopoiesis. Blood 77:34–41

Cory S, Maekawa T, McNeall J, Metcalf D 1991 Murine erythroid cell lines derived with myc retroviruses respond to leukemia inhibitory factor, erythropoietin and interleukin 3. Cell Growth & Differ 2:165–172

Ganser A, Lindemann A, Seipelt G et al 1990 Effects of recombinant human interleukin-3 in aplastic anemia. Blood 76:1287–1292

Hilton DJ, Nicola NA, Metcalf D 1991 Distribution and comparison of receptors for leukemia inhibitory factor on murine hemopoietic and hepatic cells. J Cell Physiol 146:207–215

Ishibashi T, Kimura H, Shikama Y et al 1989a Interleukin-6 is a potent thrombopoietic factor *in vivo* in mice. Blood 74:1241–1244

Ishibashi T, Kimura H, Shikama Y et al 1989b Thrombopoietic effect of interleukin (IL-6) *in vivo* in mice: comparison of the action of erythropoietin (Epo) and granulocyte colony stimulating factor (G-CSF) in combination with IL-6 on thrombocytopoiesis. Blood 74:18 (abstr)

Metcalf D, Gearing DP 1989 A fatal syndrome in mice engrafted with cells producing high levels of leukemia inhibitory factor. Proc Natl Acad Sci USA 86:5948–5952

Metcalf D, Gearing DP 1990 A myelosclerotic syndrome in mice engrafted with cells producing high levels of leukemia inhibitory factor (LIF). Leukemia (Baltimore) 3:847–852

Metcalf D, Burgess AW, Johnson GR et al 1986 *In vitro* actions on hemopoietic cells of recombinant murine GM-CSF purified after production in *Escherichia coli*: comparison with purified native GM-CSF. J Cell Physiol 128:421–431

Metcalf D, Begley CG, Nicola NA, Johnson GR 1987 Quantitative responsiveness of murine hemopoietic populations *in vitro* and *in vivo* to recombinant Multi-CSF (IL-3). Exp Hematol (NY) 15:288–295

Metcalf D, Hilton DJ, Nicola NA 1988 Clonal analysis of the action of the murine inhibitory factor on leukemic and normal murine hemopoietic cells. Leukemia (Baltimore) 2:216–221

Metcalf D, Nicola NA, Gearing DP 1990 Effects of injected leukemia inhibitory factor on hematopoietic and other tissues in mice. Blood 76:50–56

Metcalf D, Hilton D, Nicola NA 1991 Leukemia inhibitory factor can potentiate murine megakaryocyte production *in vitro*. Blood 77:2150–2153

Moreau J-F, Donaldson DD, Bennett F, Witek-Gianotti JA, Clark SC, Wong GG 1988 Leukemia inhibitory factor is identical to the myeloid growth factor human interleukin for DA cells. Nature (Lond) 336:690–692

Patchen ML, MacVittie TJ, Williams JL, Schwartz GN, Souza LM 1991 Administration of interleukin-6 stimulates multilineage hematopoiesis and accelerates recovery from radiation-induced hematopoietic depression. Blood 77:472–480

Takatsuki F, Okano A, Suzuki C et al 1990 Interleukin 6 perfusion stimulates reconstitution of the immune and hematopoietic systems after 5-fluorouracil treatment. Cancer Res 50:2885–2890

Warren MK, Conroy LB, Stelzner JD 1989 The role of interleukin-6 and interleukin-1 in megakaryocyte development. Exp Hematol (NY) 17:1095–1099

Zeidler C, Souza L, Welte K 1989 *In vivo* effects of interleukin 6 on hematopoiesis in primates. Blood 74:154 (abstr)


## DISCUSSION

*Dexter:* I was intrigued by your observation that mice with raised LIF levels (resulting from injection or from engrafting of LIF-producing cells) have reduced amounts of body fat (Metcalf & Gearing 1989). *In vitro* adipocyte cell culture systems have been developed. What happens if you add LIF to those?

*Nicola:* Mori et al (1989) showed that LIF is an inhibitor of lipoprotein lipase activity in 3T3L1 cells. Lipoprotein lipase is needed to break down the fats in lipoproteins before they can enter the cell and become re-esterified to form fatty tissues. The theory is that inhibition of lipoprotein lipase prevents accumulation of fat in cells.

*Dexter:* How quickly does the loss of fat occur?

*Metcalf:* It is evident after 3–4 days of LIF injections. It's quite amazing. The subcutaneous fat and the intraabdominal fat are reduced, but there is no change in organ weight.

*Dexter:* After treatment of primates with LIF, what happens to fatty marrow, which is a different kind of fat from the body fat?

*Metcalf:* The monkeys injected with LIF were not killed so we have no information on marrow fat. The monkeys did not lose weight, so LIF can increase platelet numbers without causing loss of body fat.

*Gearing:* IL-11 and IL-6 have been reported to be inhibitors of lipoprotein lipase in mouse 3T3-L1 pre-adipocytes (Kawashima et al 1991, Jablons et al 1990). Were there any changes in body fat in the IL-6 transgenic mice?

*Kishimoto:* I don't think so.

*Heath:* Nick Gough's work has indicated that there is no LIF transcription in most major organs, yet your results in the lung, for example, have suggested that there is a lot of protein. Is that just a difference between humans and mice?

*Metcalf:* I would guess that it is the 'inflammatory cells' in local lesions that produce the LIF. With a pleural effusion there are vastly increased numbers of cells, including macrophages, fibroblasts and activated lining cells in such cavities. The same is true with an inflamed joint. Support for such a conclusion comes from looking for transcripts of factors, such as IL-1, that are produced in higher amounts and can therefore be detected by *in situ* hybridization. All of the disease states in which levels of LIF have been found to be raised involved a cellular effusion, with the exception of amniotic fluid. Colostrum also contains a high concentration of LIF but it contains many other cytokines and it's a highly cellular fluid.

*Heath:* Have you taken the cellular component from the pleural effusion and looked for LIF expression using the polymerase chain reaction (PCR) or Northern blotting?

*Metcalf:* Not yet. One general problem complicating work with cells *in vitro* is that after they have been held in culture for a few hours large increases in mRNA for many growth factors, including LIF, occur. For example, lung tissue produces large amounts of LIF after the tissue is held in serum-free culture for 15 hours, but this does not provide definite evidence on LIF production in normal lung tissue *in vivo*.

*Aarden:* LIF can induce IL-6. Have you measured IL-6 in these situations?

*Metcalf:* I did a large survey of human sera by bioassay, but with our particular M1 clone you can't distinguish a low level of LIF from a higher level of IL-6. About 15% of the sera had detectable LIF. I checked the positives using a bioassay specific for IL-6, and they all contained high IL-6 levels.

*Aarden:* What I meant was, did you measure IL-6 in mice injected with LIF?

*Metcalf:* The mice with chronically raised LIF levels did not have increased serum IL-6.

*Williams:* IL-6 might be increased in the tissues.

*Metcalf:* There could be a local increase in the tissues. To establish this we need to look directly at stromal tissue, either by *in situ* hybridization or by the PCR, to see if there is an increase in LIF *in situ* that is not reflected in the circulation.

*Sehgal:* Have you looked for LIF in the circulation of mice after an extreme stimulus such as injection of endotoxin or live *E. coli* or a Gram-positive bacterium?

*Metcalf:* LIF is detectable in the circulation at one and at two hours after the intravenous injection of a standard 5 µg dose of endotoxin, but its rise and fall is rapid. The pattern is similar to those of IL-6, IL-1 and G-CSF.

*Baumann:* What is the half-life of LIF injected into mice?

*Metcalf:* The half-life of yeast-derived LIF injected intravenously is less than 1 min and bacterially synthesized recombinant LIF has a half-life of about 10 min. To minimize this problem, all our injections are made intraperitoneally so that we can generate an increase in LIF in the circulation that is sustained for about three hours, and we give three injections per day.

*Baumann:* Could that explain why the plasma level of LIF is so low in your system? Is it constantly consumed by the target tissue?

*Metcalf:* This is always used as an excuse for failing to find an agent in the circulation. If there is an equilibrium state a factor should still be detectable in the circulation even if its half-life is short. A short half-life is not a good explanation of why you can't detect LIF in the circulation—particularly in a patient who has a large pleural effusion containing a large amount of LIF.

*Stewart:* Have you ever tried to perfuse LIF continuously into the animal with an osmotic pump?

*Metcalf:* They have been used in rabbits to increase platelets. We feel that these pumps are too big to use with mice.

*Martin:* I understand that in mice with chronic elevations of LIF, blood clots fail to retract normally. I wonder whether there are abnormalities in the fibrinolytic system. LIF has effects on the plasminogen activator (PA) system, which is a general mechanism in various organs for the control of proteolysis. The activity of this system is correlated with a number of functions in organs that were found to be influenced in mice with high levels of LIF. The PA–plasmin system has roles in ovulation, trophoblast implantation and spermatogenesis, for example. I do not know of any effect of LIF on PAI-1 (PA inhibitor 1) formation in any cell other than the osteoblast. Do any histological changes suggest an increased tendency to thrombosis?

*Metcalf:* There is no histological evidence of bleeding or thrombotic episodes. As well as defective clot retraction there is increased erythrocyte sedimentation, which you would expect as part of the acute-phase response. Formal coagulation time tests give normal results.

*Heath:* There seems to be some problem with LIF moving between different body compartments. It has been argued that results in mice engrafted with LIF-producing tumours differ from those in injected mice because the local levels of LIF after injection are not high enough to get the effects that you see in the engrafted mice. Also, there's the difference you just mentioned between the pleural cavity and the circulation. Doesn't this suggest that there's some sink or some other control over LIF distribution in the body? On the other hand, in Nic Nicola's experiments, injected radiolabelled LIF ends up in the bone marrow. How can we reconcile all these findings?

*Metcalf:* There's no doubt that after intravenous injection of radiolabelled LIF osteoblasts are labelled in normal bone marrow. It is unclear why injections of LIF do not stimulate development of excess numbers of osteoblasts, when the levels of LIF achieved in the circulation are the same, if only for a few hours, as those measured in mice engrafted with LIF-producing cells. Perhaps this is because the increased LIF level is not sustained for the entire 24-hour period—there are only transient peaks after each injection, so one could propose that not enough LIF is being delivered into the marrow.

*Nicola:* Perhaps the barrier is in getting from the site of production into the serum. Once in the serum it can probably get to the appropriate sites.

*Heath:* LIF has to cross an endothelial cell boundary. Endothelial cells might be quite good at trapping LIF.

*Metcalf:* I don't think it is yet clear whether endothelial cells actually make LIF, although they make most other cytokines.

*Heath:* I don't think they do, but there are many different endothelial cells.

*Gough:* We and Lübbert et al (1991) have results which indicate that cultured human umbilical vein endothelial cells do express LIF mRNA, at least *in vitro*.

The excess bone formation seen in the engrafted mice but not in LIF-injected mice might be partly explained by higher local levels. However, Gideon Rodan (personal communication) did some experiments in which LIF was delivered chronically via a cannulation system directly into the bone shaft. Apparently, new bone formation did not result—if anything, there was slight net bone loss.

*Heath:* That's a very important result.

*Martin:* I am not convinced that that experiment mimics the local production of the paracrine factor well enough. I think that is better mimicked by engrafting cells in the marrow. TGF-$\beta$ is present in large amounts in the bone matrix and is produced by osteoblasts and becomes activated at specific sites; we have no idea what concentration of TGF-$\beta$ needs to be produced locally to promote bone formation, nor do we know what concentration of LIF would be required, but it could be quite high.

*Heath:* You are forced to argue that there are two different situations; Nick Gough's results suggest that if you put in high concentrations you don't see the effect and you are saying it has to be presented by the engrafted cells in some rather specific way.

*Martin:* I do suspect that it may have to be presented specifically, particularly if the matrix form you have identified has to be released from the matrix, like TGF-$\beta$, for it to act on osteoblasts. That would be mimicked better by local production than by infusion.

*Metcalf:* The study Nick Gough is referring to was a collaboration between groups 10 000 miles apart. LIF was delivered using a mini-osmotic pump; the experiment was done with materials supplied by us but we could not verify that at the time the experiment was done the material had biological activity. Also, the incorrect dose was delivered. The experiments need to be repeated and I wouldn't waste time trying to explain the outcome because there's no evidence that the experiment was successfully completed.

*Patterson:* Blocking the effects of endogenous IL-6 or LIF might be informative. Has anyone done such experiments?

*Metcalf:* I don't know of any studies involving long-term injections of antibodies against LIF.

*Gauldie:* We tried doing this with a neutralizing anti-rat IL-6 antibody and got some bizarre results. The polyclonal antibody we used inhibits IL-6 activity well *in vitro*. It will interfere with the fever response associated with IL-6 in an endotoxin-challenged rat. However, it does not block the liver response in a turpentine-induced inflammation model. I am not satisfied with that model yet, because the turpentine may have effects on its own.

*Metcalf:* The best approach is to find a mutant mouse in which the gene happens not to be functional, or to do a gene deletion experiment. The second best, I suppose, is to have an excess model, a transgenic model.

*Patterson:* The antibody experiment is the better one because you could do it locally and thereby avoid global effects. It's preferable to using a transgenic model because you would be looking at blockade of the normal factor rather than adding an exogenous factor. Also, embryos lacking the LIF gene are unlikely to develop.

*Sehgal:* There's a general problem in doing blocking experiments with anti-IL-6 antibodies. You can take a perfectly good neutralizing monoclonal antibody to IL-6 and put it into an animal, and you get a 10–100-fold higher level of biologically active IL-6 in the system. The antigen–antibody complex either blocks clearance or is a potent inducer of further IL-6.

*Metcalf:* There could be endotoxin present in your antibody.

*Sehgal:* Other investigators have had the same experience.

*Metcalf:* Did you measure the endotoxin level?

*Sehgal:* Our model was an endotoxin-injected primate, so another dose of endotoxin wouldn't have made any difference.

*Fey:* We have similar results in a rat hepatoma cell system that secretes IL-6. When we add a particular anti-peptide antibody to IL-6 in an attempt to neutralize its actions we have to physically remove the immune complex to get the neutralizing effect. If we do not do this we see an actual stimulation of IL-6 activity, suggesting that IL-6 complexed with this antibody is more effective. However, this may not be a general rule, because neutralizing antibodies against IL-6 clearly exist. Ours may be an exception.

*Lotem:* We haven't seen induction of IL-6 in serum with the injected monoclonal anti-mouse IL-6 antibody that we have used. However, irrespective of where you inject the antibody, it all ends up in the serum and stays there for a long time, which is a problem if you want to look at a local effect of the antibody.

*Gauldie:* Evidence of local activity has been obtained only when the antibody is used centrally and introduced by the intracerebral route (Rothwell et al 1991). Under these conditions anti-IL-6 inhibited fever induction and didn't spill out.

**References**

Kawashima I, Ohsumi J, Mita-Honjo K et al 1991 Molecular cloning of cDNA encoding adipogenesis inhibitory factor and identity with interleukin-11. FEBS (Fed Eur Biochem Soc) Lett 283:199–202

Jablons D, Nordan RP, McIntosh J et al 1990 Interleukin 6 inhibits adipocyte lipoprotein lipase activity in vivo and in vitro. FASEB (Fed Am Soc Exp Biol) J 4:A1713 (abstr)

Lübbert M, Mantovani L, Lindemann A, Mertelsmaan R, Hermann F 1991 Expression of leukemia inibitory factor is regulated in human mesenchymal cells. Leukemia (Baltimore) 5:361–365

Metcalf D, Gearing DP 1989 A fatal syndrome in mice engrafted with cells producing the leukemia inhibitory factor. Proc Natl Acad Sci USA 86:5948–5952

Mori M, Yamaguchi K, Abe K 1989 Purification of a lipoprotein lipase-inhibiting protein produced by a melanoma cell line associated with cancer cachexia. Biochem Biophys Res Commun 160:1085–1092

Rothwell NJ, Busbridge NJ, LeFeuvre RA, Hardwick AJ, Gauldie J, Hopkins SJ 1991 Interleukin 6 is a centrally acting endogenous pyrogen in the rat. Can J Physiol Pharmacol 69:1465–1469

# The role of interleukin 6 in plasmacytomagenesis

Toshio Hirano*, Sachiko Suematsu†, Taiji Matsusaka†, Tadashi Matsuda* and Tadamitsu Kishimoto‡

*Division of Molecular Oncology, Biomedical Research Center, Osaka University Medical School, 2-2, Yamadaoka, Suita, Osaka 565, †Institute for Molecular and Cellular Biology, Osaka University and ‡Department of Medicine III, Osaka University Medical School, 1-1-50 Fukushima, Fukushima-ku, Osaka 553, Japan

*Abstract.* Interleukin 6 (IL-6) is a polyfunctional cytokine which regulates the immune response, the acute-phase reaction and haemopoiesis. IL-6 plays a critical role in differentiation of B cells into plasma cells, and is a potent growth factor for plasmacytomas and myelomas. A relationship between IL-6 and polyclonal plasma cell abnormalities has been demonstrated. Abnormal production of IL-6 was first suggested to be related to hypergammaglobulinaemia with autoantibody production in patients with cardiac myxoma. A role of IL-6 in the generation of plasmacytoma has also been indicated. In support of these clinical and experimental observations, we demonstrated that transgenic C57BL/6 mice carrying the human IL-6 gene showed a massive polyclonal plasmacytosis with production of autoantibodies. However, the tumour was not transplantable to syngeneic animals. Susceptibility to pristane-induced plasmacytomagenesis is genetically determined—pristane can induce plasmacytomas in BALB/c but not in C57BL/6 mice. IL-6 transgenic C57BL/6 mice were backcrossed to BALB/c mice to elucidate the genetic influence on plasmacytomagenesis. Transplantable monoclonal plasmacytoma with a t(12;15) chromosomal translocation was generated in some of the backcrossed mice, indicating that IL-6 plays a key role in the multistep oncogenesis of plasma cell neoplasia.

*1992 Polyfunctional cytokines: IL-6 and LIF. Wiley, Chichester (Ciba Foundation Symposium 167) p 188–200*

There is a close relationship between both chronic inflammation and autoimmune disease and polyclonal or monoclonal B cell abnormalities. Patients with chronic inflammations show polyclonal hypergammaglobulinaemia and frequently develop plasma cell neoplasias or lymphomas (Iosbe & Osserman 1971, Isomaki et al 1978), suggesting a role for inflammatory cells in polyclonal B cell activation and malignant transformation. Potter and Boyce demonstrated that mineral oil or pristane could induce polyclonal plasmacytosis and plasmacytoma in BALB/c mice (Potter & Boyce 1962, Potter 1984). Plasmacytomas arise exclusively in the oil-induced granulomatous tissue which contains infiltrations of macrophages, neutrophils and plasma cells, suggesting that their generation is dependent on microenvironmental influences provided by the inflammatory cells (Potter et al 1985); these inflammatory cells have been found to produce interleukin 6 (IL-6), a potent growth factor for plasmacytomas (Van Damme et al 1987).

IL-6 is a polyfunctional cytokine which regulates the immune response, the acute-phase reaction and haemopoiesis. It is produced by a variety of cells including macrophages, T cells, B cells, fibroblasts, bone marrow stromal cells and endothelial cells (Sehgal et al 1989, Hirano & Kishimoto 1990, Van Snick 1990). IL-6 plays a key role in differentiation of B cells into antibody-producing plasma cells. This active factor was previously called T cell-replacing factor (TRF), B cell differentiation factor (BCDF), BCDFII and B cell-stimulating factor 2 (BSF-2) (Hirano et al 1985, Hirano & Kishimoto 1990). The abnormal production of IL-6 was first observed in patients with cardiac myxoma (Hirano et al 1987). Since then, much evidence indicating the involvement of IL-6 in polyclonal plasma cell abnormalities and plasma cell neoplasias has accumulated (Hirano 1991). In view of the possible involvement of IL-6 in chronic inflammations, autoimmune diseases and lymphoid malignancies, it is worth noting that IL-6 was identified as virus-induced $\beta$2-interferon (Weissenbach et al 1980) and as a product of infiltrated cells in the pleural effusion of patients with tuberculous pleurisy (Hirano et al 1981, Teranishi et al 1982), and that it was purified to homogeneity from the culture supernatants of a human T-cell lymphotropic virus (HTLV-I)-infected T cell line and molecularly cloned (Hirano et al 1985, Hirano et al 1986). The cloned molecule was found to be identical to other factors, called $\beta$2-interferon, 26 kDa protein, hybridoma/plasmacytoma growth factor (HPGF or IL-HP1), hepatocyte-stimulating factor (HSF) and macrophage–granulocyte inducer 2 (MGI-2). It is now called IL-6 (Sehgal et al 1989, Hirano & Kishimoto 1990, Van Snick 1990).

## IL-6 and polyclonal plasmacytosis

The first indication that IL-6 might play a role in abnormal B cell activation came from patients with cardiac myxoma, who frequently show symptoms of autoimmune disease, such as autoantibody production and an increase in acute-phase proteins. Because these symptoms disappeared after the resection of the tumour, it was supposed that myxoma cells or their product(s) induced such abnormalities. We found that myxoma cells produce a large amount of IL-6 (Hirano et al 1987). Further evidence indicating a relationship between IL-6 and polyclonal B cell abnormalities was obtained from patients with Castleman's disease; in these patients we found that activated B cells in the germinal centre of hyperplastic lymph nodes produced IL-6 (Yoshizaki et al 1989). Furthermore, IL-6 was detected in synovial fluid from the joints and serum of patients with active rheumatoid arthritis (Hirano et al 1988, Houssiau et al 1988). Because

IL-6 is not only a B cell differentiation factor, but also a major regulator of acute-phase protein synthesis and a thrombopoietic factor, the deregulation of IL-6 production may well explain the symptoms observed in patients with rheumatoid arthritis. IL-6 production was also observed in type II collagen-induced arthritis in mice (Takai et al 1989). Also, an age-associated increase in serum IL-6 was observed in MRL/*lpr* mice, which spontaneously develop autoimmune diseases with lymphoid hyperplasia associated with an infiltration of plasma cells, arthritis, hypergammaglobulinaemia and a high incidence of monoclonal or oligoclonal IgGs (Tang et al 1991). Pristane, a potent inducer of IL-6, can cause not only polyclonal plasmacytosis resulting in the generation of plasma cell neoplasias, but also arthritis in certain mouse strains (Potter & Wax 1981), further implicating IL-6 in the development of autoimmune arthritis. Pancreatic islet β-cells and thyroid cells were also found to be capable of producing IL-6 (Bendtzen et al 1989, Campbell et al 1989), suggesting that IL-6 may be involved in type I diabetes and thyroiditis, enhancing the response of auto-reactive T cells (Campbell & Harrison 1990).

All the evidence suggested that deregulated IL-6 production could cause polyclonal plasmacytosis and might be involved in the pathogenesis of autoimmune diseases. This notion was further supported by the recent observation that anti-IL-6 antibodies inhibit the development of insulin-dependent diabetes in NOD/Wehi mice (Campbell et al 1991).

## IL-6 in plasmacytomagenesis

IL-6 is a potent growth factor for murine plasmacytoma cells (Van Damme et al 1987) and human myeloma cells (Kawano et al 1988). Thus, it is possible that IL-6 is involved in the generation of plasmacytoma and myeloma (Hirano 1991). A significant association between the occurrence of plasma cell neoplasias and chronic inflammations has been reported (Isobe & Osserman 1971, Isomaski et al 1978). Plasmacytomas can be induced in BALB/c mice by intraperitoneal injection of paraffin oil or pristane, which are potent inducers of chronic inflammation and IL-6 biosynthesis (Potter & Boyce 1962, Potter 1984). The growth of primary mouse plasmacytomas *in vitro* was found to be dependent on IL-6. All this evidence suggests that IL-6 is involved in the generation of plasma cell neoplasia, although it is now known that the major producers of IL-6 in patients with multiple myeloma are bone marrow stromal cells or monocytes, rather than myeloma cells (Klein et al 1989). Where an autocrine or paracrine mechanism is functioning, it is likely that IL-6 plays an important role in the growth of myeloma cells *in vivo* and in the generation of plasma cell neoplasias. This was further supported by the following findings. The responsiveness to IL-6 *in vitro* of myeloma cells obtained from patients with multiple myeloma was directly correlated with the frequency of the S phase of the cell cycle of the tumours (Zhang et al 1989), and increased serum IL-6 levels correlated well with disease severity in patients with multiple myelomas and plasma cell leukaemias (Bataille et al 1989). Moreover, Klein et al (1991) demonstrated that treatment of patients with a mouse anti-human IL-6 monoclonal antibody can suppress growth of multiple myeloma cells *in vivo*.

### Evidence from transgenic mice carrying the human IL-6 gene

All the evidence suggests that deregulated IL-6 gene expression could be involved in polyclonal plasmacytosis and the generation of plasma cell neoplasia. In agreement with this idea, we found that a massive plasmacytosis was generated in transgenic mice (C57BL/6) carrying the human IL-6 genomic gene fused with the human immunoglobulin heavy chain enhancer (Suematsu et al 1989). The plasma cells may not have been fully transformed malignant cells, because they did not give rise to tumours after transplantation into syngeneic mice. Susceptibility to pristane-induced plasmacytoma is genetically determined; most inbred strains other than BALB/c are resistant (Potter 1984). To elucidate the genetic influences on plasmacytomagenesis we backcrossed one of the C57BL/6 IL-6 transgenic mouse lines with BALB/c mice. Some backcrossed mice (e.g. lines 7-5 and 7-2) developed ascites and showed an increased serum IgA level (50- to 200-fold), although they initially showed a preferential increase in $IgG_1$. We tested whether these plasma cells gave rise to tumours when transplanted to pristane-pretreated BALB/cA-nu/nu mice (the pristane-induced microenvironment is required for growth of primary plasmacytomas in recipient mice). The recipient nude mice developed ascites four to eight weeks after the transplantation and many tumour masses were observed in the peritoneal connective tissues. The timing of the appearance of ascites shows that the plasmacytoma is derived from the original tumour, because pristane alone takes more than 200 days to induce plasmacytoma. There were many plasma cells in the ascites, some of which were multi-nucleated. The results indicated that transplantable plasmacytomas were generated in IL-6 transgenic mice with BALB/c genetic backgrounds. To examine the clonality of these plasma cells, we did Southern blot analysis using an immunoglobulin heavy chain joining region ($J_{H4}$) probe. A rearranged monoclonal $J_{H4}$-bearing fragment was detected in DNA from both 7-5 and 7-2 mice, indicating the monoclonality of these transplantable plasmacytomas (Fig. 1). Analysis of G-banded karyotypes demonstrated that the plasmacytomas contained a t(12;15) translocation (Fig. 2). Furthermore, the c-*myc* gene was found to be rearranged.

It appears that continuous deregulated IL-6 gene expression can induce polyclonal plasmacytosis, and that abnormal production of IL-6 plays a critical role in the pathogenesis of certain autoimmune diseases (Fig. 3). Furthermore, it is conceivable that continuous polyclonal plasmacytosis may eventually result in the generation of plasma cell neoplasias if oncogenes are additionally expressed. Alternatively, IL-6 could not only expand a polyclonal normal plasma cell

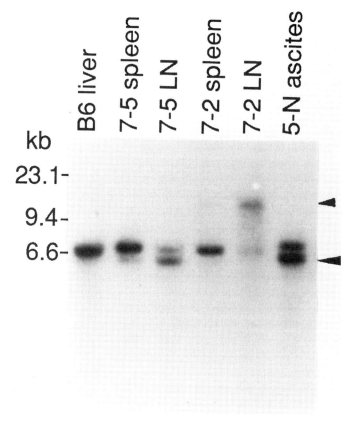

FIG. 1. Southern blot analysis of the immunoglobulin gene. DNAs (5 μg/lane) obtained from spleen or lymph node (LN) were digested with *Eco*RI (a restriction endonuclease from *Escherichia coli*) and subjected to Southern blot analysis using the immunoglobulin heavy chain joining region segment 4 ($J_{H4}$). Arrows indicate rearranged monoclonal $J_{H4}$-bearing fragments. 7-5 and 7-2 are backcrossed mice, and 5-N is a nude mouse into which plasmacytoma cells derived from 7-5 had been transplanted. (Results from Suematsu et al 1992.)

compartment but could also induce differentiation of precursors of plasma cell myeloma which already contain certain mutated oncogenes; these precommitted precursors could give rise to malignant neoplasias after further genetic changes.

## IL-6 signal transduction

The IL-6 receptor system is composed of two functional chains (Yamasaki et al 1988, Hibi et al 1990); one is an 80 kDa IL-6-binding molecule, the IL-6 'receptor' (IL-6R), and the other is a signal-transducing molecule, gp130, which can associate with IL-6R on the binding of IL-6 to IL-6R. The gp130 molecule by itself does not bind IL-6, but it is involved in the formation of a high affinity

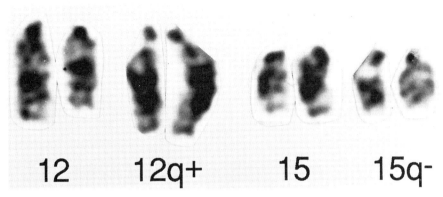

FIG. 2.  The chromosomal translocation (t12;15) in the plasmacytoma. Plasmacytoma cells from 7-2 backcrossed mice were intraperitoneally transplanted into BALB/cA-nu/nu mice. Metaphase spreads were prepared from ascites tumour cells in these BALB/cA-nu/nu mice and G-banding was done. These plasmacytoma cells were tetraploid, and one of the pairs contained a t(12;15) translocation. (Data taken from Suematsu et al 1992.)

binding site. Although the cytoplasmic region of gp130 does not contain a tyrosine kinase consensus sequence, Nakajima & Wall (1991) have demonstrated that IL-6 induces a rapid and transient phosphorylation of tyrosine residues on a 160 kDa molecule. Furthermore, they demonstrated that IL-6 induced the highly selective induction of two primary response genes, *jun*B and *tis*11, and both tyrosine kinases and H7-sensitive protein kinases appeared to be involved

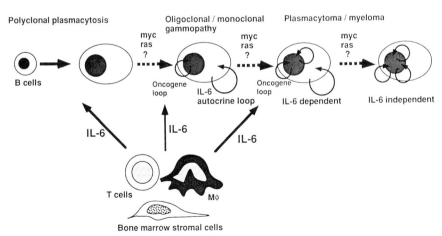

FIG. 3.  Possible involvement of IL-6 in the generation of plasmacytoma. Abnormal overexpression of the IL-6 gene could induce polyclonal plasmacytosis. IL-6 acts as a differentiation factor for normal B cells, but as an autocrine and/or paracrine growth factor for plasma cells that express oncogenes. Additional abnormal expression of oncogenes can eventually induce malignant plasmacytoma.

in the IL-6 signal transduction pathway leading to the activation of both *jun*B and *tis*11 gene transcription. We (T. Matsuda, K. Nakajima and T. Hirano) are currently investigating IL-6-mediated signal transduction leading to the expression of the *jun*B gene. We have found that tyrosine kinase activity is increased after stimulation with IL-6 in a variety of cells, including myeloma cell lines, hepatoma cell lines and myeloid leukaemic cell lines. The maximum tyrosine kinase activity was observed about 15 to 30 min after IL-6 stimulation. After gel filtration, tyrosine kinase activity was found in the fractions corresponding to molecular masses of 60 kDa and 230 kDa. Tyrosine kinase activity in both the 60 kDa and the 230 kDa fractions was increased after stimulation with IL-6, suggesting that tyrosine kinases present in both fractions could be involved in the IL-6 signal transduction pathway.

## Conclusions

The growth of B cells and their differentiation to antibody-forming plasma cells are regulated by a variety of interleukins. IL-6 plays a critical role in B cell differentiation and is a potent growth factor for plasmacytomas and myelomas. The continuous gene expression of IL-6 has been suggested to induce a polyclonal expansion of plasma cells in certain chronic inflammatory diseases, such as rheumatoid arthritis. Abnormal expression of the IL-6 gene in transgenic mice resulted in the generation of malignant plasmacytoma through polyclonal plasmacytosis. From these findings, it could be considered that continuous IL-6 gene expression plays an essential role in the multistep oncogenesis of plasma cell neoplasias.

Future studies on the molecular mechanisms involved in the abnormal expression of the IL-6 gene in myeloma patients, and on the mechanisms of signal transduction through the IL-6 receptor system should provide critical information on the oncogenesis of plasma cell neoplasia.

## Acknowledgements

The authors acknowledge their many colleagues whose work, published and unpublished, forms the basis for this chapter. The authors acknowledge in particular Dr Shinsuke Ohno, Cancer Research Institute, Kanazawa University and Dr Ken-ichi Yamamura, Institute for Medical Genetics, Kumamoto University for their collaboration in the transgenic mouse studies.

## References

Bataille R, Jourdan M, Zhang XG, Klein B 1989 Serum levels of interleukin-6, a potent myeloma cell growth factor, as a reflect of disease severity in plasma cell dyscrasias. J Clin Invest 84:2008–2011

Bendtzen K, Buschard K, Diamant M, Horn T, Svenson M 1989 Possible role of IL-1, TNF-alpha, and IL-6 in insulin-dependent diabetes mellitus and autoimmune thyroid disease. Lymphokine Res 8:335–340

Campbell IL, Harrison C 1990 A new view of the beta cell as an antigen-presenting cell and immunogenic target. J Autoimmun 3:53–62

Campbell IL, Cutri A, Wilson A, Harrison LC 1989 Evidence for IL-6 production by and effects on the pancreatic beta-cell. J Immunol 143:1188–1191

Campbell IL, Kay TW, Oxbrow L, Harrison LC 1991 Essential role for interferon-γ and interleukin-6 in autoimmune insulin-dependent diabetes in NOD/Wehi mice. J Clin Invest 87:739–742

Hibi M, Murakami M, Saito M, Hirano T, Taga T, Kishimoto T 1990 Molecular cloning and expression of an IL-6 signal transducer, gp130. Cell 63:1149–1157

Hirano T 1991 Interleukin-6 (IL-6) and its receptor: their role in plasma cell neoplasias. Int J Cell Cloning 9:166–184

Hirano T, Kishimoto T 1990 Interleukin-6. In: Sporn MB, Roberts AB (eds) Handbook of experimental pharmacology, vol 95/I: Peptide growth factors and their receptors. Springer-Verlag, Berlin, p 633–665

Hirano T, Teranishi T, Toba T, Sakaguchi N, Fukukawa T, Tsuyuguchi I 1981 Human helper T cell factor(s) (ThF). I. Partial purification and characterization. J Immunol 126:517–522

Hirano T, Taga T, Nakano N et al 1985 Purification to homogeneity and characterization of human B cell differentiation factor (BCDF or BSFp-2). Proc Natl Acad Sci USA 82:5490–5494

Hirano T, Yasukawa K, Harada H et al 1986 Complementary DNA for a novel human interleukin (BSF-2) that induces B lymphocytes to produce immunoglobulin. Nature (Lond) 324:73–76

Hirano T, Taga T, Yasukawa K et al 1987 Human B cell differentiation factor defined by an anti-peptide antibody and its possible role in autoantibody production. Proc Natl Acad Sci USA 84:228–231

Hirano T, Matsuda T, Turner M et al 1988 Excessive production of interleukin 6/B cell stimulatory factor-2 in rheumatoid arthritis. Eur J Immunol 18:1797–1801

Houssiau FA, Devogel JP, Van Damme J, De Deuxchaisnes CN, Van Snick J 1988 Interleukin-6 in synovial fluid and serum of patients with rheumatoid arthritis and other inflammatory arthritides. Arthritis Rheum 31:784–788

Isobe T, Osserman EF 1971 Pathologic conditions associated with plasma cell dyscrasias: a study of 806 cases. Ann NY Acad Sci 190:507–517

Isomaki HA, Hakulinen T, Joutsenlahti U 1978 Excess risk of lymphomas, leukemia and myeloma in patients with rheumatoid arthritis. J Chronic Dis 31:691–696

Kawano M, Hirano T, Matsuda T et al 1988 Autocrine generation and essential requirement of BSF-2/IL-6 for human multiple myelomas. Nature (Lond) 332:83–85

Klein B, Zhang XG, Jourdan M et al 1989 Paracrine rather than autocrine regulation of myeloma-cell growth and differentiation by interleukin-6. Blood 73:517–526

Klein B, Wijdenes J, Zhang XG et al 1991 Murine anti-interleukin-6 monoclonal antibody therapy for a patient with plasma cell leukemia. Blood 78:1198–1204

Nakajima K, Wall R 1991 IL-6 signals activating jun B and TIS11 gene transcription in a B cell hybridoma. Mol Cell Biol 11:1409–1418

Potter M 1984 Genetics of susceptibility to plasmacytoma development in BALB/c mice. Cancer Surv 3:247–264

Potter M, Boyce C 1962 Induction of plasma cell neoplasms in strain Balb/c mice with mineral oil and mineral oil adjuvants. Nature (Lond) 193:1086–1087

Potter M, Wax JS 1981 Genetics of susceptibility to pristane-induced plasmacytomas in BALB/cAn: reduced susceptibility in BALB/cJ with a brief description of pristane-induced arthritis. J Immunol 127:1591–1595

Potter M, Wax JS, Anderson AD, Nordan RP 1985 Inhibition of plasmacytoma development in BALB/c mice by indomethacin. J Exp Med 161:996–1012

Sehgal PB, Grieninger G, Tosata G (eds) 1989 Regulation of the acute phase and immune responses: interleukin-6. Ann NY Acad Sci vol 557 (see p ii and p 579–580)

Suematsu S, Matsuda T, Aozasa K et al 1989 IgG1 plasmacytosis in interleukin-6 transgenic mice. Proc Natl Acad Sci USA 86:7547–7551

Suematsu S, Matsusaka T, Matsuda T 1992 Generation of plasmacytomas with the chromosomal translocation t(12;15) in IL-6 transgenic mice. Proc Natl Acad Sci USA 89:232–235

Takai Y, Seki N, Senoh H et al 1989 Enhanced production of interleukin-6 in mice with type II collagen-induced arthritis. Arthritis Rheum 32:594–600

Tang B, Matsuda T, Akira S et al 1991 Age-associated increase in interleukin-6 in MRL/lpr mice. Int Immunol 3:273–278

Teranishi T, Hirano T, Arima N, Onoue K 1982 Human helper T cell factor(s) (ThF). II. Induction of IgG production in B lymphoblastoid cell lines and identification of T cell-replacing factor-(TRF) like factor(s). J Immunol 128:1903–1908

Van Damme J, Opdenakker G, Simpson RJ et al 1987 Identification of the human 26-kD protein, interferon β2 (IFNβ2), as a B cell hybridoma/plasmacytoma growth factor induced by interleukin 1 and tumour necrosis factor. J Exp Med 165:914–919

Van Snick J 1990 Interleukin-6: an overview. Annu Rev Immunol 8:253–278

Weissenbach J, Chernajovsky Y, Zeevi M et al 1980 Two interferon mRNAs in human fibroblasts: *in vitro* translation and *Escherichia coli* cloning studies. Proc Natl Acad Sci USA 77:7152–7156

Yamasaki K, Taga T, Hirata Y et al 1988 Cloning and expression of the human interleukin-6 (BSF-2/IFNβ2) receptor. Science (Wash DC) 241:825–828

Yoshizaki K, Matsuda T, Nishimoto N et al 1989 Pathological significance of interleukin 6 (IL-6/BSF-2) in Castleman's disease. Blood 74:1360–1367

Zhang XG, Klein B, Bataille R 1989 Interleukin-6 is a potent myeloma-cell growth factor in patients with aggressive multiple myeloma. Blood 74:11–13

## DISCUSSION

*Baumann:* The *jun*B gene is known to be activated by many cytokines. Does the regulatory element that you described for IL-6 also function with TNF, IL-1 or colony-stimulating factor?

*Hirano:* We have never tested whether TNF or IL-1 activate the IL-6 response element.

*Lotem:* Is the biological response that follows the increase in *jun*B also sensitive to H-7?

*Hirano:* We did not check the sensitivity of the biological activity to H-7.

*Sehgal:* The tyrosine kinase purification results are reminiscent of information published by Wong & Goldberg (1984). They fractionated rat liver extracts using a similar methodology, and looked for cellular tyrosine kinases. Their profiles of elution of tyrosine kinase activity were like those you described. They purified to homogeneity a tyrosine kinase from rat liver of about 65–70 kDa.

*Nicola:* Professor Hirano, did the 230 kDa fraction you described also contain gp130?

*Hirano:* We have not yet examined that.

*Nicola:* Was the material you put on the gel filtration columns a cytoplasmic extract or a whole-cell extract including membrane extracts?

*Hirano:* We used a cytoplasmic extract.

*Aarden:* Have you looked at the tyrosine kinase activity using different stimuli that produce the same ultimate result?

*Hirano:* No.

*Ciliberto:* Were there any higher molecular mass fractions containing tyrosine kinase activity?

*Hirano:* We detected some tyrosine kinase activity in a 440 kDa fraction. However, this activity was not altered by IL-6 stimulation.

*Ciliberto:* The 230 kDa activity could arise by formation of multimers from the 60 kDa species.

*Fey:* You said that if you did the backcross to BALB/c mice malignant plasmacytomas resulted, whereas with other genetic backgrounds malignancy did not result. Why does one genetic background give a malignant plasmacytoma whereas others do not? You also mentioned that this particular strain carried a t(12;15) translocation. Was that a pre-existing component of the genetic background of those mice before you brought the IL-6 gene in?

*Hirano:* I think the translocation was a secondary event.

*Metcalf:* There are two mouse strains that traditionally develop plasmacytomas—BALB/c and NZB. Professor Hirano was backcrossing to the BALB/c strain. The translocation is a secondary event—it's not present in BALB/c mice. The reason why BALB/c mice are susceptible to plasmacytoma is not clear, though there have been suggestions that they have a defective capacity to control normal immune responses.

*Romero:* Professor Hirano, what is the reproductive efficiency of the IL-6 transgenic mice?

*Hirano:* It is almost the same as that of normal C57BL/6 mice.

*Ciliberto:* What is the average length of survival of your backcrossed mice?

*Hirano:* Backcrossed mice usually developed ascites around 30 weeks of age and died around 40 weeks.

*Ciliberto:* In these mice, where there is chronic production of IL-6, do you observe other effects, such as rheumatoid arthritis or psoriasis?

*Hirano:* The mice developed a mesangial cell proliferative glomerulonephritis, but not rheumatoid arthritis.

*Dexter:* I am still unclear how you think IL-6 is involved in the generation of plasmacytoma, and of myeloma, when we have heard earlier that IL-6 does not appear to be a proliferative stimulus for pre-B cells or B precursor cells. In Fig. 3 you showed an autocrine proliferative mechanism. Was your scheme speculative, or do you have results indicating that IL-6 acts as a proliferative stimulus for pre-B or B precursor cells *in vivo*?

*Hirano:* That figure does not indicate any growth effect of IL-6 on either pre-B or B precursor cells. It indicates that IL-6 induces proliferation by either an autocrine or a paracrine mechanism in plasma cells which already contain abnormalities in certain proto-oncogenes. I have no evidence indicating that IL-6 induces proliferation of pre-B or B precursor cells. IL-6 can act on activated B cells and induce their differentiation into plasma cells.

*Dexter:* The question is whether or not IL-6 is a proliferative stimulus for B cells that are actually capable of proliferating. I thought there was no evidence for this, and that IL-6 acts as a stimulus for the production of immunoglobulin, rather than as a proliferative stimulus. Am I wrong?

*Kishimoto:* I think IL-6 is a growth factor for normal plasma cells.

*Dexter:* But what is the evidence for that?

*Kishimoto:* The IL-6 transgenic mice of the B6 strain showed plasmacytosis, which suggests that IL-6 is a growth factor for non-malignant plasma cells.

*Dexter:* That is a circular argument. You need to treat normal mice, BALB/c mice for example, with IL-6 and show conclusively that it causes proliferation of normal precursor cells. Studies in transgenic mice can give misleading information about the true biology of cytokines.

*Aarden:* Vink et al (1988) have shown that for normal mouse splenic B cells, IL-6 in combination with IL-1 is a potent growth stimulator.

*Hirano:* IL-6 alone has no effect on B cell growth *in vitro*.

*Gauldie:* Didn't Professor Hirano imply that the autocrine stimulation occurs only after a plasmacytoma has been generated?

*Hirano:* That is correct. Fig. 3 indicates that autocrine growth stimulation occurs only after the expression of certain oncogenes in plasma cells.

*Lotem:* Is it known whether IL-6 affects the viability of plasma cells, if not their multiplication?

*Dexter:* That is an important question. We have been trying to reproduce some of the results from myeloma cells, in which it has been claimed that IL-6 acts as a growth factor *in vitro*, and have been unable to reproduce the results using primary myeloma cells. Cell lines, as Professor Metcalf has stressed, can give aberrant responses.

*Metcalf:* Do you have a positive control? Does anything else make those cells proliferate?

*Dexter:* We suspect that IL-7, perhaps in combination with the c-*kit* ligand (stem cell factor), may do so.

*Kishimoto:* We have studied myeloma cells from the bone marrow of 24 patients. IL-6 augments the proliferation of these cells and antibodies against IL-6 inhibit their growth. Not all myeloma cells produce IL-6, but stromal cells from some patients with myeloma produce large amounts of IL-6. Myeloma cells may produce IL-1 or TNF, to stimulate IL-6 production; IL-6 would then augment the proliferation of myeloma cells. In IL-6 transgenic mice of the B6 strain there was proliferation only of plasma cells, but in BALB/c mice monoclonal plasmacytomas were generated. Continuous proliferation of plasma cells in the BALB/c mouse may induce the t(12;15) translocation, which includes the c-*myc* gene. Alternatively, the translocation could occur at the pre-B stage or earlier; IL-6 could then bring about the expansion of this cell to generate a monoclonal plasmacytoma. In the IL-6 transgenic mouse, the only difference is the over-production of IL-6; this shows that abnormal production of IL-6 in a certain genetic environment can generate plasmacytoma.

*Dexter:* How did you measure the proliferation in response to IL-6 of myeloma cells from your patients?

*Kishimoto:* By measuring thymidine uptake.

*Dexter:* That is not adequate.

*Hirano:* We also measured the increase in plasma cell number.

*Gough:* Do you know whether the t(12;15) translocation is in fact from the BALB/c background? Does the *myc* gene or the immunoglobulin locus from the BALB/c chromosome have a high propensity to rearrange? Is the t(12;15) translocation that you see in the F1 mice always from the BALB/c?

*Hirano:* We do not know.

*Gough:* By analysing either partner in the translocation, the *myc* gene or the immunoglobulin locus, using an appropriate restriction fragment length polymorphism, you could determine whether or not the translocation always involves the BALB/c-derived chromosome.

*Hirano:* These mice were produced by backcrossing the F1 mice with BALB/c mice.

*Gough:* It is nevertheless possible that in all cases the t(12;15) translocation is from the BALB/c partner.

*Lotem:* Expression of the *bcl*-2 proto-oncogene has been shown to maintain the viability of cells that would otherwise die in the absence of growth factors. The cells do not multiply, they just live longer, which might allow events eventually leading to a malignancy to occur. What is known about the effect of IL-6 on the viability or the life-span of a plasma cell?

*Kishimoto:* In lymph nodes of patients with Castleman's disease there is massive proliferation of mature plasma cells and abnormal production of IL-6. In some patients we can detect a chromosomal translocation which includes the *bcl*-2 gene. Perhaps during proliferation this translocation occurs in plasma cells, resulting in lymphoma.

*Dexter:* In the absence of expression of the *bcl*-2 gene, do B precursors or B cells live longer in the presence of IL-6 than they do in the absence of IL-6?

*Kishimoto:* When we put the affected lymph node cells from patients with Castleman's disease into culture, with or without IL-6, they can survive for longer than normal B cells.

*Aarden:* Can the transplantable plasmacytomas from these mice be grown easily *in vitro*?

*Hirano:* It is very difficult to maintain them *in vitro*. We tried, but failed, to establish a plasmacytoma cell line.

*Aarden:* Mike Potter had the same problem with his original plasmacytomas. One of the solutions was IL-6, but *in vitro* they were also dependent on serum factors.

*Metcalf:* Most plasmacytomas cell lines were established with great difficulty. The cells almost died *in vitro*, then were rescued by transplantation, and this process was continued until a mutant cell line that would grow continuously *in vitro* emerged.

*Sehgal:* In the transplantable plasmacytoma model is the t(12;15) translocation the only genetic alteration? Have you looked for other alterations?

*Hirano:* We detected only the t(12;15) translocation.

*Sehgal:* So there could be other alterations which might contribute to plasmacytomagenesis.

*Hirano:* That's correct. I am not saying that this translocation is solely responsible.

*Metcalf:* That is the typical translocation in mouse plasmacytomas.

*Gauldie:* Professor Kishimoto has indicated that some IL-6 may come from stromal cells, not from the plasmacytoma. Is IL-6 necessary, but not sufficient, to cause plasmacytosis?

*Hirano:* From our results we can say that over-production of IL-6 is sufficient to cause plasmacytosis, but the genetic background of the BALB/c mouse is additionally required for generation of plasmacytoma.

*Gauldie:* Your scheme implied more than that, that the release of IL-6 becomes autoregulatory.

*Hirano:* The major source of IL-6 could be bone marrow stromal cells; nevertheless, an autocrine mechanism could also be functioning.

*Stewart:* Have you generated IL-6 transgenic mice using a different promoter? This would enable you to determine whether the effects you have described can be attributed to the B cells themselves synthesizing IL-6, with the IL-6 acting in an autocrine manner.

*Hirano:* We obtained the same results using the H-2Ld promoter.

## References

Vink A, Coulie PG, Wouters P, Nordan RP, Van Snick J 1988 B-cell growth and differentiation activity of interleukin HP1 and related murine plasmacytoma growth factors: synergy with interleukin-1. Eur J Immunol 18:607–612

Wong TW, Goldberg AR 1984 Purification and characterization of the major species of tyrosine kinase in rat liver. J Biol Chem 259:8505–8512

# General discussion II

## Interleukin 6, leukaemia inhibitory factor and granulocyte colony-stimulating factor

*Metcalf:* In mice injected with IL-6 there is no real increase in white cell levels. One of the differences between IL-6 and LIF is that IL-6, but not LIF, will stimulate granulocyte colony formation in a culture of mouse bone marrow cells. You cannot tell the difference between a culture stimulated by G-CSF and one stimulated by IL-6. If you combine IL-6 with the c-*kit* ligand, stem cell factor (SCF), there is a large enhancement of proliferation, which includes the formation of blast cell colonies, though most colonies are composed of granulocytes. *In vitro*, the response to a combination of SCF and G-CSF is similar to the response to SCF plus IL-6. When injected *in vivo*, G-CSF induces high levels of neutrophils and so the lack of stimulation of neutrophils in animals injected with IL-6 is puzzling. The IL-6 experiments in mice were done with human IL-6, which is a possible worry. Human IL-6 does bind to the mouse IL-6 receptor, but there may be an abnormally rapid clearance of human IL-6 in mouse urine or problems with access of the agent to various tissue compartments. I would be happier if somebody produced recombinant mouse IL-6 so that we could check some of these things.

*Gauldie:* I have used recombinant rat IL-6. In rats, it caused no alteration in white cell counts. The half-life in rats is about 17–19 min. This is similar to the half-life of human IL-6 in the rat that Professor Heinrich mentioned—the second phase, that is.

*Heinrich:* One should still compare the biological activity of recombinant rat IL-6 with that of the native, glycosylated material.

*Gauldie:* Rat IL-6 has only a small amount of carbohydrate.

*Heinrich:* It is *O*-glycosylated. The negative charges of the sialic acid residues may be important; they affect clearance rates.

*Metcalf:* The problem is that there are some responses to IL-6 in mice that can be observed, such as an increase in platelet levels. Failure of IL-6 to bind to cellular receptors cannot be the explanation.

*Williams:* There is a great difference in the expansion of haemopoietic cells derived from primitive precursors and late precursors. With stem cell factor plus IL-6 there is a bias towards more primitive cells, with only some feedout into late progenitors. Stem cell factor plus G-CSF gives the reverse result—there's enormous expansion of late precursors with minimal maintenance of more primitive cells (Williams et al 1992).

*Metcalf:* In our hands the two treatments give essentially identical results.

*Williams:* The experiments I referred to were done on defined primitive cell populations.

*Lotem:* The similarity between the effects of G-CSF and IL-6 on colony formation depends on the assay. In the colony formation assay with bone marrow cells cultured with 12-*O*-tetradecanoylphorbol-13-acetate (TPA) alone, if you go below a certain concentration of cells per plate, you get no colonies, but when the cell number is increased two-fold, you can get 50–100 granulocyte/macrophage colonies, and that's not because TPA acts directly as a colony-stimulating factor. Similarly, we find that if we seed about $10^4$ bone marrow cells/ml, with IL-6 we get do not get any colonies whereas with G-CSF we do. In addition, when we seed isolated bone marrow myeloid precursors in suspension at about $10^5$ cells/ml with IL-6, the number of cells remains the same for four days, and there may even be some decrease in cell number. When we add G-CSF instead of IL-6, after four days we get $2 \times 10^6$ cells/ml. The fact that G-CSF, but not IL-6, is a strong stimulant of granulocyte formation *in vivo* does not surprise me.

*Metcalf:* In our laboratory the effects of IL-6 and G-CSF in cultures of mouse bone marrow cells are essentially identical. I agree that the observed effect of a factor in a culture of bone marrow cells does depend on the way you do the experiment. As Dr Williams pointed out, if you analyse the actions of IL-6 and G-CSF on more ancestral haemopoietic target cells, you begin to see differences. The point about G-CSF *in vivo* is that some of its effects probably depend on interactions with SCF, because G-CSF, when injected into *Steel* mice or $W^v$ mutant mice, is a relatively poor stimulus for granulocyte formation. However, that doesn't explain the puzzling difference between the effects of injected IL-6 and injected G-CSF. It would help to have more information on white cell and platelet counts in the IL-6 transgenic mice.

*Heinrich:* I would like to return to the use of heterologous IL-6. Human IL-6 is internalized by human hepatoma cells (HepG2). With mouse IL-6 we did not observe internalization. We have speculated on the importance of IL-6 internalization. The internalized IL-6–IL-6 receptor complex could be processed intracellularly to generate a soluble form of the receptor which could be expelled again, interact with its ligand, IL-6, bind to gp130 and generate a signal.

*Metcalf:* Do you know whether mouse IL-6 binds to human cells? Human IL-6 will bind to mouse cells, but will mouse IL-6 bind to human cells?

*Heinrich:* Mouse IL-6 does not stimulate acute-phase protein synthesis in human HepG2 cells.

*Kishimoto: In vivo*, recombinant mouse IL-6 and human IL-6 have the same effects on platelet induction and on acute-phase protein induction.

*Metcalf:* Are there effects on white cell counts?

*Kishimoto:* Human IL-6 stimulates production of antibodies in mice, because IL-6 is a B cell differentiation factor.

*Ciliberto:* Is there weight loss in mice injected with mouse IL-6?

*Kishimoto:* There is an increase in platelet counts and an acute-phase reaction, but there is no weight loss and no fever.

*Metcalf:* What dose were you injecting?

*Kishimoto:* 10 µg per mouse.

*Sehgal:* In our hands, baculovirus-expressed mouse IL-6 can stimulate $\alpha_1$-antichymotrypsin synthesis in human Hep3B cells, though more is needed than with human IL-6. I think these are grey zones rather than black and white distinctions between species.

*Heinrich:* Human IL-6 is recognized by human as well as by mouse IL-6 receptors, whereas mouse IL-6 seems to exert its action only on mouse IL-6 receptors. We therefore asked ourselves which domain or domains in human IL-6 is or are important for the interaction of the ligand with the human receptor. We replaced various regions of human IL-6 by mouse sequences and assayed a series of chimeras in the mouse B9 cell proliferation assay and in the $\gamma$-fibrinogen induction assay in human HepG2 cells. We found that the second (amino acids 36–90) and fourth (135–184) domains of IL-6 are responsible for the species specificity and also for receptor recognition.

*Ciliberto:* We have developed an *in vitro* binding assay using the human soluble receptor. With this we can study competition with wild-type human IL-6. Mouse IL-6 cannot displace bound human IL-6 from the receptor.

*Heinrich:* We find the same.

*Metcalf:* If it can't bind to the receptor, it can't work—that's a quite unambiguous situation, not a grey area.

*Sehgal:* My batch of Hep3B cells could differ from other people's.

## The effects of leukaemia inhibitory factor and interleukin 6 on vascular function

*Williams:* We haven't yet discussed the systemic effects of IL-6 or LIF, whether they affect blood volume. There have been reports of IL-6 being produced in smooth muscle cells responding to it. What vascular effects do these cytokines have?

*Metcalf:* Most of the results so far are from mice, and this species does not provide good models of cardiovascular disease.

*Gauldie:* The studies at Sandoz with IL-6 on primates involved exhaustive monitoring. There were no apparent major changes in haemodynamics (P. Meyer, personal communication).

*Dexter:* Our first patients are now being treated with IL-6 in Manchester. The trials are still at Phase I, but to my knowledge there are no dramatic vascular effects.

*Gauldie:* From what I have heard so far, it appears that IL-6 has systemic actions, but I am not sure that I have heard that LIF does.

*Metcalf:* I think you are asking whether there are disease states or experimental perturbations in which LIF can be detected in the circulation. There are situations in humans in which increased local levels can be detected, but increased circulating LIF has not been observed so far.

I am rather confused about IL-6. Does everyone in this room have detectable IL-6 in the circulation, or is it circulating only in people with certain disease states?

*Sehgal:* I think that although we might disagree about the exact concentrations, we would agree that in certain disease states there are systemic elevations of IL-6. We can clearly detect IL-6 in normal subjects.

*Heinrich:* Marathon runners have increased serum IL-6 levels.

*Gauldie:* That's been found in several studies, but it's not surprising.

*Aarden:* In our extensive experience with serum and plasma, very low amounts of B9 cell growth-supporting activity are found in normal individuals, and this activity cannot be inhibited with anti-IL-6. However, any increased growth-supporting activity in patients can be inhibited by antibodies against IL-6. The results from the B9 assay are in agreement with those from our ELISA, which can detect 1 pg/ml IL-6 in serum and doesn't detect any IL-6 in normal healthy individuals.

*Patterson:* Eve Wolinsky and I found that normal adult rat serum contains a cholinergic differentiation activity (Wolinsky & Patterson 1985). This factor was not, however, demonstrated to be authentic CDF/LIF.

## References

Williams N, Bertoncello I, Kavnoudias H, Zsebo K, McNiece I 1992 Recombinant rat stem cell factor stimulates the amplification and differentiation of fractionated mouse stem cell populations. Blood 79:58–64

Wolinsky E, Patterson PH 1985 Rat serum contains a developmentally regulated cholinergic inducing activity. J Neurosci 5:1509–1512

# Interleukin 6 determination in the detection of microbial invasion of the amniotic cavity

Roberto Romero*†, Waldo Sepulveda†, John S. Kenney‡, Linda E. Archer‡, Anthony C. Allison‡ and Pravinkumar B. Sehgal°

†Department of Obstetrics and Gynecology, Yale University School of Medicine, New Haven, CT 06510, ‡Institute of Immunology and Biological Sciences, Syntex Research, Palo Alto, CA 94304 and °Departments of Microbiology, Immunology and Medicine, The New York Medical College, Valhalla, NY 10595, USA

*Abstract.* A growing body of evidence suggests a role for cytokines in the mechanisms responsible for preterm parturition associated with intrauterine infection. Interleukin 6, a polyfunctional cytokine that is secreted by tissues in the feto–maternal interface in response to microbial products, has been implicated in the host response to intrauterine infection. The purpose of this study was to establish whether measurement of amniotic fluid concentrations of interleukin 6 could be of value in the diagnosis of microbial invasion of the amniotic cavity. Fluid was obtained by transabdominal amniocentesis from patients with preterm labour and intact chorioamniotic membranes and cultured for aerobic and anaerobic bacteria and mycoplasmas. Interleukin 6 concentrations were determined by an ELISA validated for human amniotic fluid. An interleukin 6 concentration above 11.2 ng/ml had a 93.7% sensitivity and a 92.3% specificity in the diagnosis of intra-amniotic infection. Moreover, patients with an amniotic fluid interleukin 6 level above 11.2 ng/ml and a negative amniotic fluid culture failed to respond to tocolysis, delivered a preterm infant and showed histological evidence of chorioamnionitis, and their neonates were at risk for congenital infections.

*1992 Polyfunctional cytokines: IL-6 and LIF. Wiley, Chichester (Ciba Foundation Symposium 167) p 205–223*

Accumulating evidence indicates that cytokines are key mediators in the pathogenesis of inflammatory diseases. This evidence consists of the demonstration of: (1) enhanced cytokine gene expression and translation in diseased tissues; (2) increased concentrations of cytokines in the plasma/serum or other biological fluids of patients with specific disease states; and (3) clinical

---

*To whom correspondence should be addressed. *Present address*: Wayne State University, Department of Obstetrics and Gynecology, 4707 St Antoine Boulevard, Hutzel Hospital, Detroit, MI 48201, USA

manifestation of disease in animal models in response to local (i.e., intra-articular) or systemic administration of cytokines and the blockade of these biological effects by specific neutralizing antibodies to cytokines or by other cytokine antagonists.

The potential participation of cytokines in these disease states has given rise to the expectation that measurement of cytokine levels may have diagnostic or prognostic value and that the administration of these mediators or their antagonists may have therapeutic value. However, even if there is a change in the concentrations of cytokines during the course of disease states, this does not automatically mean that the measurement of these analytes produces information that is of diagnostic or prognostic value. Here, we briefly discuss the principles involved in the assessment of the performance of a laboratory test. These principles will be applied to the analysis of interleukin 6 (IL-6) determination in the management of a specific condition—premature labour with intact chorioamniotic membranes.

## Interleukin 6 and human disease

IL-6 has been implicated in the pathogenesis of several disease states (Shimizu et al 1988, Van Oers et al 1988, Biondi et al 1989, Horii et al 1989, Breen et al 1990, Jourdan et al, 1990, Andus et al 1991, Bauer & Herrmann 1991, Molyneux et al 1991, Nijsten et al 1991; Table 1). In the context of acute and chronic inflammatory processes (such as meningitis, intra-amniotic infection, trauma, burns and rheumatoid arthritis), IL-6 is considered to be a mediator of inflammation (Nijsten et al 1987, Hirano et al 1988, Guerne et al 1989a,b,

**TABLE 1   Clinical disorders in which interleukin 6 may play a pathogenic role**

*Acute inflammatory disorders*
Meningitis
Intra-amniotic infection
Acute transplant rejection
Inflammatory bowel disease
Burns
Surgical trauma

*Chronic disease*
Rheumatoid arthritis
Mesangial proliferative glomerulonephritis
Psoriasis
Castleman's disease
Alzheimer's disease

*Neoplasia*
Multiple myeloma
Solid tumours (e.g., renal cell carcinoma)
Leukaemia

Hermann et al 1989, Sehgal et al 1989, Gauldie et al 1990, Van Snick 1990). In neoplasia and other chronic diseases, IL-6 seems to be an autocrine growth factor (Bataille et al 1989, Freeman et al 1989, Oster et al 1989). Raised concentrations of IL-6 have been demonstrated in the cerebrospinal fluid of patients with bacterial meningitis (Houssiau et al 1988a, Helfgott et al 1989, Waage et al 1989), in the amniotic fluid of women with microbial invasion of the amniotic cavity (Romero et al 1990), in the plasma of patients with sepsis (Hack et al 1989) and in the synovial fluid of patients with rheumatoid arthritis (Houssiau et al 1988b). Plasma concentrations of IL-6 also increase after surgery and after injection of endotoxin or tumour necrosis factor (TNF) into normal volunteers (Shenkin et al 1989, Cruickshank et al 1990, Pullicino et al 1990). In psoriasis, a disease characterized by epidermal inflammation and hyperplasia, IL-6 may be both an inflammatory mediator and an autocrine growth factor. Indeed, keratinocytes can synthesize IL-6 and also grow in response to this cytokine (Grossman et al 1989). Plasma concentrations of IL-6 are increased during the active phase of the disease and decrease after treatment.

A role for IL-6 in neoplasia has also been postulated. Constitutive IL-6 production has been demonstrated in several tumour cell lines (Kirnbauer et al 1989, Utsumi et al 1990). However, the effect of this cytokine on tumour growth is unclear. Although IL-6 may have autocrine growth-promoting effects in renal cell carcinoma (Miki et al 1989), it has been shown to have some anti-proliferative effects on other tumours (including breast and cervical cancer, and melanoma) (Morinaga et al 1989, Mule et al 1990, Levy et al 1991). A role for IL-6 in the pathophysiology of multiple myeloma is supported by the demonstration of: (1) increased plasma levels of this cytokine in patients affected by the disease; (2) a correlation between IL-6 plasma concentrations and the activity/severity of the disease; (3) a constitutive increase in mRNA for IL-6 in the bone marrow of patients with multiple myeloma; and (4) inhibition of spontaneous proliferation of plasmacytoma cells by a neutralizing anti-IL-6 antibody (Kawano et al 1988, Klein et al 1989, Klein et al 1990). IL-6 has also been implicated in the pathogenesis of Castleman's disease (Yoshizaki et al 1989). The multicentric variety of this disease is a lymphoproliferative disorder characterized by enlargement of lymph nodes, hepato-splenomegaly, anaemia, hypoalbuminaemia and polyclonal hypergammaglobulinaemia. Examination of lymph nodes reveals diffuse plasmacytosis. Yoshizaki et al (1989) suggested that IL-6 is a key mediator in this condition and that the germinal centre of affected lymph nodes is the source of IL-6 detected in the peripheral circulation of these patients. Further evidence supporting this view is that transplantation of congenitally anaemic mice with retrovirus-infected bone marrow cells bearing IL-6-coding sequences produced a syndrome that closely resembles Castleman's disease (Brandt et al 1990).

**The clinical problem of preterm labour**

Prematurity is the leading cause of perinatal morbidity and mortality worldwide. One third of all preterm deliveries occur after spontaneous preterm labour in which the chorioamniotic membranes are intact; another third result from premature rupture of the membranes and the remaining third occur as a consequence of maternal or fetal complications which indicate the need for preterm delivery. The standard therapeutic approach to preterm labour with intact membranes is pharmacological inhibition of labour (tocolysis), which generally requires the use of potent drugs such as β-adrenergic agents. Therapy with these agents is associated with significant cardiovascular side-effects (such as tachycardia and cardiac arrhythmia), pulmonary oedema, myocardial ischaemia, cerebral vasospasm and even maternal death (Benedetti 1983). A desirable clinical goal is to identify the patient who may benefit from this potentially life-threatening treatment as well as the patient who will not. The latter group could be spared the risk associated with ineffective therapy and could be the subject of different therapeutic approaches.

In cases of premature labour with intact membranes, 20% of the preterm neonates are born to women with microbial invasion of the amniotic cavity (Romero et al 1989a). This condition is often subclinical in nature and its diagnosis requires microbiological studies of amniotic fluid (Romero et al 1988a). The early identification of microbial invasion of the amniotic cavity is a desirable clinical goal because neonates born to mothers with intra-amniotic infections are at high risk of both infectious and non-infectious complications (Romero et al 1989a). Moreover, patients with microbial invasion of the amniotic cavity are refractory to tocolysis (Romero et al 1989a) and are at increased risk of developing pulmonary oedema during treatment.

Despite the importance of microbial invasion of the amniotic cavity in preterm labour, an early diagnosis is difficult to accomplish in practice; clinical signs of infection (fever, uterine tenderness, etc.) occur late and are present in only 12% of these patients. Results of amniotic fluid culture, considered the gold standard (definitive test) for diagnosis, are not immediately available and may take several days. Tests allowing rapid examination of amniotic fluid for microorganisms, such as the Gram stain (Romero et al 1988b) and the acridine orange stain (Romero et al 1989b), can detect only 50–60% of all cases with a positive amniotic fluid culture. Therefore, there is a need for sensitive and rapid tests for the identification of microbial invasion of the amniotic cavity.

**Cytokines in premature labour associated with microbial invasion of the amniotic cavity**

We have proposed that cytokines, including IL-6, play a role in the mechanism of preterm parturition associated with infection. The evidence for this includes that: (1) interleukin 1 (IL-1) and TNF stimulate prostaglandin production by human amnion, chorion and decidua (Romero et al 1989c,d,e,f); (2) bacterial products (such as endotoxin) stimulate human decidua to produce IL-1 (Romero et al 1989g), TNF (Romero et al 1991a) and IL-6 (Romero et al 1990); (3) in preterm labour associated with intra-amniotic infection cytokines not normally present in amniotic fluid (IL-1 and TNF, for example) can be detected (Romero et al 1989c,e, 1991b, 1992a) and there is a dramatic increase in the concentration of IL-6 (which is a normal constituent of amniotic fluid) (Romero et al 1990); (4) extracorporeal perfusion of human uteri with IL-1 and TNF can induce regular myometrial contractions (R. Romero & C. Bulletti, unpublished observations); (5) systemic administration of IL-1 and TNF can induce preterm parturition in mice (Romero et al 1991c); and (6) pre-treatment with the natural IL-1 receptor antagonist—a novel cytokine that binds to the IL-1 receptor and blocks the biological effects of IL-1 (Eisenberg et al 1991)—prevents IL-1-induced preterm parturition in mice (Romero et al 1992b) and blocks IL-1-induced prostaglandin production by intrauterine tissues (Romero et al 1992c).

The interest in the potential diagnostic value of measurement of amniotic fluid IL-6 in the management of premature labour arose from the observation that intra-amniotic infection was associated with a dramatic increase in the concentration and biological activity (measured with both the hepatocyte-stimulating assay and the B9 assay) of this cytokine in amniotic fluid (Romero et al 1990, Santhanam et al 1991). Moreover, patients with raised IL-6 concentrations but without demonstrable infection showed histological evidence of chorioamnionitis and/or failed to respond to tocolysis. These observations justify a formal examination of the diagnosic and prognostic value of amniotic fluid IL-6 determination in patients in preterm labour with intact membranes.

## Assessment of the performance of a laboratory test

The importance of diagnostic testing in clinical medicine is self-evident— appropriate treatment and accurate prognosis are not possible without a precise diagnosis. Clinicians are called on to make assessments of the usefulness of an ever-increasing number of laboratory tests. How should this be decided?

The main objective of a clinical laboratory test is to provide information that will alter the management of patients. The information may help differentiate between healthy and sick individuals, provide an indication of the severity of the disease, and predict the clinical course (prognosis) and the response to therapy. A test should become part of the clinical diagnostic armamentarium only if it is the best source of the required information and if the cost of the test can be justified when compared with that of other alternatives (Zweig et al 1987).

Two elements are considered in the analysis of the performance of a clinical laboratory test—accuracy and efficiency. A test's accuracy is its ability to discriminate between alternative clinical states (infection or no infection, for example). It is evaluated by determining the sensitivity and specificity, and is represented by receiver operating characteristic (ROC) curves. A test's efficiency refers to its actual practical value and takes into account the benefits, the costs and the risks of acting on the results of a test. The formal tool of clinical decision analysis is used to evaluate efficiency (Beck 1982, Robertson et al 1983, Zweig et al 1987).

Table 2 shows guidelines which can be used in the evaluation of the performance of a laboratory test. We shall examine these guidelines in the context of the assessment of the usefulness of measuring IL-6 concentrations in amniotic fluid in the management of women with premature labour.

## Selection of the study sample

A general principle in the design of any study is that the sample must be representative of the larger population of interest. Every effort must be made to eliminate biases in sample selection, because they limit extrapolation of results.

The ideal method for evaluating a laboratory test is a prospective study of consecutive patients who meet the criteria of the clinical group of interest. Inclusion criteria must be defined to reflect the real diagnostic problem. Enrollment of patients who do not represent a diagnostic dilemma weakens the validity of the study. All patients meeting the eligibility criteria must be kept in the study and accounted for in the final analysis; cases of inadequate sampling, analytical error, non-follow-up, or death should be included, because exclusion of these patients creates biases and may affect the interpretation of the results.

Prospective studies of consecutive patients are difficult to conduct, expensive and time-consuming. For these reasons, alternative approaches are often used in the evaluation of laboratory tests. The second-best study design is a random selection of patients meeting the eligibility criteria. Non-random selection is likely to introduce biases.

**TABLE 2   Guidelines for the evaluation of a laboratory test**

1. Define the clinical problem
2. Select a representative study sample
3. Establish the clinical state of each subject according to a 'gold standard'
4. Perform the laboratory test blindly (without knowing the results of the 'gold standard')
5. Compare the results of the performance of the new diagnostic tests with other methods available to answer the clinical question
6. Define the inter- and intra-assay variability of the test
7. Define the medical and financial costs and benefits of the test (efficiency or cost–benefit analysis)

The sample size is critical: the larger the sample, the smaller the sampling error and the better the estimates of diagnostic accuracy. However, in practice, this advantage must be balanced against other factors such as the duration and cost of the study. A practical approach is to decide *a priori* the confidence limits for the desired sensitivity and specificity. These figures can then be used to calculate the sample size necessary.

To evaluate the diagnostic and prognostic value of amniotic fluid IL-6 determination in patients with preterm labour, we conducted a study of consecutive patients who were admitted with this condition and who underwent amniocentesis for microbiological evaluation of the amniotic cavity. Preterm labour was defined as the presence of regular uterine contractions occurring before the 37th week of gestation.

### *Establishing the 'true' clinical state—'the gold standard'*

For assessment of the accuracy of a laboratory test, two pieces of information are required—the results of the test and the true state of the patient (the 'gold standard'). The gold standard is a definitive diagnosis attained by biopsy, autopsy, surgery, microbiological studies, long term follow-up, or other means. In some cases, an adequate gold standard is not available and clinicians use the logical consequences of the disease (term constructs) for analytical purposes.

To determine the value of amniotic fluid IL-6 assay in the diagnosis of microbial invasion of the amniotic cavity, we have chosen the results of amniotic fluid culture for microorganisms as the gold standard. Amniotic fluid obtained by transabdominal amniocentesis is normally sterile, and, therefore, the recovery

| | Disease | |
|---|---|---|
| | Present | Absent |
| Positive | A | B |
| Negative | C | D |

**Sensitivity = A ÷ (A + C)**
**Specificity = D ÷ (B + D)**
**Positive Predictive Value = A ÷ (A + B)**
**Negative Predictive Value = D ÷ (C + D)**

FIG. 1. Contingency table and definition of sensitivity, specificity and positive and negative predictive values (see text p 214).

of any microorganism from this fluid is considered abnormal. Microbiological studies included the traditional cultures for aerobic and anaerobic bacteria. Cultures for mycoplasmas were also done, because recent studies indicate that *Ureaplasma urealyticum* is the microorganism most commonly isolated from the amniotic fluid of women with preterm labour (Romero et al 1989a).

The use of microbial cultures as a gold standard is not perfect. False-positive results may occur if the specimen is contaminated with skin flora at the time of amniocentesis or during transport or manipulation of the specimen. False-negative results can occur if microorganisms do not grow in culture because of a small inoculum size, loss of viability during transport, or inappropriate culture techniques. Despite these limitations, it is believed that microbiological studies are the best method currently available for diagnosing microbial invasion of the amniotic cavity.

The other clinical outcome of interest in our study was the response to pharmacological inhibition of labour. To assess the ability of measured amniotic fluid IL-6 concentrations to predict response to tocolysis, we used the amniocentesis-to-delivery interval.

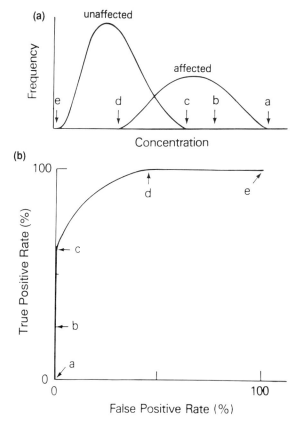

*Blinding of the results of the laboratory test*

It is essential that individuals carrying out or interpreting the results of the test under study are ignorant of the clinical outcome of interest (i.e., amniotic fluid culture result). The ideal situation is for the test to be performed before the outcome variable is known. This will avoid the temptation of selectively repeating tests or rejecting data when the results do not fit the expected or desired outcome. An alternative method is for the test and extraction of clinical data to be done by separate teams blinded to each other's results. In our study, IL-6 concentrations in amniotic fluid were assayed by a team unaware of amniotic fluid culture results and the clinical outcome. Similarly, results of IL-6 determination were unavailable to the clinicians managing the patients,

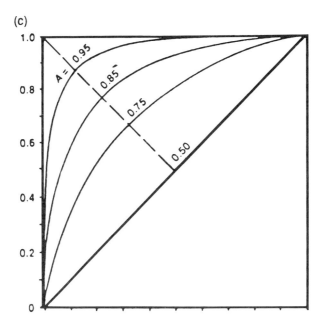

FIG. 2. (a) Hypothetical frequency distribution; the curve on the left represents results from unaffected individuals and the curve on the right represents results from affected individuals. (b) Receiver operating characteristic curve, generated by varying the decision level according to the cut-off points a–e shown in (a) and then plotting the resulting pairs of true- and false-positive rates. (c) Receiver operating characteristic curve in which the proportion of true-positives is plotted against the proportion of false-positives for various possible settings of the decision criterion. The idealized curves shown correspond to the indicated values of the accuracy measure A, which represents the area under the curve. (a) and (b) are reprinted with permission from Zweig & Robertson 1982; (c) is reprinted with permission from Swets 1988, copyright 1988 by the AAAS.

and, thus, these results could not influence the decision to continue or discontinue tocolysis or any other intervention which might result in preterm delivery.

### Evaluation of the performance of a new laboratory test

The performance of a test is assessed on the basis of its ability to discriminate between healthy and sick patients according to the results of the gold standard. This generally involves calculating the diagnostic indices. Figure 1 (p 211) illustrates a standard table for these calculations and the formulae for sensitivity, specificity, and positive and negative predictive values.

The following definitions are helpful:

*Sensitivity*, or the detection rate, is the percentage of affected individuals yielding a positive result.

The *false-negative* rate is the percentage of affected individuals yielding a negative result and can be calculated by the formula *100 – sensitivity*.

*Specificity* is the percentage of healthy individuals yielding a negative result.

The *false-positive* rate is the percentage of healthy individuals with a positive result and can be calculated with the formula *100 – specificity*.

*Positive predictive value* is the proportion of patients with a positive test who are affected by the disease.

*Negative predictive value* is the proportion of individuals with a negative test who are not affected by the disease.

Although sensitivity and specificity are generally considered to be intrinsic properties of a test, the positive and negative predictive values are critically dependent on the prevalence of the disease in the population.

Receiver operating characteristic (ROC) curves are a tool with which to assess the performance of a test. They display the relationship between the sensitivity and false-positive rate of a test at different cut-off points. Figure 2a shows a hypothetical frequency distribution of the concentration of an analyte in both an affected and an unaffected population. Figure 2b shows an ROC curve generated by varying decision levels according to the cut-off points labelled in Fig. 2a. It is possible to calculate the area under the curve (denoted as A), and this is considered a good index of the accuracy of a test. Figure 2c illustrates various ROC curves with different A values. A curve with an A value of 0.5 lies along the major diagonal, indicative of an equal sensitivity and specificity. A test with this performance has no more discriminating value than chance alone. The greater the area under the curve, the better is the accuracy of the test. Perfect discrimination would be achieved by a test with an A value of 1.0 (Swets 1988). Statistical techniques have been developed to compare the area under two ROC curves, and thus the performance of two laboratory tests.

**Performance of amniotic fluid IL-6 determination in the diagnosis of microbial invasion of the amniotic cavity**

Concentrations of IL-6 in the amniotic fluid were determined in 146 patients admitted consecutively with a diagnosis of preterm labour and intact membranes who underwent amniocentesis for evaluation of the microbiological state of the amniotic cavity. The prevalence of positive amniotic fluid cultures was 10.9% (16/146). IL-6 levels in amniotic fluid were measured by a two-site ELISA (enzyme-linked immunosorbent assay) using two mouse monoclonal antibodies which bind non-overlapping epitopes on human IL-6 molecules. The monoclonal antibodies to IL-6 were developed as previously described (Kenney et al 1987). ELISA conditions were optimized and the assays performed as previously described (Kenney et al 1990). ROC curve analysis of the performance of amniotic fluid IL-6 measurement in the identification of microbial invasion of the amniotic cavity clearly demonstrated that amniotic fluid IL-6 concentrations provide discrimination between patients with positive amniotic fluid cultures and patients with negative cultures (data not shown). The diagnostic indices for two cut-off values (amniotic fluid IL-6 equal to or above 11.2 ng/ml or 6.7 ng/ml) are displayed in Table 3. An IL-6 concentration above 6.7 ng/ml identified 93.7% of patients with microbial invasion of the amniotic cavity and had a false-positive rate of 11.6%. However, the false-positive results may be more apparent than real. Neonates born to two women with amniotic fluid IL-6 concentrations above 6.7 ng/ml but a negative amniotic fluid culture for microorganisms had congenital syphilis. Moreover, all patients with amniotic IL-6 concentrations above 6.7 ng/ml and negative amniotic fluid cultures delivered a preterm neonate, and all placentas available for study showed severe chorioamnionitis, acute inflammatory lesions of the chorioamniotic membranes. Chorioamnionitis and acute inflammatory lesions of the umbilical cord (funisitis) are well-recognized risk factors for neonatal sepsis and poor perinatal outcome in preterm births. However, this information is not currently used in clinical decision-making, because results are not available shortly after delivery. These results suggest that an increased IL-6 concentration in amniotic fluid identifies a group of patients with an acute inflammatory process that was frequently associated with documented microbial invasion of the amniotic cavity. False-positive results may be attributed to infections which escape detection by microbiological techniques used in this study, or to an inflammatory process of non-infectious aetiology.

**TABLE 3   Sensitivity and specificity of amniotic fluid interleukin 6 assay in the diagnosis of microbial invasion of the amniotic cavity**

| IL-6 cut-off value (ng/ml) | Sensitivity (%) | Specificity (%) |
|---|---|---|
| 6.7 | 93.7 | 88.4 |
| 11.2 | 93.7 | 92.3 |

**Comparison of amniotic fluid IL-6 assay with other currently available tests for the rapid detection of microbial invasion of the amniotic cavity**

The Gram stain is universally used for the rapid detection of infection in biological fluids. In this study, Gram staining of amniotic fluid as a test for infection had a sensitivity of 62% (10/16), a specificity of 100% (130/130), a positive predictive value of 100% (10/10) and a negative predictive value of 95.5% (130/136). The low sensitivity of the Gram stain may be due to its inability to detect mycoplasmas (Romero et al 1988b). Indeed, four of the six patients with false-negative Gram stain results had microbial invasion limited to mycoplasmas. The sensitivity of amniotic fluid IL-6 assay (cut-off $>11.2$ ng/ml) was higher than that of the Gram staining method (93.7% vs. 62%; $P > 0.05$ but $< 0.10$). However, the false-positive rate of amniotic fluid IL-6 testing was greater than that of the Gram stain (7.7% vs. 0%; $P < 0.05$), but, as previously indicated, this may be due to the shortcomings of using amniotic fluid culture as the gold standard.

In our study, amniotic fluid IL-6 assay had a higher sensitivity than amniotic fluid culture results in the prediction of chorioamnionitis (identified histologically) (80% vs. 53%; $P < 0.05$). Moreover, the three cases with early congenital infections had dramatically raised amniotic fluid IL-6 concentrations. Because results of IL-6 determinations could be available before birth, this information could be used to start early antibiotic treatment. The timing of antibiotic administration is critical, because there is recent evidence that antepartum treatment is more effective than postnatal treatment in reducing the rate of neonatal sepsis (Gibbs et al 1988). Moreover, administration of antibiotics has been shown to delay preterm delivery in patients with chorioamnionitis. Although the increased sensitivity of amniotic fluid IL-6 assay in the detection of chorioamnionitis is associated with a higher rate of false-positive results (16% vs. 7%; $P < 0.05$), this is not considered a problem, because it is more serious to delay antibiotic treatment in affected cases than to treat a few normal newborns.

**Performance of amniotic fluid IL-6 assays in predicting response to tocolysis**

The relationship between amniotic fluid concentrations of IL-6 and the admission-to-delivery interval was examined in patients who received intravenous tocolysis and had a negative Gram stain for microorganisms. ROC curve analysis indicated that amniotic fluid IL-6 measurement could discriminate between the patient who responded and the patient who did not respond to tocolysis. No patient with an amniotic fluid IL-6 level above 100 ng/ml had a pregnancy prolongation beyond 24 hours. This information can be used to avoid the risk of using tocolytic agents in patients unlikely to respond and to attempt other forms of therapy.

**Conclusion**

Amniotic fluid IL-6 measurement is of value in the diagnosis of microbial invasion of the amniotic cavity. Patients with an increased amniotic fluid IL-6 level, despite negative amniotic fluid cultures, are at risk for histological chorioamnionitis and preterm delivery.

*Acknowledgements*

This work was supported by a grant from the Walter Scott Foundation for Medical Research. R. Romero is a recipient of a Physician Scientist Award from the NICHD. P. B. Sehgal is supported by NIH grant AL 16262.

**References**

Andus T, Gross V, Casar I et al 1991 Activation of monocytes during inflammatory bowel disease. Pathobiology 59:166–170

Bataille R, Jourdan M, Zhang X-G, Klein B 1989 Serum levels of interleukin 6, a potent myeloma cell growth factor, as a reflect of disease severity in plasma cell dyscrasias. J Clin Invest 84:2008–2011

Bauer J, Herrmann F 1991 Interleukin-6 in clinical medicine. Ann Hematol 62:203–210

Beck JR 1982 The role of new laboratory tests in clinical decision-making. Clin Lab Med 2:751–777

Benedetti TJ 1983 Maternal complications of parenteral β-sympathomimetic therapy for premature labor. Am J Obstet Gynecol 145:1–6

Biondi A, Rossi V, Bassan R et al 1989 Constitutive expression of the interleukin-6 gene in chronic lymphocytic leukemia. Blood 73:1279–1284

Brandt SJ, Bodine DM, Dunbar CE, Nienhuis AW 1990 Dysregulated interleukin 6 expression produces a syndrome resembling Castleman's disease in mice. J Clin Invest 86:592–599

Breen EC, Rezai AR, Nakajima K et al 1990 Infection with HIV is associated with elevated IL-6 levels and production. J Immunol 144:480–484

Cruickshank AM, Fraser WD, Burns HJ, van Damme J, Shenkin A 1990 Response of serum interleukin-6 in patients undergoing elective surgery of varying severity. Clin Sci (Lond) 79:161–165

Eisenberg SP, Brewer MT, Verderber E, Heimdal P, Brandhuber BJ, Thompson RC 1991 Interleukin 1 receptor antagonist is a member of the interleukin 1 gene family: evolution of a cytokine control mechanism. Proc Natl Acad Sci USA 88:5232–5236

Freeman GJ, Freedman AS, Rabinowe SN et al 1989 Interleukin 6 gene expression in normal and neoplastic B cells. J Clin Invest 83:1512–1518

Gauldie J, Northemann W, Fey GH 1990 IL-6 functions as an exocrine hormone in inflammation: hepatocytes undergoing acute phase responses require exogenous IL-6. J Immunol 144:3804–3808

Gibbs RS, Dinsmoor MJ, Newton ER, Ramamurthy RS 1988 A randomized trial of intrapartum versus immediate postpartum treatment of women with intra-amniotic infection. Obstet Gynecol 72:823–828

Grossman RM, Krueger J, Yourish D et al 1989 Interleukin 6 is expressed in high levels in psoriatic skin and stimulates proliferation of cultured human keratinocytes. Proc Natl Acad Sci USA 86:6367–6371

Guerne P-A, Zuraw BL, Vaughan JH, Carson DA, Lotz M 1989a Synovium as a source of interleukin 6 in vitro: contribution to local and systemic manifestations of arthritis. J Clin Invest 83:585–592

Guerne P-A, Vaughan JH, Carson DA, Terkeltaub B, Lotz M 1989b Interleukin-6 (IL-6) and joint tissues. Ann NY Acad Sci 557:558–561

Hack CE, de Groot ER, Felt-Bersma RJF et al 1989 Increased plasma levels of interleukin-6 in sepsis. Blood 74:1704–1710

Helfgott DC, Tatter SB, Santhanam U et al 1989 Multiple forms of IFN-$\beta_2$/IL-6 in serum and body fluids during acute bacterial infection. J Immunol 142: 948–953

Hermann E, Fleischer B, Mayet W-J, Poralla T, Meyer zum Büschenfelde K-H 1989 Correlation of synovial fluid interleukin 6 (IL-6) activities with IgG concentrations in patients with inflammatory joint disease and osteoarthritis. Clin Exp Rheumatol 7:411–414

Hirano T, Matsuda T, Turner M et al 1988 Excessive production of interleukin 6/B cell stimulatory factor-2 in rheumatoid arthritis. Eur J Immunol 18:1797–1801

Horii Y, Muraguchi A, Iwano M et al 1989 Involvement of IL-6 in mesangial proliferative glomerulonephritis. J Immunol 143:3949–3955

Houssiau FA, Bukasa K, Sindic CJM, van Damme J, van Snick J 1988a Elevated levels of the 26K human hybridoma growth factor in cerebrospinal fluid with acute infection of the central nervous system. Clin Exp Immunol 71:320–323

Houssiau FA, Devogelaer J-P, van Damme J, de Deuxchaisnes CN, van Snick J 1988b Interleukin-6 in synovial fluid and serum of patients with rheumatoid arthritis and other inflammatory arthritides. Arthritis Rheum 31:784–788

Jourdan M, Bataille R, Seguin J, Zhang X-G, Chaptal PA, Klein B 1990 Constitutive production of interleukin-6 and immunologic features in cardiac myxomas. Arthritis Rheum 33:398–402

Kawano M, Hirano T, Matsuda T et al 1988 Autocrine generation and requirement of BSF-2/IL-6 for human multiple myelomas. Nature (Lond) 332:83–85

Kenney JS, Masada MP, Eugui EM, DeLustro BM, Mulkins MA, Allison AC 1987 Monoclonal antibodies to human recombinant interleukin 1 (IL-1) β: quantitation of IL-β and inhibition of biological activity. J Immunol 138:4236–4242

Kenney JS, Masada MP, Allison AC 1990 Development of quantitative two-site ELISAs for soluble proteins. In: Zola H (ed) Laboratory methods in immunology. CRC Press, Boca Raton, FL, vol 1:232–240

Kirnbauer R, Kock A, Schwartz T et al 1989 IFN-β2, B cell differentiation factor 2, or hybridoma growth factor (IL-6) is expressed and released by human epidermal cells and epidermoid carcinoma cell lines. J Immunol 142:1922–1928

Klein B, Zhang X-G, Jourdan M et al 1989 Paracrine rather than autocrine regulation of myeloma-cell growth and differentiation by interleukin-6. Blood 73:517–526

Klein H, Becher R, Lubbert M et al 1990 Synthesis of granulocyte colony-stimulating factor and its requirement for terminal divisions in chronic myelogenous leukemia. J Exp Med 171:1785–1790

Levy Y, Tsapis A, Brouet J-C 1991 Interleukin-6 antisense oligonucleotides inhibit the growth of human myeloma cell lines. J Clin Invest 88:696–699

Miki S, Iwano M, Miki Y et al 1989 Interleukin-6 (IL-6) functions as an in vitro autocrine growth factor in renal cell carcinomas. FEBS (Fed Eur Biochem Soc) Lett 250:607–610

Molyneux ME, Taylor TE, Wirima JJ, Grau GE 1991 Tumour necrosis factor, interleukin-6, and malaria. Lancet 337:1098

Morinaga Y, Suzuki H, Takatsuki F et al 1989 Contribution of IL-6 to the anti-proliferative effect of IL-1 and tumor necrosis factor on tumor cell lines. J Immunol 143:3538–3542

Mule JJ, McIntosh JK, Jablons DM, Rosenberg SA 1990 Antitumor activity of recombinant interleukin 6 in mice. J Exp Med 171:629–636

Nijsten MWN, de Groot ER, ten Duis HJ, Klasen HJ, Hack CE, Aarden LA 1987 Serum levels of interleukin-6 and acute phase responses. Lancet 333:921

Nijsten MW, Hack CE, Helle M, ten Duis HJ, Klasen HJ, Aarden LA 1991 Interleukin-6 and its relation to the humoral immune response and clinical parameters in burned patients. Surgery (St Louis) 109:761–767

Oster W, Cicco NA, Klein H et al 1989 Participation of the cytokines interleukin 6, tumor necrosis factor-$\alpha$, and interleukin 1-$\beta$ secreted by acute myelogenous leukemia blasts in autocrine and paracrine leukemia growth control. J Clin Invest 84:451–457

Pullicino EA, Carli F, Poole S, Rafferty B, Malik STA, Elia M 1990 The relationship between the circulating concentrations of interleukin-6 (IL-6), tumor necrosis factor (TNF) and the acute phase response to elective surgery and accidental injury. Lymphokine Res 9:2–6

Robertson EA, Zweig MH, van Steirteghem AC 1983 Evaluating the clinical efficacy of laboratory tests. Am J Clin Pathol 79:78–86

Romero R, Mazor M 1988a Infection and preterm labor. Clin Obstet Gynecol 31:553–584

Romero R, Emamian M, Quintero R et al 1988b The value and limitations of the Gram stain examination in the diagnosis of intraamniotic infection. Am J Obstet Gynecol 159:114–119

Romero R, Sirtori M, Oyarzun E et al 1989a Infection and labor. V. Prevalence, microbiology and clinical significance of intraamniotic infection in women with preterm labor and intact membranes. Am J Obstet Gynecol 161:817–824

Romero R, Emamian M, Quintero R et al 1989b Diagnosis of intraamniotic infection: the acridine orange stain. Am J Perinatol 6:41–45

Romero R, Brody DT, Oyarzun E et al 1989c Infection and labor. III. Interleukin-1: a signal for the onset of parturition. Am J Obstet Gynecol 160:1117–1123

Romero R, Durum S, Dinarello C, Oyarzun E, Hobbins JC, Mitchell MD 1989d Interleukin-1 stimulates prostaglandin biosynthesis by human amnion. Prostaglandins 37:13–22

Romero R, Manogue KR, Mitchell MD et al 1989e Infection and Labor. IV. Cachectin-tumor necrosis factor in the amniotic fluid of women with intraamniotic infection and preterm labor. Am J Obstet Gynecol 161:336–341

Romero R, Mazor M, Wu YK et al 1989f Bacterial endotoxin and tumor necrosis factor stimulate prostaglandin production by human decidua. Prostaglandins Leukotrienes Essent Fatty Acids 37:183–186

Romero R, Wu YK, Brody D, Oyarzun E, Duff GW, Durum SK 1989g Human decidua: a source of IL-1. Obstet Gynecol 73:31–34

Romero R, Avila C, Santhanam U, Sehgal PB 1990 Amniotic fluid IL-6 in preterm labor: association with infection. J Clin Invest 85:1392–1400

Romero R, Mazor M, Manogue K, Oyarzun E, Cerami A 1991a Human decidua: a source of cachectin-tumor necrosis factor. Eur J Obstet Gynecol Reprod Biol 41:123–127

Romero R, Mazor M, Brandt F, Sepulveda W, Avila C, Dinarello CA 1991b Interleukin-1$\alpha$ and interleukin-1$\beta$ in human preterm and term parturition. Am J Reprod Immunol, in press

Romero R, Mazor M, Tartakovsky B 1991c Systemic administration of interleukin-1 induces preterm parturition in mice. Am J Obstet Gynecol 165:969–971

Romero R, Mazor M, Sepulveda W, Avila C, Copeland D, Williams J 1992a Tumor necrosis factor in preterm and term labor. Am J Obstet Gynecol 166:1576–1587

Romero R, Mazor M, Tartakovsky B 1992b The natural interleukin-1 receptor antagonist prevents interleukin-1-induced preterm parturition. 12th Annual Meeting of the Society of Perinatal Obstetricians. Orlando, Florida. February 1992. Am J Obstet Gynecol 166:358 (abstr)

Romero R, Sepulveda W, Mazor M, Dinarello C, Mitchell M 1992c Natural interleukin-1 receptor antagonist blocks interleukin-1-induced prostaglandin production by intra-uterine tissues: the basis for a novel approach to the treatment of preterm labor in the setting of infection. 12th Annual Meeting of the Society of Perinatal Obstetricians. Orlando, Florida. February 1992. Am J Obstet Gynecol 166:274 (abstr)

Santhanam U, Avila C, Romero R et al 1991 Cytokines in normal and abnormal parturition: elevated amniotic fluid interleukin-6 levels in women with premature rupture of membranes associated with intrauterine infection. Cytokines 3:155–163

Sehgal PB, Grieninger G, Tosato G (eds) 1989 Regulation of the acute phase and immune responses: interleukin-6. Ann NY Acad Sci vol 557

Shenkin A, Fraser WD, Series J et al 1989 The serum interleukin-6 response to elective surgery. Lymphokine Res 87:123–127

Shimizu S, Hirano T, Yoshioka R et al 1988 Interleukin-6 (B-cell stimulatory factor-2)-dependent growth of a Lennert's lymphoma-derived T-cell line (KT-3). Blood 72:1826–1828

Swets J 1988 Measuring the accuracy of diagnostic systems. Science (Wash DC) 240:1285–1293

Van Oers MHJ, van der Heyden AAPAM, Aarden LA 1988 Interleukin-6 (IL-6) in serum and urine of renal transplant recipients. Clin Exp Immunol 71:314–319

Van Snick J 1990 Interleukin-6: an overview. Annu Rev Immunol 8:253–278

Waage A, Halstensen A, Shalaby R, Brandtzaeg P, Kierulf P, Espevik T 1989 Local production of tumor necrosis factor $\alpha$, interleukin 1, and interleukin 6 in meningococcal meningitis. Relation to the inflammatory response. J Exp Med 170:1859–1867

Utsumi K, Takai Y, Tada T, Ohzeki S, Fujiwara H, Hamaoka T 1990 Enhanced production of IL-6 in tumor-bearing mice and determination of cells responsible for its augmented production. J Immunol 145:397–403

Yoshizaki K, Matsuda T, Nishimoto N et al 1989 Pathogenic significance of interleukin-6 (IL-6/BSF-2) in Castleman's disease. Blood 74:1360–1367

Zweig MH, Robertson EA 1982 Why we need better test evaluation. Clin Chem 28:1272–1276

Zweig MH, Beck JR, Collinnsworth WL et al 1987 Assessment of clinical sensitivity and specificity of laboratory tests. National Committee for Clinical Laboratory Standards (NCCLS). Publication GP10-P. Villanova, PA

## DISCUSSION

*Metcalf:* Are you implying that IL-6 has a causative role to play in this syndrome, or are you simply using IL-6 as an indicator, in which case you could perhaps have used IL-1 or some other cytokine?

*Romero:* I believe that IL-6 plays a role in the process of infection-induced preterm parturition and in the host response to intrauterine infection. Indeed, IL-6 stimulates production of prostaglandins by intrauterine tissues and thus can participate in signalling parturition (Mitchell et al 1991). Induction by IL-6 of acute-phase plasma proteins may play a role in host defence. However, you are correct to point out that we are using IL-6 as an indicator and that other cytokines may be as informative as IL-6. Comparative studies of the diagnostic value of different cytokines are in progress.

*Dexter:* Some women have a history of premature labour. Have you managed to look at any patients who have gone through two or three premature labours, to see if there is any consistency in IL-6 levels?

*Romero:* The single most important risk factor for preterm delivery is having delivered preterm in a previous pregnancy. Our studies have focused on measurement of IL-6 in the amniotic fluid of women presenting with an acute episode of preterm labour. Serial amniocenteses in women at risk have not been performed because amniocentesis is an invasive procedure which carries a small risk of pregnancy loss, about 0.5–1%.

In women with an acute episode of preterm labour we have examined IL-6 levels in maternal plasma, amniotic fluid and fetal plasma. Amniotic fluid IL-6 levels correlated best with amniotic fluid culture results (Santhanam et al 1991). Because intrauterine infection is generally a localized process, I doubt that IL-6 maternal plasma concentrations could be a sensitive index of impending preterm labour. Other approaches which are being considered are measurement of cervical length by ultrasound and the detection of oncofetal fibronectin in vaginal secretions (Lockwood et al 1991).

*Dexter:* Is there always a history of infection in patients at risk?

*Romero:* No. However, we have some results suggesting that patients who have had premature labour associated with infection in their first pregnancy tend to have premature labour associated with infection in subsequent pregnancies.

*Lotem:* What is the degree of correlation between the concentration of IL-6 in the amniotic fluid and the severity of the infection itself?

*Romero:* There is a very weak correlation between inoculum size and amniotic fluid IL-6.

*Fey:* Do you detect IL-6 only after infection with Gram-positive bacteria?

*Romero:* No; we find raised amniotic fluid IL-6 concentrations in cases of microbial invasion with Gram-positive and Gram-negative microorganisms, with fungi and with mycoplasmas.

*Stewart:* Is it clear that all the IL-6 in the amniotic fluid is maternal in origin, or is some derived from the fetus?

*Romero:* Actually, we do not know the source of amniotic fluid IL-6. We have assessed IL-6 concentrations in maternal plasma, fetal plasma and amniotic fluid in a limited number of cases. Increased IL-6 in the amniotic fluid has been detected in the absence of detectable IL-6 in maternal and fetal plasma. Therefore, we believe that most of the IL-6 is generated within the uterine cavity. We have demonstrated that tissues in the feto–maternal interface (decidua and chorion) can produce IL-6 (Romero et al 1990).

*Gauldie:* Are there viral infections associated with preterm labour?

*Romero:* There has not been a careful prospective examination of the potential role of viruses in the aetiology of preterm labour. I suspect that a fraction of cases are due to undiagnosed viral infections.

*Gauldie:* Did you check urine, to see if IL-6 was detectable?

*Romero:* We have checked urine for IL-1 and TNF, but were unable to detect either. We haven't tested urine for IL-6.

*Gauldie:* If this test is to be of clinical relevance, the result will need to be produced quickly enough for it to have an impact on the clinical decision. Some of your results indicated that women give birth within three to four hours of the onset of labour.

*Romero:* You are correct. For a test to be valuable, the results need to be available in time to change clinical management. This places a real limitation on the use of bioassays. Immunoassays are used very often in clinical medicine. For example, pregnancy testing is now routinely done with immunoassays which have been configured to produce results within two minutes. If amniotic fluid IL-6 concentrations are an adequate discriminator of infection and if this information is helpful, I think it's just a matter of configuring an immunoassay to provide the appropriate answer.

*Gough:* If I understood correctly, the test must not only be rapid, but also highly quantitative. You aren't dealing with a plus or minus situation, but with a precise cut-off value. Rapid immunoassay tests will give you a plus/minus answer, not a quantitative one.

*Romero:* We could configure an immunoassay with two separate cut-off values, one with a high cut-off—11.2 ng/ml—for the detection of microbial invasion of the amniotic cavity and one with a lower cut-off to identify patients who will fail to respond to tocolysis. This again is a problem of assay configuration.

*Heinrich:* There are organs other than the liver in which acute-phase proteins can be synthesized, such as the choroid plexus and placenta. The uterus may also be a site of acute-phase protein synthesis.

*Fey:* We have found mRNA for acute-phase proteins ($\alpha_2$-macroglobulin) in pregnant rat uterus (Northemann et al 1988), and Dr Sakaki's group in Japan have similar results.

*Heinrich:* So IL-6 could have a local function also in the uterus.

*Fey:* Not necessarily; it was not established that IL-6 was responsible for the induction of the $\alpha_2$-macroglobulin gene in the uterus. As you know, this gene is induced in the ovary and testis by agents other than IL-6, placental lactogens.

# References

Lockwood CJ, Senyei AE, Dische MR et al 1991 Fetal fibronectin in cervical and vaginal secretions as a predictor of preterm delivery. N Engl J Med 325:669–674

Mitchell MD, Dudley DJ, Edwin SS, Schiller SL 1991 Interleukin-6 stimulates prostaglandin production by human amnion and decidual cells. Eur J Pharmacol 192:189–191

Northemann W, Shiels BR, Braciak TA, Hanson RW, Heinrich PC, Fey GH 1988 Structure and acute phase regulation of the rat $\alpha_2$-macroglobulin gene. Biochemistry 27:9194–9203

Romero R, Avila C, Santhanam U, Sehgal PB 1990 Amniotic fluid interleukin 6 in preterm labor: association with infection. J Clin Invest 85:1392–1400

Santhanam U, Avila C, Romero R et al 1991 Cytokines in normal and abnormal parturition: elevated amniotic fluid interleukin-6 levels in women with premature rupture of membranes associated with intrauterine infection. Cytokines 3:155–163

# General discussion III

### The multimeric nature of human interleukin 6

*Sehgal:* When you take sera from patients with sepsis, and purify the IL-6 using an immunoaffinity purification scheme, Western blots show that the IL-6 found in the peripheral circulation, in denaturing conditions, is in low molecular weight form, 20–30 kDa (Helfgott et al 1989). Everyone agrees with this. When we began studying purified fibroblast-derived IL-6 we found that under non-denaturing, non-reducing, conditions most of the IL-6 is in the form of a complex of about 85 kDa (May et al 1991). There are additional bands from 45 to 120 kDa. When the 85 kDa band is eluted and SDS added, it breaks up into the usual 25 and 30 kDa monomeric IL-6 species. The 25 kDa species is $O$-glycosylated, and the 30 kDa species is $N$- and $O$-glycosylated.

On gel filtration of fibroblast-derived IL-6 on a G-200 column we observe a fairly broad distribution of IL-6 protein from 25 to 85 kDa. We can immunoprecipitate IL-6 in fractions eluted from the column and show that the high molecular mass material consists of the two 25 and 30 kDa species. However, when we test the 50–85 kDa material in the B9 assay and compare it with the 25–45 kDa material we find that the latter is very active but the former is essentially inactive. The 50–85 kDa and 25– 45 kDa IL-6 fractions show equal ability to stimulate $\alpha_1$-antichymotrypsin synthesis in the Hep3B assay.

In collaboration with Tony Allison and John Kenney at Syntex Research we raised some mouse monoclonal antibodies to human IL-6. Dr Romero has been using an ELISA with two antibodies, 4IL6 and 5IL6. 4IL6 is used as the capture antibody and biotinylated 5IL6 as the reporter antibody. In collaboration with Dr N. Ida and colleagues at Toray, Japan, we have also used another assay, which involves a monoclonal antibody called Ig61 as the capture antibody, and the same reporter antibody, 5IL6. None of these antibodies recognizes denatured IL-6 protein. All these monoclonal antibodies immunoprecipitate the various fibroblast-derived forms of IL-6. The Ig61–5IL6 ELISA fails to recognize the high molecular mass fibroblast-derived IL-6 eluted from the gel filtration column, whereas the 4IL6–5IL6 ELISA preferentially recognizes the high molecular mass material. The Ig61–5IL6 ELISA is an extremely sensitive assay with recombinant IL-6.

With serum taken from a patient 20 days after bone marrow transplant, the conventional B9 bioassay gave an activity of 300 units per ml. After gel filtration of the serum, the B9 assay indicated that all the IL-6 activity was in low molecular mass material, about 20 kDa. The Ig61–5IL6 assay, which works extremely well with recombinant IL-6 or IL-6 derived from cell culture, gave the traditional

224

result—a serum IL-6 concentration of about 200 pg/ml, with all the IL-6 in a low molecular mass peak. However, the 4IL6–5IL6 ELISA picked up a 100 kDa peak and a 400–500 kDa peak in this serum sample, in plasma samples from patients with epidermolysis bullosa and psoriasis and also in my own plasma. Even in my plasma the levels detected were in 1000s of pg/ml. These fractions show absolutely no activity in the B9 assay; they do not interfere with IL-6 in the B9 assay, and do not show activity in the B9 assay after heating.

Is the high molecular mass material simply cross-reactive material, or is it really IL-6? We have pooled fractions in the 100 kDa and the 400–500 kDa peaks and have purified the IL-6 in each of the pools using a 5IL6 immunoaffinity column. Western blotting showed that the 400 kDa material is enriched in the 30 kDa IL-6 species; the 100 kDa complex contains both the 25 and the 30 kDa monomeric forms. There are huge amounts of IL-6 in this serum, perhaps up to 1 μg/ml as judged by amino acid sequencing of the IL-6, yet it showed no activity in the B9 assay.

After immunoaffinity purification through a 5IL6 column, but not before, this material becomes active in the B9 assay. Coomassie blue staining of the 'purified' IL-6 reveals several protein bands. Sequencing of three of these bands verifies that they are IL-6. Additionally, the proteins 'associated' with IL-6 include fragments of complement C3 and C4, C-reactive protein and serum albumin. The albumin band is so thick that it's possible that the soluble receptor is buried underneath it.

My essential point is that depending on the bioassay or ELISA you use, you can detect different amounts of IL-6. Perhaps the bulk of the IL-6 in the bloodstream, in the ng/ml range, is transparent to most conventional assays.

*Gough:* Were the plasma fractions which were inactive in the B9 assay active in the Hep3B assay?

*Sehgal:* Yes, with $\alpha_1$-antichymotrypsin synthesis as the assay.

*Gough:* Do you think there is an intrinsic difference in the presentation of the IL-6 to the IL-6 receptor in Hep3B cells, or does the Hep3B cell convert the IL-6 to an active form?

*Sehgal:* Either is possible. We are not yet in a position to guess which is the case.

*Gearing:* If the binding protein is the soluble IL-6 receptor, then, according to Professor Kishimoto's group, the complex might be stimulatory.

*Sehgal:* Because there is a large amount of serum albumin in the 70 kDa region I cannot exclude the possibility that buried in the serum albumin is the soluble receptor. However, we have found no evidence for the soluble receptor in this complex.

*Aarden:* Did you try to block the activity in the HepG2 assay with neutralizing antibodies?

*Sehgal:* Using a polyclonal antiserum against IL-6 we found, as Georg Fey has also described, that there was even greater synthesis of $\alpha_1$-antichymotrypsin. We have not yet physically removed the antigen–antibody complex and assayed the soluble part in Hep3B cells.

*Gearing:* You said that the high molecular mass complexes are inactive, but become active after immunoaffinity purification. Is it the way that you get the complex off the column that releases the IL-6; that is, can you take the 400 kDa fraction, treat it in the same way and release IL-6 activity?

*Sehgal:* We have thought about this. We elute the column by using glycine to produce a pH drop. We have tried a simple acid treatment, but this did not yield material active in the B9 assay.

*Gearing:* Do you think that you are removing an inhibitor in the immunoaffinity purification?

*Sehgal:* That's a possibility.

### References

Helfgott D, Tatter SB, Santhanam U et al 1989 Multiple forms of IFN-β2/IL-6 in plasma and body fluids during acute bacterial infection. J Immunol 142:948–953

May LT, Santhanam U, Sehgal PB 1991 On the multimeric nature of natural human interleukin-6. J Biol Chem 266:9950–9955

# Distribution and binding properties of receptors for leukaemia inhibitory factor

Douglas J. Hilton*, Nicos A. Nicola and Donald Metcalf

*The Walter and Eliza Hall Institute of Medical Research and Co-operative Research Centre for Cellular Growth Factors, Post Office, Royal Melbourne Hospital, 3050 Victoria, Australia*

*Abstract.* The pleiotropic biological actions of leukaemia inhibitory factor (LIF) on haemopoietic cells (macrophages and megakaryocytes), hepatocytes, osteoblasts, pre-adipocytes, embryonic stem cells, myoblasts and neuronal cells must be mediated through the interactions of LIF with specific cellular receptors. The demonstration by equilibrium binding analysis and autoradiography of LIF receptors on all of the above cells and cell lines suggests that each of these pleiotropic effects of LIF is mediated by direct interactions with the responding cells rather than by the indirect release of secondary cytokines. Despite the differing biological effects of LIF on these cells, equilbrium binding, kinetic analyses and receptor internalization studies have all suggested that these cells display essentially identical high affinity LIF receptors. Nevertheless, there is evidence on some cell types (granulocyte-macrophage colony-stimulating factor [GM-CSF] transgenic peritoneal cells and F9 embryonal carcinoma cells) for a second class of low affinity LIF receptors ($K_d = 1.5$ nM versus $K_d = 30$ pM for high affinity receptors) which differ from the high affinity receptors only in kinetic dissociation rate. Moreover, the evidence suggests that low and high affinity receptors are structurally related and interconvertible, because detergent solubilization of LIF receptors from any cell type results in the quantitative conversion of high affinity receptors into low affinity receptors. As is the case for other related cytokine receptors, these data suggest that high affinity LIF receptors may be composed of two protein subunits—one responsible for LIF-specific low affinity binding and the other responsible for affinity conversion and cell signalling by the receptor. Such a model provides a possible explanation for the pleiotropy of LIF's biological actions.

*1992 Polyfunctional cytokines: IL-6 and LIF. Wiley, Chichester (Ciba Foundation Symposium 167) p 227–244*

Although leukaemia inhibitory factor (LIF) was first described as a glycoprotein that induced terminal differentiation and suppression of proliferation of mouse M1 myeloid leukaemic cells (Tomida et al 1984, Hilton et al 1988a) it is now clear that this cytokine has a wide range of biological actions on disparate tissues

---

*Present address*: The Whitehead Institute for Biomedical Research, 9 Cambridge Center, Cambridge, MA 02142, USA

and organs. These actions include the direct or synergistic proliferative stimulation of certain haemopoietic cells, inhibition of the differentiation of embryonic stem cell lines, stimulation of the release of acute-phase proteins by hepatocytes, inhibition of the lipoprotein lipase activity of adipocytes, stimulation of osteoblast bone-forming activity and, in certain circumstances, an indirect stimulation of osteoclast bone-resorbing activity, stimulation of myoblast proliferation and an influence on the survival and neurotransmitter choice of certain neurons (Gough & Williams 1989, Hilton & Gough 1991, Hilton 1992). These wide-ranging and, in some cases, contradictory actions raise several questions concerning the molecular nature of LIF-induced signalling. LIF receptors are presumably expressed at the very earliest stages of development in the inner cell mass (embryonic stem cells) of the blastocyst. Are these the same receptors as those subsequently selectively expressed on differentiated tissues? Are the different actions of LIF (on survival, proliferation, differentiation, secretion) on developmentally unrelated cell types mediated by the same receptor and the same signalling mechanisms? How are the synergistic actions of LIF mediated at the receptor level? It is difficult to envisage physiological situations where all these different actions of LIF would need to be expressed coordinately; how are the actions of LIF localized to specific sites and times? Is this localization of action regulated only by controlling LIF production and clearance rates, or can cellular receptors also be modulated according to needs?

## The cellular distribution of LIF receptors

As expected from the pleiotropic biological actions of LIF, radioligand binding assays indicate that receptors for LIF have a broad distribution, being found on various cell types. However, only specific subsets of cells in various tissues and certain cell lines are receptor positive. Of a very large set of mouse, rat and human cell lines tested (Table 1) only a few displayed LIF receptors. These included monocytoid haemopoietic cell lines, oncogene-transformed haemopoietic cell lines, hepatic cell lines, neural crest, neuroepithelial and neuronal cell lines, osteoblastic cell lines, pre-adipocyte cell lines and embryonal carcinoma (EC) and embryonic stem (ES) cell lines. However, in each class (with the exception of ES and EC cell lines) some cell lines with similar characteristics were receptor negative (Table 1), suggesting that LIF receptor expression is not strongly correlated with cellular phenotype.

Amongst normal haemopoietic populations, LIF binding is restricted to monocyte/macrophages, megakaryocytes and certain lymphocyte-like cells (Hilton et al 1988b, 1991a). However, even within these populations LIF binding is not uniform. In mouse bone marrow populations less than half of the promonocytes and monocytes were labelled whereas in peritoneal populations more than 90% of macrophages were labelled (as assessed by autoradiography). 'Activated' macrophages from the peritoneal cavities of mice injected with thioglycollate or *Mesocestoides corti* or from granulocyte-macrophage colony-

**TABLE 1   Distribution of LIF receptors on cell lines**

| Cell type | Receptor-positive | Receptor-negative |
|---|---|---|
| Monoctyic (M) | M1, HA-15, AC-5, NFS-60 | PU5-1.8, J774, WEHI-3B, AC-8, FDCP-1, 32D, NFS-61, WR19, WEHI-265, DA-1, DA-3 |
| Monocytic (H) | — | U-937, HL-60, K6-1, K-562, HEL, DU-528 |
| Oncogene-transformed | Zen-37, 2-Mes/raf, 3-Mes/raf | 1-Bra/raf, J2EG11 |
| Hepatic (H) | HepG2, Hep3B | Mahlavu, PLC |
| Embryonal (M) | PCC3-A1, P19, NG-2, F9, PC-13, D3 | — |
| Osteoblastic (R) | UMR-106, RCT-1 | UMR-201 |
| Pre-adipocyte (M) | 3T3-L1 | |
| Neural crest (M) | 14.4.4, 14.4.3G | 14.9.10, 14.4.6 |
| Neuroepithelial (M) | NZen-40, NZen-37, OLF-442 | 2.3D |
| B,T lymphoid (M) | — | LB3, EL4, E9.D4 |
| (H) | | Raji, BALL, LILA |
| Fibroblast (M) | — | L929, 3T3 |
| Epithelial (H) | — | A 431, CT-20, HeLa |

M, mouse; H, human; R, rat.

stimulating factor (GM-CSF) transgenic mice tended to display higher numbers of LIF receptors per cell, although considerable heterogeneity in receptor levels was seen. A small sub-population of lymphoid-like cells (2–6%) in the bone marrow, spleen and peritoneal cavity were also labelled with LIF, but most lymphoid cells were negative.

In human haemopoietic populations, bone marrow and peripheral blood monocytes displayed only low levels of receptors, and patients with acute or chronic myeloid leukaemia had LIF receptor-positive cells only occasionally (generally M5 leukaemias in the FAB [French–American–British] classification). However, peritoneal macrophages from patients undergoing dialysis for renal disease or from patients undergoing abdominal surgery were routinely LIF receptor-positive.

Primary cultures or preparations of mouse fetal and adult hepatocytes, myoblasts and osteoblasts (but not osteoclasts), and of fetal neurons and their progenitors from the neural crest and the dorsal root ganglia were also shown to be LIF receptor-positive by cell autoradiography (Allan et al 1990, Hilton et al 1991a,b, Hendry et al 1992).

For a molecule with pleiotropic effects like LIF these receptor distribution studies have been useful for two reasons. Firstly, they have suggested potential target cells and effects of LIF that might not otherwise have been considered, such as LIF's actions on megakaryocyte and platelet formation (Metcalf et al 1991), hepatocyte acute-phase protein release (Baumann & Wong 1989), sensory neuron survival and function (Murphy et al 1991) and an as yet undetermined role in monocyte/macrophage functioning. Secondly, they have helped to define which actions of LIF are likely to be mediated directly rather than through the induced release of secondary mediators. For example, LIF receptors were clearly present on osteoblasts and their precursors, but not on osteoclasts. This finding is consistent with the bone-forming effects of LIF (Metcalf & Gearing 1989), and suggests that the effects of LIF in stimulating bone resorption by osteoclasts are indirectly mediated through a LIF–osteoblast interaction.

### *In vivo* distribution of injected LIF

To complement the above *in vitro* studies, which determine *potential* cellular targets of LIF, we have determined the distribution of injected $^{125}$I-labelled LIF *in vivo* to define the accessibility of different tissues to LIF (Hilton et al 1991b). LIF injected intravenously displayed a biphasic clearance pattern, with an $\alpha$-phase of $t_{1/2} = 6$–8 min and a $\beta$-phase of $t_{1/2} = 3$–7 h, depending on the glycosylation status of the LIF. Accumulation of $^{125}$I-LIF was predominant in the liver and kidney, with relatively high levels also being found in the lung, spleen, pancreas and thymus. Autoradiographs showed LIF binding to hepatocytes, lung and thymic endothelial cells, renal glomeruli, spleen red pulp and marginal zones, intestinal crypt cells and the inner layer of the adrenal cortex. Bone sections from several sites revealed labelling of osteoblasts and megakaryocytes, but, despite the accumulation of radioactivity in the pancreas, no specific labelling of pancreatic cells was observed.

There was clear evidence that some structures are inaccessible to LIF. After injection of $^{125}$I-LIF into pregnant mice the labyrinthine trophoblasts were heavily labelled at sites in contact with the maternal circulation but not at sites in contact with the fetal circulation. Consistent with this observation, LIF was not detected in the fetal circulaton or in fetal organs such as the liver, and, similarly, the amniotic and yolk sac membranes were not labelled. On the other hand, after injection of $^{125}$I-LIF into the embryonic exocoelum *in utero*, trophoblasts were labelled only on the fetal side, and the fetal membranes, in addition to fetal liver, mesonephros, retina and endothelial cells, were specifically labelled. This suggests that there is a two-way barrier to LIF transit between the maternal and fetal circulations, an observation of some relevance given the effects of LIF on embryonic stem cells and developing fetal organs. Similarly, in adult mice LIF did not appear to cross the blood–brain barrier. The only brain structure labelled was the choroid plexus, which is believed to determine the composition of the cerebrospinal fluid.

In other studies, LIF injected *in vitro* has been shown to undergo retrograde axonal transport in sensory neurons (Hendry et al 1992). Injection of [125]I-labelled LIF in the foot pad led to accumulation of radiolabel at the point of ligation of the sciatic nerve. With unligated nerves, labelled LIF appeared in the cell bodies of sensory neurons in the dorsal root ganglia but not in motor neurons of the spinal cord. This suggests that LIF, like nerve growth factor, may act as a true neurotrophic factor for sensory neurons.

## Low and high affinity receptors for LIF

The binding of LIF to various cell types is specific, and interleukin 6 (IL-6), interleukin 1 (IL-1), tumour necrosis factor (TNF) and the colony-stimulating factors, despite having some similar biological activities, do not compete with LIF for its binding sites (Hilton et al 1988b). Scatchard analyses of saturation binding curves have revealed, on most cell types, a single class of high affinity receptor, present at low density (Hilton et al 1988b, 1991a). Normal mouse bone marrow, spleen, thymus and peritoneal cells had a mean of 1–170 receptors per cell showing an equilibrium dissociation constant ($K_d$) of 30–60 pM at 4 °C. Even after excluding receptor-negative cells in calculation of the average, to allow for the small numbers of positive cells in these populations (as determined by cell autoradiography), monocytes and macrophages displayed only 150–400 receptors per cell. Normal fetal and adult mouse hepatocytes also expressed one class of receptor, $K_d = 100$ pM, but at greater levels (about 4000 per cell). Various other cell lines, including the mouse myelomonocytic cell line M1, osteoblastic cell lines such as UMR-106, ES cell lines such as D3 and the pre-adipocytic cell line 3T3-L1, displayed a single class of high affinity ($K_d = 30$–70 pM) receptor at relatively low densities (100–2000 receptors per cell).

Kinetic analyses of the binding of LIF to cells at 4 °C revealed a single fast kinetic association rate ($k_{on} = 4$–$9 \times 10^8$ M$^{-1}$ min$^{-1}$) and a single very slow kinetic dissociation rate at 4 °C ($k_{off} < 0.01$ min$^{-1}$). This nearly irreversible binding at 4 °C is common to many cytokine–receptor interactions (Nicola 1991) and presents some problems in interpretation. Irreversible binding implies that at low ligand concentrations all ligand will be bound, given sufficient time and adequate receptor numbers. This, however, is never seen. There is always a significant fraction of ligand remaining free in solution and it can be shown that this free ligand is capable of binding to fresh cells. For the interaction of M-CSF with macrophages it has been postulated that this is due to the selective loss of unoccupied M-CSF receptors from the cell surface at 4 °C, such that the number of free receptors, rather than the amount of free ligand, becomes limiting (Stanley & Guilbert 1981). However, an alternative interpretation might be that ligand binding involves two kinetic steps—reversible binding of LIF to its cell surface receptor might be followed by a rapid kinetic conversion of the ligand–receptor complex to a form from which LIF does not dissociate. In this model, the proportion of LIF bound is determined by the 'reversible' equilibrium constant, but, because the conversion of the ligand–receptor complexes is fast relative to the rate of ligand dissociation from the reversible complex the measured kinetic dissociation rate will be that for the converted complex rather than the reversible complex. In this case the primary binding interaction is not in fact irreversible, so only a fraction of the ligand will be bound.

In a few cases, the pattern of binding of LIF to its cellular receptor was not consistent with a single class of receptor (Fig. 1). With activated peritoneal macrophages, such as those found in transgenic mice expressing the mouse GM-CSF gene (GM-CSF transgenic peritoneal cells), and F9 embryonal

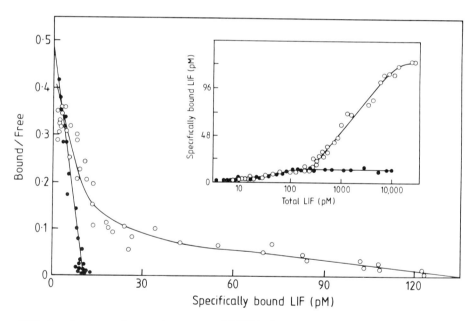

FIG. 1.   Analyses of saturation curves of LIF binding to (C57BL × SJL)F$_2$ granulocyte-macrophage colony-stimulating factor (GM-CSF) transgenic peritoneal cells and RCT-1 cells at equilibrium. $5 \times 10^6$ GM-CSF transgenic peritoneal cells (○) or $1.5 \times 10^6$ RCT-1 pre-osteoblast cells (●) were incubated with various concentrations (5 pM–20 nM) of $^{125}$I-emLIF (iodinated E. coli-derived mouse LIF) with or without excess unlabelled LIF for 16 hours on ice. The concentrations of $^{125}$I-emLIF bound to cells and free in solution were determined, and the latter was adjusted for the bindable fraction (see Hilton et al 1988b). Scatchard transformation of the data is shown in the major graph, and a plot of specifically bound LIF versus the total concentration of LIF on a logarithmic scale is shown in the inset.

carcinoma cells LIF binding data at 4 °C consistently gave rise to curvilinear Scatchard transformations. Figure 1 shows that, whereas binding of $^{125}$I-LIF to osteoblastic RCT-1 cells was clearly saturated above 100 pM LIF, binding to GM-CSF transgenic peritoneal cells was not saturated until LIF concentrations reached about 10 nM, and there was clear evidence for two classes of binding site (about 300 sites with $K_d = 50$ pM and about 5000 sites with $K_d = 1.4$ nM). F9 cells showed two classes of binding site with the same dissociation constants as transgenic cells, but there were fewer (200) low affinity than high affinity sites (300).

Measurements of kinetic association rates did not reveal any major differences between cells with only high affinity receptors and cells displaying significant numbers of low affinity receptors. However, there were dramatic differences in kinetic dissociation rates between such cells (Fig. 2). LIF bound to cells with only high affinity receptors (bone marrow, resident peritoneal cells, hepatocytes, M1 and 3T3-L1 cells) dissociated in a monoexponential manner, with less than 50% dissociation after 20 h at 4 °C. However, dissociation of LIF from F9 or GM-CSF transgenic peritoneal cells was clearly biphasic, with a rapid component ($t_{1/2} = 2$ min) and a slow component ($t_{1/2} > 20$ h) corresponding to the relative numbers of low affinity and high affinity receptors, respectively. These results suggest that the difference between high and low affinity receptors resides entirely in the kinetic rates of ligand dissociation.

Dissociation kinetics with transgenic cells which had been incubated with $^{125}$I-LIF for various times at 4 °C (2 min to 24 h) revealed identical biphasic dissociation patterns (Fig. 2B). This suggests either that the two affinity sites are independent or that if conversion of low affinity to high affinity receptors occurs after ligand association, it is very fast relative to the ligand dissociation rate, and, consequently, the two receptor classes would behave as if they were independent even though they are interconvertible. The dissociation of bound $^{125}$I-LIF from such cells was not affected by inclusion of excess unlabelled LIF in the dissociation medium (Fig. 2C), indicating that re-binding of dissociated LIF was not occurring in these experiments.

## The characteristics of solubilized LIF receptors

To understand better the relationship between high and low affinity LIF receptors, and their complex kinetic behaviour, we have developed methods to study LIF receptors in isolated cell membranes and in detergent solution. By using biologically active recombinant LIF produced as a non-glycosylated protein in *E. coli*, we could separate labelled LIF bound to glycosylated solubilized receptors from unbound labelled LIF by adsorption onto concanavalin A beads. After optimization, this technique was used to study the equilibrium binding and kinetic characteristics of detergent-solubilized LIF receptors from a variety of cells. Figure 3 shows that the specific binding of LIF to Triton

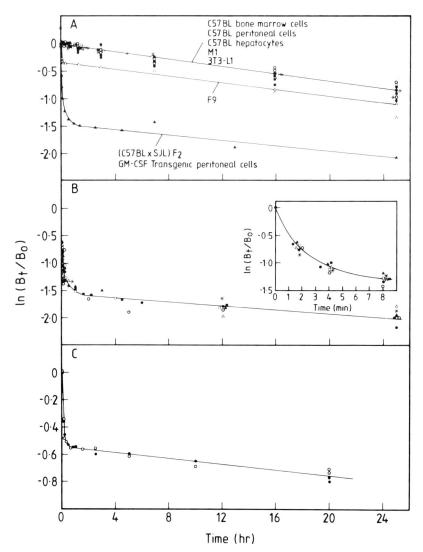

FIG. 2. Analysis of the dissociation of LIF from receptors on various cells. (A) 400 pM
$^{125}$I-emLIF was incubated with $4 \times 10^8$ bone marrow cells (○), $2 \times 10^8$ M1 cells (●),
$2 \times 10^8$ resident peritoneal cells (□), $4 \times 10^7$ 3T3-L1 cells (■), $2 \times 10^7$ adult parenchymal
hepatocytes (*), $8 \times 10^7$ F9 EC cells (△) or $2 \times 10^8$ (C57BL × SJL)F$_2$ GM-CSF transgenic
peritoneal cells (▲) for two hours on ice, with or without excess emLIF, in a volume
of 4.0 ml. The cells were centrifuged and then resuspended in 4.0 ml of medium containing
100 ng/ml unlabelled LIF. At the indicated times 100 μl aliquots were removed and the
amount of $^{125}$I-emLIF remaining bound to cells was determined by counting after
centrifugation through fetal calf serum. The natural log of the ratio of the amount of
$^{125}$I-emLIF bound specifically after a given time of dissociation (B$_t$) to that at the

X-100-solubilized receptors was relatively insensitive to pH (over the range 4–9), ionic strength (0.03–0.15 M NaCl) and the concentration of Triton X-100 in the binding medium (0.1–0.5% v/v). Moreover, provided that a sufficient quantity of concanavalin A beads was used (20 μl), the specific binding of $^{125}$I-LIF was linearly dependent on the concentration of solubilized LIF receptors.

The behaviour of LIF receptors in intact transgenic peritoneal cells, purified membranes and in Triton X-100-solubilized form is shown in Fig. 4. The Scatchard transformations of the equilibrium binding data showed that whereas both high and low affinity receptors were detected in intact cells and purified membranes, only low affinity receptors were found in detergent solution. Because the detergent-solubilized receptors were recovered in good yield from the membranes and showed the same affinity ($K_d \approx 1.5$ nM) as low affinity receptors in intact cells, this result suggests that high affinity receptors might be converted to low affinity receptors by detergent solubilization. This was confirmed by examining the behaviour of LIF receptors from cells which display only high affinity receptors in the intact state (Fig. 5). Mouse adult hepatocytes had a single class of high affinity receptor ($K_d = 35$ pM) whereas both high and low affinity receptors were found in membranes and detergent-solubilized receptors displayed only low-affinity receptors ($K_d = 1$ nM). Several different detergents, including Triton X-100, β-octyl glucoside, digitonin, CHAPS (3-[(3-cholamidopropyl)dimethylammonio]-1-propane sulphonate) and 3.12 (N-decyl-N,N-dimethyl-3-ammonio-1-propane sulphonate), were capable of extracting LIF receptors in an active (LIF-binding) form and each produced solubilized LIF receptors of identical low affinity (Fig. 5), so it is unlikely that the affinity conversion of the LIF receptor is induced by detergent. Rather, it is probable that LIF receptors can exist in either of two interconvertible forms (low affinity and high affinity) in intact cells, and that the interactions required for the maintenance of the high affinity state are lost during detergent solubilization. These results are reminiscent of the behaviour of GM-CSF receptors on solubilization (Nicola 1990) and suggest that similar mechanisms may be operating in these two receptor systems.

---

beginning of dissociation (B$_0$) is plotted versus time. (B) $2 \times 10^8$ transgenic peritoneal cells were incubated for various times from 1.66 minutes to 24 hours with 1.0 nM $^{125}$I-emLIF in the presence or absence of an excess of emLIF. After the given time, dissociation was carried out as for panel A. Association times were 1.66 min (■), 6.0 min (□), 18 min (●), 45 min (○), 2 h (▲), 6 h (△) and 24 h (∗). (C) Binding was carried out as for panel A, with $4 \times 10^8$ GM-CSF transgenic peritoneal cells, and dissociation was carried out in 10 ml medium with (○) or without (●) 19 nM emLIF.

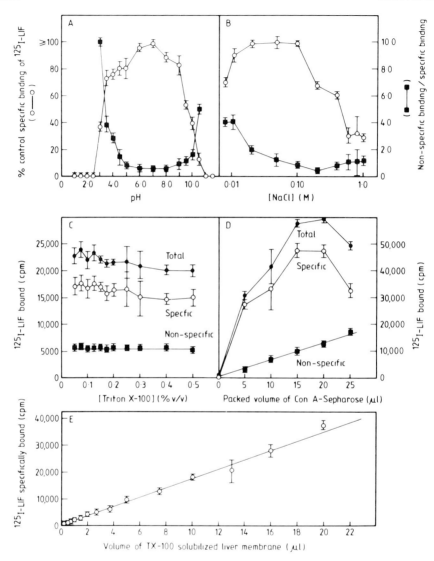

FIG. 3.   Development of an assay for detergent-solubilized LIF receptors: *Top*: Binding
of [125]I-LIF to solubilized receptors in buffers of various pH (A) or ionic strength (B) is
presented in terms of the percentage of specific binding in 100 mM Na acetate at pH 6.0 (O)
and as the ratio of non-specific to specific binding (■). *Middle*: Binding in the presence
of various amounts of Triton X-100 (C) and Con A–Sepharose (D) was assayed using
100 mM Na acetate at pH 6.0. Total (●), specific (O) and non-specific binding of [125]I-LIF
(■) are shown. *Bottom*: E shows specific binding of [125]I-LIF to various amounts of
solubilized extract, again in 10 mM Na acetate at pH 6.0, with optimal amounts of Triton
X-100 and Con A–Sepharose. All incubations were for 15 h at 4 °C and, unless otherwise
stated, 15 µl packed Con A beads were used. Error bars show SEM values; $n = 2$.

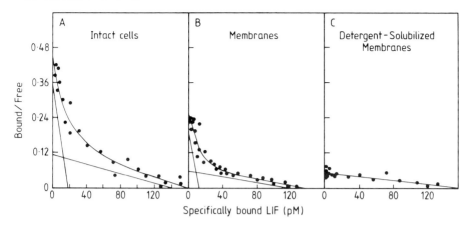

FIG. 4.   Scatchard analysis of saturation binding curves of LIF bound at equilibrium to (A) $5 \times 10^6$ peritoneal cells from (C57BL × SJL)$F_2$ GM-CSF transgenic mice in 40 µl, (B) 40 µl of peritoneal cell membranes or (C) 40 µl of peritoneal cell membranes solubilized in 1% (v/v) Triton X-100, incubated with increasing concentrations of $^{125}$I-emLIF (5–8000 pM), with or without 80 nM unlabelled LIF, in 100 mM Na acetate (pH 6.0) for 16 h on ice. LIF bound to soluble receptors was separated from unbound LIF by adsorption to 15 µl Con A–Sepharose.

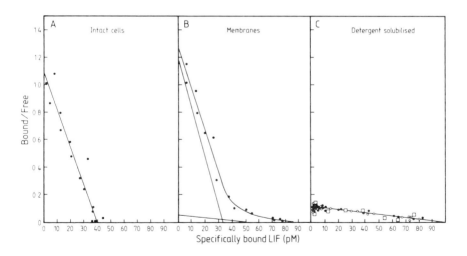

FIG. 5.   Scatchard analyses of saturation binding curves of LIF bound at equilibrium to (A) $4 \times 10^5$ adult C57BL/6 hepatocytes in 40 µl, (B) 40 µl of liver membranes and (C) 40 µl of liver membranes solubilized in 1% (w/v) β-octyl glucoside (●), 1% (v/v) Triton X-100 (□) or 1% (w/v) 3.12 (N-decyl-N,N-dimethyl-3-ammonio-1-propane sulphonate) (■). Binding of $^{125}$I-emLIF (5–8000 pM) in the presence or absence of 80 nM unlabelled LIF was assayed as described in Fig. 4.

**Conclusions**

The interconversion between high and low affinity states is a characteristic of LIF receptors shared with other cytokine receptors, including those for GM-CSF, IL-2, IL-6 and IL-3 (Nicola 1991). These receptors have been demonstrated to exist as heterotypic oligomers, in at least dimeric form. At least one subunit behaves as a low affinity ligand-specific binding protein while a second subunit (which may or may not bind ligand by itself) can form a complex with the ligand-occupied first subunit, producing a high affinity receptor complex. The association of the second subunit is dependent on binding of ligand to the first subunit and results in a greatly reduced rate of ligand dissociation from the complex. All these properties are consistent with the kinetic binding properties of LIF described here, provided that the ligand-induced conversion of receptor affinity is fast. The LIF receptor has been recently cloned (Gearing et al 1991) and shown to belong to the same family of receptors as those mentioned above; moreover, it forms only a low affinity receptor when expressed in COS-7 cells. It is likely, therefore, that this LIF receptor can also form a heterotypic high affinity complex.

The above model of the LIF receptor provides a plausible basis for the mechanism of the redundancy and pleiotropy of LIF's biological actions. The affinity-converting second subunits of other cytokine receptors are responsible for signal transduction (Nicola 1991, Kitamura et al 1991); thus, the occurrence of common signalling subunits for different ligand-binding chains could explain the redundancy of biological actions of different cytokines (LIF and IL-6, LIF and G-CSF). Similarly, if different cells produced differing receptor subunits, each able to interact with a common LIF-binding chain, this arrangement could permit qualitatively differing signalling and provide the mechanism mediating the pleiotropic actions of LIF on different cells.

*Acknowledgments*

The authors' original work was supported by the National Health and Medical Research Council, Canberra; the Anti-Cancer Council of Victoria; the Co-operative Research Centre's Programme of the Australian Government; the National Institutes of Health, Bethesda, MD, Grant CA-22556 and AMRAD Corporation, Melbourne.

**References**

Allan EH, Hilton, DJ, Brown MA et al 1990 Osteoblasts display receptors for and responses to leukemia inhibitory factor. J Cell Physiol 145:110–119
Baumann H, Wong GG 1989 Hepatocyte-stimulating factor-III shares structural and functional identity with leukemia-inhibitory factor. J Immunol 143:1163–1167
Gearing DP, Thut CJ, VandenBos T et al 1991 Leukaemia inhibitory factor receptor is structurally related to the IL-6 signal transducer, gp130. EMBO (Eur Mol Biol Organ) J 10:2839–2848

Gough NM, Williams RL 1989 The pleiotropic actions of leukemia inhibitory factor. Cancer Cells 1:77–80

Hendry I, Murphy M, Hilton DJ, Nicola NA, Bartlett PF 1992 Binding and retrograde transport of leukemia inhibitory factor by the sensory nervous system. J Neurosci, in press

Hilton DJ 1992 LIF: lots of interesting functions. Trends Biochem Sci 17:72–76

Hilton DJ, Gough NM 1991 Leukemia inhibitory factor: a biological perspective. J Cell Biochem 46:21–26

Hilton DJ, Nicola NA, Metcalf D 1988a Purification and characterization of a murine leukemia inhibitory factor from Krebs ascites cells. Anal Biochem 173:359–367

Hilton DJ, Nicola NA, Metcalf D 1988b Specific binding of murine leukemia inhibitory factor to normal and leukemic monocytic cells. Proc Natl Acad Sci USA 85:5971–5975

Hilton DJ, Nicola NA, Metcalf D 1991a Distribution and comparison of receptors for leukemia inhibitory factor on murine hemopoietic and hepatic cells. J Cell Physiol 146:207–215

Hilton DJ, Nicola NA, Waring PM, Metcalf D 1991b Clearance and fate of leukemia inhibitory factor (LIF) after injection into mice. J Cell Physiol 148:430–439

Kitamura T, Hayashida K, Sakamaki K, Yokota T, Arai K, Miyajima A 1991 Reconstruction of functional receptors for human granulocyte/macrophage colony stimulating factor (GM-CSF): evidence that the protein encoded by AIC2B cDNA is a subunit of the murine GM-CSF receptor. Proc Natl Acad Sci USA 88:5082–5086

Metcalf D, Gearing DP 1989 A fatal syndrome in mice engrafted with cells producing high levels of leukemia inhibitory factor (LIF). Proc Natl Acad Sci USA 86:5948–5952

Metcalf D, Hilton DJ, Nicola NA 1991 Leukemia inhibitory factor (LIF) can potentiate murine megakaryocyte production in vitro. Blood 77:2150–2153

Murphy M, Reid K, Hilton DJ, Bartlett PF 1991 Generation of sensory neurons is stimulated by leukemia inhibitory factor. Proc Natl Acad Sci USA 88:3498–3501

Nicola NA 1990 Characteristics of soluble and membrane-bound forms of haemopoietic growth factor receptors: relationships to biological function. In: Molecular control of haemopoiesis. Wiley, Chichester (Ciba Found Symp 148) p 110–126

Nicola NA 1991 Structural and functional characteristics of receptors for colony-stimulating factors. In: Quesenberry PJ, Asano S, Saito K (eds) Hematopoietic growth factors. Excerpta Medica, Amsterdam, p 101–120

Stanley ER, Guilbert LJ 1981 Methods for the purification, assay, characterization and target cell binding of a colony stimulating factor (CSF-1). J Immunol Methods 42:253–284

Tomida M, Yamamoto-Yamaguchi Y, Hozumi M 1984 Purification of a factor inducing differentiation of mouse myeloid leukemic M1 cells from conditioned medium of mouse fibroblast L929 cells. J Biol Chem 259:10978–10982

## DISCUSSION

*Heath:* Your model is in agreement with current dogma; however, it's probably worth pointing out that many receptors—the EGF receptor, for example—lose affinity on solubilization. That the affinity lowers on solubilization is not hard evidence that the model is correct.

*Nicola:* It's consistent with the model, though I agree it does not prove it. There are many ways in which affinity can be lost; with the EGF receptor the reduction in affinity is thought to be due to loss of EGF receptor–EGF receptor dimerization, and that's also thought to be true for G-CSF receptors. Such a mechanism is a variation on this model—intermolecular interactions between two proteins are still required, but in this case the two molecules are the same. I don't think the results can be explained by loss of affinity on solubilization, because several detergents give soluble receptors with absolutely identical affinities, and the affinity in solution is identical to that of the low affinity receptor in intact transgenic cells in terms of kinetic association and dissociation rates.

*Heath:* Under this intermolecular dimerization model, why are there low affinity receptors on some cell types but not on others? If the LIF receptor is analogous to the EGF receptor, you would expect roughly the same proportion of high and low affinity sites wherever it is expressed.

*Nicola:* I don't think the LIF receptor high affinity complex is a homodimer. There's significant evidence against that. For example, homotypic dimerization should be a concentration-dependent phenomenon, and David Gearing has shown that when LIF receptors are overexpressed in COS cells high affinity complexes are not detected (Gearing et al 1991).

*Stewart:* Can you take the detergent-solubilized receptors which have low affinity and convert them back to high affinity, by adding the putative second subunit, such as gp130?

*Nicola:* We would like to be able to do that but we haven't been able to. It can't be done simply by concentrating the solution containing the low affinity receptor and any potential second chain. We are trying to do it by reconstitution into membrane vesicles, but we haven't got any results yet.

*Ciliberto:* Do you assume that cells which carry only the high affinity receptors have a large quantity of the converter protein? Have you tried to do fusion experiments between the two types of cells?

*Nicola:* That's a good question. I think this cannot be done in solution. We need to turn to membrane reconstitution experiments to address that question.

*Heinrich:* When we did our study on the plasma clearance and target organs of iodinated human IL-6 we were warned that there might be active dehalogenases or halogen transferases, enzymes which remove the [125]I or even transfer it to other polypeptides. Can you be sure that you are really detecting LIF in your distribution studies?

*Nicola:* There are clearly some examples where the detected radiolabel is free iodine—the accumulation in the thyroid, for example. However, iodinated ovalbumin was used as a parallel control, so a positive result in the LIF-injected mice is one not also seen in ovalbumin-injected mice.

*Metcalf:* That's true, but an internal label such as tritium or [35S]methionine would be far superior. With radiolabelled iodine, all that silver grains tell you is that there is radiolabelled iodine present.

*Heinrich:* From most organs you can extract the radiolabelled protein and run it on an SDS gel to show that it contains the intact labelled molecule. In our experiments we found that five hours after injection most of the labelled IL-6 was found in the skin, but we were unable to extract the radioactive material from skin and show that it was intact IL-6.

*Nicola:* We observed a similar effect; there was specific accumulation of LIF, relative to ovalbumin, in the pancreas, but we were unable to show specific labelling of any pancreatic structures. I assume this was due to accumulation of degraded LIF which couldn't be fixed in the sections. It might be labelled tyrosine, or there might have been degradation of LIF after binding. You do have to be careful.

*Gauldie:* LIF is more heavily glycosylated than IL-6. Although the carbohydrate doesn't appear to have an impact on function, I would expect it to have a dramatic impact on distribution.

*Nicola:* It doesn't. We have compared iodinated native LIF purified from Kreb's ascites with *E. coli*-derived LIF. There is actually little difference in clearance rates. There was some evidence from TCA precipitation comparisons that native glycosylated LIF is degraded slightly more slowly than unglycosylated LIF, so slightly higher levels are maintained. There was no difference in organ distributions.

*Gauldie:* That leads me to question further the function of the carbohydrate.

*Nicola:* The glycosylation clearly protects the molecule to some extent from degradation. I wouldn't be surprised if it increased solubility. There are few examples of carbohydrate on glycoproteins having a specific biological effect other than one of general stabilization.

*Heath:* The localization of labelled LIF in the kidney is interesting in the light of two reports of LIF activity in the kidney. One describes inhibition of $Na^+$-dependent hexose transport in renal epithelial cells (Tomida et al 1990); the other showed that LIF inhibits the induction of differentiation in kidney development (Bard & Ross 1991). That accumulation could well be physiologically significant.

*Nicola:* I think you are right. However, LIF is obviously cleared through the kidney, and its not easy to discriminate specific binding structures from a general clearance mechanism. Doug Hilton ground up organs to see whether specific binding of LIF could be measured; it could in liver, lung and spleen, but not in kidney (Hilton 1990). That may reflect the fact that in the kidney there are greater numbers of specific binding sites, so that a huge excess of cold LIF would be needed to compete with the bound labelled LIF. It is still an open question whether the kidney contains specific LIF receptor sites.

*Patterson:* Does LIF bind to subpopulations of neural crest cells in culture?

*Nicola:* LIF seems to bind to almost all the cells that come out of the neural crest. It certainly binds to some cells which are bipotential—binding is not restricted to neuronal cells in the neural crest.

*Patterson:* Did you look at tissue distribution of injected LIF in young animals or embryos?

*Nicola:* Doug Hilton looked at the distribution of LIF after injection into the exocoelum of embryos *in situ*. In some experiments there was labelling of structures in the peritoneum, the mesonephros and the developing liver. There was also some labelling in the brain area around the eye (Hilton 1990).

*Lotem:* You have shown that LIF labels mature megakaryocytes. Do you see any labelling of earlier precursors of megakaryocytes?

*Nicola:* Obviously, we can detect only morphologically recognizable megakaryocytes, and I don't think there were many in our cultures.

*Metcalf:* We haven't looked at acetylcholinesterase-positive cells. It is possible that we could use this stain on autoradiographs.

*Lotem:* Is anything known about the actions of LIF on myoblast cell lines?

*Nicola:* Laurie Austin has some results indicating that LIF can stimulate myoblast proliferation *in vitro* (Austin & Burgess 1991).

*Stewart:* Is it possible to determine the affinity of the receptors that you have looked at *in vivo*? For example, are the receptors in the placenta low affinity receptors or high affinity?

*Nicola:* That can't easily be established *in vivo*. Problems arise with potential affinity conversion upon tissue damage. For example, intact hepatocytes appear to have only high affinity receptors, but when you make hepatocyte membranes you get both high and low affinity components, which probably results from a loss of the ability to maintain the high affinity state. With megakaryocytes the competition by cold LIF is not good; with most cell types the number of grains is reduced to background levels, but there's always still significant labelling of megakaryocytes. It is possible that megakaryocytes have only low affinity receptors, but we can't get enough cells to do Scatchard analysis.

*Stewart:* It has been pointed out that once a cell is moved from its *in vivo* environment to an *in vitro one* there is often rapid induction of growth factor/cytokine synthesis by the explanted cells. Does this happen with LIF receptors?

*Nicola:* Most of the cell types that we expect to have LIF receptors do have them *in situ*, at the time of injection of the radiolabel. When you inject the radiolabel, there are sufficiently high concentrations to occupy receptors only in the first few minutes, so I presume that these receptors were present *in vivo*, and were not just expressed after preparing a bone marrow smear, for example. There's no evidence for rapid up-regulation by *in vitro* cultivation for these cells, but it is a possibility.

*Martin:* You seem to favour the idea that the basis of the pleiotropic effects is the action of different second messengers. I am not sure why you need to suggest that. There are plenty of examples of hormones acting through the same specific second messenger, where the actual effect in a particular cell depends on the differentiated properties of that cell.

*Nicola:* There could be a system in which the only thing needed is the appropriate receptor, because the cell is genetically programmed to respond to activation of that receptor in a particular way. Having said that, I am trying to think of an example of one cell type responding differently to two different signals. There are certain cells which can proliferate in response to one cytokine, such as IL-3, and differentiate in response to another, G-CSF. In that case there must be two independent signals coming through two different receptors.

*Metcalf:* Are you postulating a need for differing signalling cascades?

*Nicola:* Professor Martin is suggestting that two different signalling cascades are not needed. The cells just need to be programmed to respond differently to one signal cascade. Personally, I don't think this is always the case; in some situations such a mechanism might be sufficient. There are also examples of cells giving different responses to different signals, where there must be different types of signal transduction through the receptors.

*Fey:* You suggested that in the liver there is selective or exclusive binding of LIF to hepatocytes. Which cell types in the liver did you try to discriminate? One class of cell that is frequently mentioned is the 'oval' cell, which is considered to be the progenitor responsible for liver regeneration. We have seen mRNA for LIF in regenerating liver, so I would expect the oval cells to have LIF receptors.

*Metcalf:* In mice injected with radiolabelled LIF essentially every cell in the liver was labelled. There was variability between hepatocytes in the level of labelling, but we did not subdivide the hepatocytes on the basis of their morphology.

*Baumann:* There seemed to be a substantial number of receptors in adult cells. The concentration of LIF is low, but the off rate of LIF from the receptor seems also to be very low. Does that mean that only a minor fraction of the receptors need to be occupied to get the maximum response?

*Nicola:* That is a very difficult question to answer. We have done studies of receptor internalization, and have determined the kinetic constants for internalization, degradation and re-expression of the receptors to try and address that question. One should bear in mind the difference in kinetic dissociation rate between the low and the high affinity receptors. For the low affinity receptor, ligand molecules go on and off quickly. With a certain rate of internalization, the chance of the ligand being internalized through a low affinity receptor is actually small. We have been able to distinguish the internalization of low affinity and high affinity receptors in transgenic cells where there are both; there's little internalization through the low affinity receptors but with the high affinity receptor there is a high chance that the ligand and receptor will be internalized. Removal of the reaction 'product' by internalization means that the receptor occupancy level is increased, so although the high affinity receptor *equilibrium* dissociation constant is about 50 pM, the actual calculated *steady state* affinity for these receptors is about 5 pM, which is much closer to the concentration of LIF required for biological effects.

*Baumann:* The biological effect you measure is a late event though. Is it proportional to the immediate early response that follows binding of LIF to the receptor?

*Nicola:* Internalization is very fast. There is more LIF–receptor complex inside the cell than outside in certain cell types within 3–4 min at 37 °C.

## References

Austin L, Burgess AW 1991 Stimulation of myoblast proliferation in culture by leukaemia inhibitory factor and other cytokines. J Neurol Sci 10:193–197

Bard JBL, Ross ASA 1991 LIF, the ES cell inhibition factor reversibly blocks nephrogenesis in cultured mouse kidney rudiments. Development 113:193–198

Gearing DP, Thut CJ, Vandenbos T et al 1991 Leukaemia inhibitory factor receptor is structurally related to the IL-6 signal transducer, gp130. EMBO (Eur Mol Biol Organ) J 10:2839–2848

Hilton DJ 1990 Characterization of leukaemia inhibitory factor and its cell surface receptor. PhD thesis, University of Melbourne, Australia

Tomida M, Yamamoto-Yamaguchi Y, Hozumi M, Holmes W, Lowe D, Goeddel DV 1990 Inhibition of $Na^+$ dependent hexose transport in renal epithelia LLC-PK1 cells by differentiation-stimulating factor for myeloid leukemia cells, leukaemia inhibitory factor. FEBS (Fed Eur Biochem Soc) Lett 268:261–264

# Reconstitution of high affinity leukaemia inhibitory factor (LIF) receptors in haemopoietic cells transfected with the cloned human LIF receptor

David P. Gearing, Tim VandenBos, M. Patricia Beckmann, Catherine J. Thut, Michael R. Comeau, Bruce Mosley and Steven F. Ziegler

*Immunex Research and Development Corporation, 51 University Street, Seattle WA98101, USA*

*Abstract.* cDNA clones encoding the human leukaemia inhibitory factor (hLIF) receptor were isolated by screening a placental cDNA expression library in COS-7 cells with [125]I-hLIF. The cloned LIF receptor is a member of the haemopoietin receptor family and comprises a signal sequence (44 amino acids), an extracellular region of two haemopoietin receptor domains and three fibronectin type III domains (789 amino acids), a transmembrane domain (26 amino acids) and a cytoplasmic domain (238 amino acids). The LIF receptor is expressed in COS-7 cells as a 190 kDa glycoprotein that specifically binds human LIF with low affinity, but does not bind mouse LIF. Clones encoding a soluble form of the homologous mouse LIF receptor have been isolated, suggesting complex interactions between the various forms of LIF ligand and receptor *in vivo*. The LIF receptor is most related to the gp130 signal-transducing component of the IL-6 receptor, a feature that may provide a molecular basis for the intertwined biologies of LIF and IL-6 in the absence of obvious structural similarily between the ligands. Mouse B9 plasmacytoma cells transfected with the human LIF receptor display novel high affinity LIF receptors that are presumed to consist of transfected receptors in association with endogenous mouse high affinity-converting subunits. Unlike the low affinity human LIF receptor, the mixed species high affinity receptor is capable of binding mouse LIF.

*1992 Polyfunctional cytokines: IL-6 and LIF. Wiley, Chichester (Ciba Foundation Symposium 167) p 245–259*

In recent years it has become apparent that the biologies of two polyfunctional cytokines, interleukin 6 (IL-6) and leukaemia inhibitory factor (LIF), are inter-twined. LIF was initially discovered in a search for molecules that could induce the differentiation of certain mouse myeloid leukaemic cell lines, an activity which IL-6, as macrophage/granulocyte differentiation inducer type 2 (MGI-2), and granulocyte colony-stimulating factor (G-CSF) were also known to show. At that stage, it was clear that the structures of the IL-6 and G-CSF polypeptides were related, but LIF was not obviously structurally related to either. As more information accumulated on the range of functions attributable to LIF, IL-6 and G-CSF, it became evident that LIF and IL-6 were very closely related functionally. To date, the systems in which LIF and IL-6 have both been implicated include the differentiation of myeloid leukaemic cells, the growth and development of haemopoietic stem cells and megakaryocyte progenitors, bone and calcium metabolism, induction of the acute-phase response in hepatocytes, inhibition of adipogenesis, regulation of nerve differentiation, proliferation of myoblasts, regulation of kidney development and control of the early embryo (reviewed in Hirano et al 1990, Gearing 1991, Hilton & Gough 1991, and elsewhere in this volume). The actions of LIF and IL-6 may not be identical in each case, but the range of systems they affect illustrates how closely the activities of the two cytokines correlate.

The molecular basis for the interaction of IL-6 with its receptor has been largely elucidated. Two components of the IL-6 receptor have been cloned—a low affinity IL-6-binding subunit (Yamasaki et al 1988) and a signal-transducing subunit, gp130 (Taga et al 1989, Hibi et al 1990) that alone does not bind IL-6, but binds to the ligand-occupied binding subunit, to produce a high affinity complex. Both subunits are members of the recently defined haemopoietin receptor family (Cosman et al 1990); the relationship between these two subunits forms a paradigm for receptor subunit interactions within other members of this family. Studies of the LIF receptor have been limited to determination of the number and affinities of receptor types on responsive cells. Most responsive cells were found to display small numbers of receptors of relatively high affinity (150–400 per cell, $K_d = 10$–200 pM; Yamamoto-Yamaguchi 1986, Hilton et al 1988, 1991), although receptors of a lower affinity class have also been reported (Hilton et al 1991). Direct demonstration of the size of the LIF receptor by cross-linking was hampered by its low abundance. Thus, we attempted to determine the molecular basis for the similarities between the biologies of LIF and IL-6 by studying the nature of the LIF receptor and its associated signal transduction apparatus. As a first step, we tried to clone LIF-binding polypeptides by expression screening of cDNA libraries using [125]I-labelled LIF as a probe and a sensitive microscopic autoradiographic approach (Gearing et al 1989).

### Expression cloning of the human leukaemia inhibitory factor receptor

We initially demonstrated that placental membrane preparations bound approximately 60 fmol of [125]I-labelled human (h) LIF per mg of membrane protein, about four times less than the amount of G-CSF bound. Because clones for the G-CSF receptor had previously been isolated from a placental library (Larsen et al 1990), we chose to screen the same library for LIF receptors. Using a cDNA pool size of 2400 recombinants per transfection, we detected a single cDNA clone that was able to confer [125]I-hLIF-binding activity to COS-7 cells

```
         0                                             49
Hulifr   MRTASNFQWL LSTFILLYLM NQVNSQKKGA PHDLKCVTNN LQVWNCSWKA
Rbprlr   .......... .......... ....QSPPGK PFIFKCRSPE KETFTCWWRP
Cons.    .......... .......... ........G. P...KC.... .....C.W..

         50                                            99
Hulifr   PSGTGRGTDY EV........ ....CIENRS ...RSCYQLE K.TSIKIPAL
Rbprlr   GADGGLPTNY TLTYHKEGET ITHECPDYKT GGPNSCYFSK KHTSI.....
Cons.    ....G..T.Y .......... ....C..... ....SCY... K.TSI.....

         100                                           149
Hulifr   SHGDYEITIN SLHDFGSSTS KFTLNEQNVS LIPDTPEILN LSA...DFST
Rbprlr   .WTIYIITVN ATNQMGSSVS DPRYVDVTYI VEPDPPVNLT LEVKHPEDRK
Cons.    ....Y.IT.N .....GSS.S .......... ..PD.P..L. L.........

         150                                           199
Hulifr   STLYLKWNDR GSVFPHRSNV IWEIKVLRKE SMELVKLVTH NTTLNGKDTL
Rbprlr   PYLWVKWLPP TLVDVRSGWL TLQYEIRLKP EKAAEWETHF AGQQTQFKIL
Cons.    ..L..KW... ..V....... ........K. .......... .........L

         200                                           249
Hulifr   HHWSWASDMP LECAIHFVEI RCYIDNLHFS GLEE.WSDWS PVKNNSWIPD
Rbprlr   SLYPGQK... .......... ..YLVQVRCK PDHGFWSVWS PESSIQ.IPN
Hugp130                                                    ELLD
Cons.    .......... .......... ..Y....... .....WS.WS P......ipd

         250                                           299
Hulifr   SQTKVFPQDK VILVGSDITF CCVSQEKVLS ALIGHTNCPL IHLDGENV.A
Hugp130  PCGYISPESP VVQLHSNFTA VCVLKEKCMD YFHVNANYIV WKTNHFTI.P
Hgcsfr   ECGHISVSAP IVHLGDPITA SCIIKQNCSH LDPEP.Q.IL WRLGAE.LQP
Rbprlr   DFTMKD
Cons.    .c..isp..p vv.lgs.iTa .Cv.kekc.. ......n.il w.l..e...p

         300                                           349
Hulifr   .IKIRN.ISV SASSG.TNVV FTTEDNIF.. .......... ...GTVIFAG
Hugp130  .KEQYTIINR TASSV.TFTD IASLNIQLTC NILTFGQLEQ NVYGITIISG
Hgcsfr   GGRQQRLSDG TQESIITLPH LNHTQAFLSC C.LNWGNSLQ ILDQVELRAG
Cons.    ...q...i.. tasS..T... .......l.c ..L..G...Q ...g..i.aG

         350                                           399
Hulifr   YPPDTPQQLN CETHDLKEI. ICSWNPGRVT ALVGPRATSY TLVESFSGKY
Hugp130  LPPEKPKNLS CIVNEGKKM. RCEWDGGRET HLETNF.TLK SEWATHKFAD
Hgcsfr   YPPAIPHNLS CLMNLTTSSL ICQWEPGPET HLPTSF.TLK SFK.SRGNCQ
Cons.    yPP..P.nLs C..n..k... iC.W.pGreT hL.t.f.Tlk s...s.....

         400                                           449
Hulifr   VRLKRAEAPT NESYQL.... LFQMLPNQE. ..IYNFTLNA HNPLGRSQST
Hugp130  CKAKR.DTPT SCTVDY.... STVYFVNIE. ..VWVEAENA LGKVT.SDH.
Hgcsfr   TQGDSILDCV PKDGQSHCCI PRKHLLLYQN MGIWVQAENA LGTSM.SPQ.
Cons.    ...kr...pt ....q..... ....l.n.e. ..iwv.aeNA lg....S...
```

```
          450                                                          499
Hulifr    ILVNITEKVY PHTPT..SFK V.KDI...NS TAVKLSWHLP GNFAKINFLC
Hugp130   INFDPVYKVK PNPPHNLSVI NSEEL....S SILKLTWTNP SIKSVIILKY
Hgcsfr    LCLDPMDVVK LEPPMLRTMD PSPEAAPPQA GCLQLCWEPW QPGLHINQKC
Cons.     i..dp..kVk p.pP...s.. .s.e.....s ..lkL.W..p .....In.kc

          500                                                          549
Hulifr    EIEIK.KSNS VQEQRNVTIQ GVEN.SSYLV ALDKLNPYTL YTFRIRCSTE
Hugp130   NIQYR.TKDA STWSQIPPED TASTRSSFTV ..QDLKPFTE YVFRIRCMKE
Hgcsfr    ELRHKPQRGE ASWALVGPLP LEALQ..YEL ..CGLLPATA YTLQIRCIRW
Cons.     ei..k..... ..w....p.. .....ssy.v ....L.P.T. YtfrIRC..e

          550                                                          599
Hulifr    .TFWKWSKWS NKKQHLTTEA SPSKGPDTW. REWSSD...G KNLIIYWKPL
Hugp130   DGKGYWSDWS EEASGITYED RPSKAPSFWY KIDPSHTQGY RTVQLVWKTL
Hgcsfr    PLPGHWSDWS PSLELRTTER APTVRLDTWW RQRQLD...P RTVQLFWKPV
Cons.     ...g.WSdWS ......TtE. .Psk.pdtW. r...sd.... rtvql.WKpl

          600                                                          649
Hulifr    PINEANGKIL SYNVS.CSSD EETQSLSEIP DPQHKAEIRL DKNDYIISVV
Hugp130   PPFEANGKIL DYEVTLTRWK SHLQNYTVNA T...KLTVNL TNDRYLATLT
Hgcsfr    PLEEDSGRIQ GYVVSWRPSG QAGAILPLCN TTELSCTFHL PSEAQEVALV
Cons.     P..EanGkIl .Y.Vs...s. ...q.l.... t...k.t..L ....y...lv

          650                                                          699
Hulifr    AKNSVGSSPP SKIASMEIP. NDDLKIEQVV GMGKG...IL LTWHYDPNMT
Hugp130   VRNLVGKSDA AVLTIPACD. FQATHPVMDL KAFPKDNMLW VEWTTPRESV
Hgcsfr    AYNSAGTSRP TPVVFSESRG PALTRLHAMA RDPHS...LW VGWEPPNPWP
Cons.     a.NsvG.S.p ......e... ...t...... ........lw v.W..p....

          700                                                          749
Hulifr    CDYVIKWCNS SRSEPCLMD. WRKVPSNSTE TVIESDEFRP GIRYNFFLYG
Hugp130   KKYILEWCVL SDKAPCITD. WQQEDGTVHR TYLRGNLAES KC.YLITVTP
Hgcsfr    QGYVIEWGLG PPSASNSNKT WRMEQNGRAT GFLLKENIRP FQLYEIIVTP
Cons.     ..YvieWc.. s.sapc..d. Wr.e...... t.l.....rp ...Y.i.vtp

          750                                                          799
Hulifr    CRNQGYQLLR SMIGYIEELA PIVAPNFTVE DTSADSILVK WEDIPVEELR
Hugp130   VYADGPGSPE SIKAYLKQAP PSKGPTVRTK KVGKNEAVLE WDQLPVDVQN
Hgcsfr    LYQDTMGPSQ HVYAYSQEMA PSHAPELHLK HIGKTWAQLE WVPEPPELGK
Cons.     .y.dg.g... s..aY..e.a Ps.aP....k ..gk..a.le W...Pve...

          800                                                          849
Hulifr    GFLRGY.LFY FGKGERDTSK MRVLESGRSD IKVKNITDIS QKTLRIADLQ
Hugp130   GFIRNYTIFY RTIIGNETAV ..NVDSSHTE Y......... ..TLSSLTSD
Hgcsfr    SPLTHYTIFW TNA.QNQSFS AILNASSRGF VLHGLEPASL YHIHLMAASQ
Cons.     gflr.YtiFy .....n.t.. .....Ssr.. .......... ..tl..a.sq

          850                                                          899
Hulifr    GKTSYHLVLR AYTD.GGV.G PEKSMYVVTK ENSVGLIIAI LIPVAVAVIV
Hugp130   ..TLYMVRMA AYTDEGGKDG PEFT..FTTP KFAQGEIEAI VVPVCLAFLL
Hgcsfr    AGATNSTVLT LMTLTP..EG SELH...... .IILGLFGLL LLLTCLCGTA
Cons.     ..t.y..vl. ayTd.gg..G pE......t. ....Gli.ai l.pvcla...
```

```
            900                                                   949
  Hulifr   .GVVTSILCY RKREWIKETF YPDIPNPENC KALQFQKSVC EGSSALKTLE
  Hugp130  TTLLGVLFCF NKRDLIKKHI WPNVPDPSKS HIAQWSPHTP PRHNFNSKDQ
  Hgcsfr   WLC.....CS PNR...KNPL WPSVPDPAHS SLGSWVP... .TIMEEDAFQ
  Cons.    ........C. .kR..iK... wP.vPdP..s ...qw.p... .........q

            950                                                   999
  Hulifr   MNPCT.PNNV EVLETRSAFP KIEDTEIISP VAERPEDRSD AEPENHVVVS
  Hugp130  MYSDGNFTDV SVVEIEANDK KPFPEDLKSL DLFKKEKINT EGHSSGIGGS
  Hgcsfr   LPGLGTPPIT KLTVLEEDEK KPVPWESHNS SETCGLPTLV QTYVLQGDPR
  Cons.    m...g.p..v .v.e.e...k Kp.p.e..s. .....e.... .........s

            1000                                                  1049
  Hulifr   YCPPIIEEEI PNPAADEAGG TAQVI....Y I.DVQSMYQP QAKPE.....
  Hugp130  SCMSSSRPSI SSS..DENES SQNTSSTVQY STVVHSGYRH QVPSVQVFSR
  Hgcsfr   AVSTQPQSQS GTS..DQAGP PRRSA....Y FKDQIML.HP APPNGLLCLF
  Cons.    .c.......i ..s..Deag. .........Y ..dv.s.y.p q.p.......

            1050                                                  1099
  Hulifr   .......... EEQENDPV.. ....GGAGYK .......... ......PQMH
  Hugp130  SESTQPLLDS EERPEDLQLV DHVDGGDGIL PRQQYFKQNC SQHESSPDIS
  Hgcsfr   PITSVL*
  Cons.    .......... EE...D.... ....GG.G.. .......... ......P...

            1100                                                  1149
  Hulifr   LPINSTVEDI AAEEDLDKTA GYRPQANVNT WNLVSPDSPR SIDSNSEIVS
  Hugp130  HFERSKQVSS VNEEDFVRLK QQISDHISQS CGSGQMKMFQ EV.SAADAFG
  Cons.    ....S..... ..EED..... .......... .......... ...S......

            1150                              1188
  Hulifr   FGSPCSINSR Q..FLIPPKD EDSPKSNGGG WSFTNFFQNKPND*
  Hugp130  PGTEGQVERF ETVGMEAATD EGMPKSYLPQ TVRQGGYM..PQ*
  Cons.    .G........ .........D E..PKS.... ..........P..
```

FIG. 1.   Alignment of the amino sequences of the human LIF receptor (Hulifr) with the human G-CSF receptor (Hgcsfr) and the human IL-6 receptor signal-transducing subunit, gp130 (Hugp130). The first haemopoietin receptor domain of the LIF receptor is aligned, for convenience, with another member of the same family, the rabbit prolactin receptor (Rbprlr), to illustrate the positions of the conserved residues in the haemopoietin receptor family. The numbers refer to the aligned consensus sequence (Cons.), not to any of the receptors. The consensus shows identities as upper case and residues shared by two receptors out of the three (Hulifr, Hgcsfr and Hugp130) as lower case. Underlined sequences show the predicted transmembrane domains.

(SV40-transformed African Green monkey kidney cells). Binding activity was observed as accumulation of silver grains in photographic emulsion placed over the transfected COS-7 cell monolayer (Gearing et al 1991).

Sequence analysis of the cDNA insert from the positive clone revealed a single long open reading frame of 971 amino acids that had no in-frame termination signal, terminating instead in vector sequences. Several other cDNA clones were isolated from this library and another was isolated from a human hepatoma (SKHep) cDNA library, and in all cases the cDNAs had 3' ends at the same site. We concluded the the oligo-dT used to prime the cDNA synthesis reactions in both libraries had spuriously primed the LIF receptor clones from an adenine-rich sequence in the region of the mRNA encoding the cytoplasmic domain of the LIF receptor. To identify sequences encoding the remainder of the cytoplasmic tail of the LIF receptor we turned to a human genomic library for overlapping sequences. The sequence of one such genomic clone extended the cDNA sequence by 111 amino acids, until a stop codon was encountered. Using primers from the cDNA and genomic sequences we used the polymerase chain reaction (PCR) to amplify products from placental cDNA and not genomic DNA that confirmed the assignment of the genomic sequences as part of the LIF receptor coding region (Gearing et al 1991).

The cloned LIF receptor is a member of the haemopoietin receptor family, showing extensive sequence similarity to the IL-6 receptor signal-transducing subunit gp130 and to the G-CSF receptor (Fig. 1). Its domain structure is typical of this class of receptor; following the 44 amino acid residue signal sequence the LIF receptor has an extracellular domain of 789 residues made up of two copies of the haemopoietin receptor domain (cysteine-rich region to WSXWS box) and three copies of a fibronectin type III domain, followed by a 26 residue hydrophobic transmembrane domain and a 238 residue cytoplasmic domain rich in serine, threonine, proline and acidic residues. Confirmation that the sequence from the genomic clone was correct also came from comparisons with the IL-6 receptor gp130 subunit. Alignment of the transmembrance and cytoplasmic domains of the two proteins revealed a close similarity throughout, and placed the predicted stop codon in the LIF receptor one amino acid apart from the stop codon in gp130.

## Transfection studies with the cloned leukaemia inhibitory factor receptor

Transfection of the hLIF receptor cDNA into COS-7 cells resulted in expression of a low affinity ($K_d$ about 1 nM) LIF receptor on the cell surface, with a size of about 190 kDa (Gearing et al 1991). The large size of the LIF receptor may account in part for some of the difficulties in its detection on responsive cells by standard cross-linking methods, because most gel systems do not resolve proteins adequately in this size range. The low affinity binding to the cloned receptor was characteristic of the type observed on macrophages and macrophage cell lines; however, because most LIF receptors detected on responsive cells are of a higher affinity type, we wished to determine whether the cloned LIF receptor could interact with another membrane-bound subunit to form a high affinity complex. To examine this possibility we used a mouse plasmacytoma cell line, B9, that displays no LIF-binding sites (Baumann & Wong 1989), into which we transfected the cloned human LIF receptor (hLIFR-65; Gearing et al 1991). We had hoped that the B9 cells would become LIF-dependent after transfection of the hLIFR, but both hLIFR cDNA transfectants and control plasmid transfectants became growth factor-independent during the course of the electroporation and selection. Nevertheless, the expected high number of low affinity hLIF receptors was observed (5800 per cell, $K_d = 2.2 \times 10^{-9}$ M) on the B9/hLIFR transfectants, as well as a lower number of high affinity hLIF receptors (430 per cell, $K_d = 1.6 \times 10^{-10}$ M) (Fig. 2). This result suggests that B9 cells normally express an endogenous non-LIF-binding high affinity-converting subunit that is capable of cross-species interaction with the cloned human LIF receptor. Moreover, because the high affinity sites number less than the low affinity sites, the affinity converter would be limiting. Such a model is consistent with other related haemopoietin-type receptor systems that are composed of multiple affinity classes. For example, the human granulocyte-macrophage colony-stimulating factor (GM-CSF) receptor system is made up of a low affinity ($K_d = 6.6$ nM) GM-CSF-binding $\alpha$ subunit that can complex with a $\beta$ subunit, which by itself cannot bind GM-CSF, to form a high affinity complex ($K_d = 120$ pM; Hayashida et al 1990). The IL-6 receptor system, described above, also has this design; in this case an IL-6-binding $\alpha$ subunit complexes with a non-binding $\beta$ subunit (gp130), to form a high affinity complex (Taga et al 1989, Hibi et al 1990).

We had previously observed that the binding of [125]I-hLIF to the cloned hLIFR expressed in COS-7 cells was not competitively inhibited by mouse (m) LIF (Gearing et al 1991), despite the fact that it shares extensive amino acid identity with hLIF. Furthermore, mLIF does not bind to high affinity hLIF receptors on human cell lines (N. A. Nicola, personal communication). We therefore expected a similar lack of binding of mLIF to the cloned receptor on B9/hLIFR cells. However, when [125]I-hLIF binding to B9/hLIFR cells was assessed in the presence of constant excess mLIF, we observed selective competition for the high affinity sites, such that only the low affinity sites (6660 per cell, $K_d = 2.6 \times 10^{-9}$ M) were detected (Fig. 2). The binding of mLIF to the mixed-species high affinity receptor on B9/hLIFR cells suggested that mLIF may interact weakly with the low affinity hLIF receptor and that this interaction can be stabilized through association with the putative mouse affinity-converting subunit.

In summary, these results support the hypothesis that a high affinity LIF receptor complex can be assembled in non-LIF-responsive mouse B9 cells from a low affinity hLIFR and an endogenous mouse affinity-converting chain (Fig. 3). Alternative hypotheses to account for the affinity conversion in B9/hLIFR cells include post-translational modification of the hLIFR subunit, resulting in a second species capable of high affinity hLIF binding as well as mLIF binding. However, because of the similarity to the related IL-6 and GM-CSF receptor

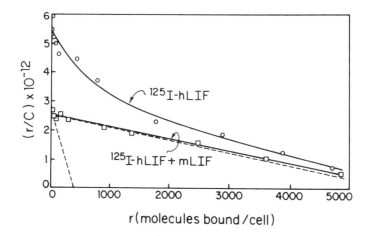

FIG. 2. Scatchard analysis of [125]I-labelled human (h) LIF binding to B9/hLIFR cells (mouse plasmacytoma cells expressing the cloned human LIF receptor). Direct binding of [125]I-hLIF (circles) is presented. Binding of [125]I-hLIF in the presence of excess (2 µg per point) unlabelled mouse LIF is also shown (squares). For comparison, the computer-generated contributions of high and low affinity [125]I-hLIF binding in the absence of competitor are shown as broken lines. Human LIF was radiolabelled to a specific activity of $0.58$–$1.0 \times 10^{16}$ c.p.m./mmol as described in Gearing et al (1991), and binding was assayed essentially as described by Benjamin & Dower (1990). Non-specific binding was determined using a 200-fold molar excess of hLIF.

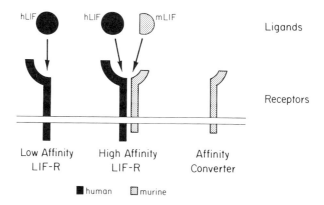

FIG. 3. Interpretation of LIF binding to B9/hLIFR transfectants. Human LIF binds to both the low affinity transfected hLIFR and to a novel mixed-species high affinity receptor complex. The mouse affinity-converting subunit does not bind human LIF alone. Mouse LIF does not bind to the low affinity hLIFR on its own, but can compete for binding of human LIF to the high affinity mixed-species receptor complex.

systems we favour the 'second affinity-converting subunit' hypothesis. Accordingly, we predict that the purely high affinity LIF receptors observed on most responsive cells (in the absence of low affinity receptors) consist of low affinity ligand-binding subunits in the presence of an excess of affinity-converting subunits.

An initial observation that followed the comparison between the LIF receptor and the IL-6 receptor β subunit (gp130) was that the extensive sequence similarity of the cytoplasmic domains of these two molecules provided a potential basis for the similarity between the biological effects of their ligands (Gearing et al 1991). Of particular note was that the sequences of the two cytoplasmic domains could be aligned to within one amino acid in length of each other. Although we do not yet have evidence of signal transduction by the cloned LIF receptor, the truncated form encoded by clone pHLIFR-65 was apparently capable of interacting with another subunit in B9 cells to generate a high affinity complex. Therefore, the carboxy-terminal 110 amino acids missing from the cytoplasmic domain of clone pHLIFR-65 are not required for formation of the high affinity complex. This observation is similar to that made in the IL-6 receptor system, where a soluble IL-6 receptor α subunit can transduce a signal through the gp130 β subunit when the α subunit is complexed with IL-6, with subunit interaction occurring between the extracellular domains. We are attempting to determine which part(s) of the cytoplasmic domain of the LIF receptor is required for signal transduction.

The relationship of the cloned LIF receptor to the gp130 β subunit of the IL-6 receptor raises the possibility that the IL-6 receptor α subunit might also interact with the LIF receptor to form a high affinity complex. However, co-transfection experiments with the LIF receptor and the IL-6 receptor α subunit in COS-7 cells did not reveal altered affinity of LIF binding, nor did excess hIL-6 influence the binding of LIF to the receptors on B9/hLIFR cells (D. P. Gearing, M. R. Comeau & B. Mosely, unpublished work). Thus, the subunits of the IL-6 receptor appear to be independent of the LIF receptor.

Recently, a number of reports have documented some of the intracellular consequences of binding of LIF and IL-6 to responsive cells. Tyrosine phosphorylation of a 160 kDa protein was an immediate early event in M1 cells after addition of LIF or IL-6 (Lord et al 1991), as was induction of a series of so-called myeloid differentiation primary response genes including interferon regulatory factor 1 (Abdollahi et al 1991). In a separate study, phosphorylation of the heat-shock protein hsp27 was observed in M1 cells similarly treated with LIF or IL-6 (Michishita et al 1991). Activation of hepatocytes by IL-6 is thought to be associated with transcriptional activation through NF-IL6 (Akira et al 1992, this volume). It will be interesting to see whether LIF also acts through NF-IL6, and whether the cloned LIF receptor transfected into appropriate target cells can activate the same intracellular components.

*Acknowledgements*

We are grateful to S. Dower and C. Maliszewski for critical comments on this manuscript.

## References

Abdollahi A, Lord KA, Hoffman-Liebermann B, Liebermann DA 1991 Interferon regulatory factor 1 is a myeloid differentiation primary response gene induced by interleukin-6 and leukemia inhibitory factor: role in growth inhibition. Cell Growth & Differ 2:401–407

Akira S, Isshiki H, Nakajima T et al 1992 Regulation of expression of the IL-6 gene: structure and function of the transcription factor NF-IL6. In: Polyfunctional cytokines: IL-6 and LIF. Wiley, Chichester (Ciba Found Symp 167) p 47–67

Baumann H, Wong GG 1989 Hepatocyte-stimulating factor-III shares structural and functional identity with leukemia-inhibitory factor. J Immunol 143:1163–1167

Benjamin D, Dower SK 1990 Human B cells express two types of interleukin-1 receptors. Blood 75:2017–2023

Cosman D, Lyman SD, Idzerda RL et al 1990 A new cytokine receptor superfamily. Trends Biochem Sci 15:265–270

Gearing DP, King JA, Gough NM, Nicola NA 1989 Expression cloning of a receptor for human granulocyte-macrophage colony-stimulating factor. EMBO (Eur Mol Biol Organ) J 8:3667–3676

Gearing DP 1991 Leukemia inhibitory factor; does the cap fit? Ann NY Acad Sci 628:9–18

Gearing DP, Thut CJ, VandenBos T et al 1991 Leukemia inhibitory factor receptor is structurally related to the IL-6 signal transducer, gp130. EMBO (Eur Mol Biol Organ) J 10:2839–2848

Hayashida K, Kitamura T, Gorman DM, Arai K, Yokota T, Miyajima A 1990 Molecular cloning of a second subunit of the receptor for human granulocyte-macrophage colony-stimulating factor (GM-CSF): reconstitution of a high-affinity GM-CSF receptor. Proc Natl Acad Sci USA 87:9655–9659

Hibi M, Murakami M, Saito M, Hirano T, Taga T, Kishimoto T 1990 Molecular cloning and expression of an IL-6 signal transducer, gp130. Cell 63:1149–1157

Hilton DJ, Gough NM 1991 Leukemia inhibitory factor: A biological perspective. J Cell Biochem 46:21–26

Hilton DJ, Nicola NA, Metcalf D 1988 Specific binding of murine leukaemia inhibitory factor to normal and leukaemic monocytic cells. Proc Natl Acad Sci USA 85:5971–5975

Hilton DJ, Nicola NA, Metcalf D 1991 Distribution and comparison of receptors for leukemia inhibitory factor on murine hemopoietic and hepatic cells. J Cell Physiol 146:207–215

Hirano T, Akira S, Taga T, Kishimoto T 1990 Biological and clinical aspects of interleukin 6. Immunol Today 11:443–449

Larsen A, Davis T, Curtis BM et al 1990 Expression cloning of a human granulocyte colony-stimulating factor receptor: a structural mosaic of hematopoietin receptor, immunoglobulin, and fibronectin domains. J Exp Med 172:1559–1570

Lord KA, Abdollahi A, Thomas SM et al 1991 Leukemia inhibitory factor and interleukin-6 trigger the same immediate early response, including tyrosine phosphorylation upon induction of myeloid leukemia differentiation. Mol Cell Biol 11:4371–4379

Michishita M, Satoh M, Yamaguchi M, Hirayoshi K, Okuma M, Nagata K 1991 Phosphorylation of the stress protein hsp27 is an early event in murine myelomonocytic leukemic cell differentiation induced by leukemia inhibitory factor/D-factor. Biochem Biophys Res Commun 176:979–984

Taga T, Hibi M, Hirata Y et al 1989 Interleukin-6 triggers the association of its receptor
    with a possible signal transducer, gp130. Cell 58:573–581
Yamasaki K, Taga T, Hirata Y et al 1988 Cloning and expression of the human
    interleukin-6 (BSF-2/IFNβ) receptor. Science (Wash DC) 241:825–828
Yamamoto-Yamaguchi Y, Tomida M, Hozumi M 1986 Specific binding of a factor
    inducing differentiation to mouse myeloid leukemic M1 cells. Exp Cell Res 164:97–102

## DISCUSSION

*Gough:* You alluded to the existence of a high affinity-converting chain. Do
you have any direct evidence for such a converter?

*Gearing:* When there is excess expression of the LIF receptor in COS cells
you don't see high affinity receptors. Therefore, one would presume that the
cloned LIF receptor is not capable of forming a dimer to give high affinity
receptors. However, when LIF is transfected into B9 cells a high affinity LIF
receptor complex forms. I prefer the idea that there is another chain in B9 cells
which converts the affinity. An alternative explanation might be that in B9 cells
post-translational modification of the LIF receptor produces a variant form
which is capable of binding human LIF with high affinity. The fact that there
is competition for binding between mouse and human LIF increases the
likelihood that there is a mouse high affinity converter. We are looking for direct
evidence for a converter. [See note added in proof by D. P. Gearing.]

*Martin:* What is the relationship, on a molar basis, of competition
between mouse LIF and human LIF for the high affinity specific receptors?
Is mouse LIF less effective than human in the competition for binding to
transfected B9 cells?

*Gearing:* That's difficult to answer because we use human LIF as the cold
competitor for Scatchard analysis. The affinity is probably similar, because we
used only 2 μg mouse LIF yet we see effective blockade of the high affinity site.

*Lee:* It's still possible that the mouse converting chain binds mouse LIF, but
that the affinity is so low that you can't detect it.

*Gearing:* Exactly—the same is true with human LIF.

*Ciliberto:* When you transfect the LIF receptor into B9 cells do the cells show
a growth response to LIF?

*Gearing:* The aim of the experiment was to get signal transduction. Both
control cells and the LIF-transfected cells became factor independent during
the course of the experiment. We are constructing a retrovirus encoding the
human LIF receptor to try a softer approach. We have also tried BAF/B03 cells;
in these, transfected G-CSF receptors transduce a proliferative signal, but to
date we have observed no effect with transfected LIF receptors.

*Ciliberto:* Have you tried to transfect Hep3B cells, for example, which don't
carry LIF receptors?

*Gearing:* I discovered that only yesterday. That would be a good idea.

*Ciliberto:* Have you done cross-linking experiments to see if heterodimers can be formed between human LIF and mouse LIF?

*Gearing:* We have been unable to do cross-linking experiments, because the LIF that I have managed to iodinate is a high molecular mass form made in yeast. I have tried to iodinate *E. coli*-derived LIF without success. The yeast LIF gives a smear on an SDS gel over 100–200 kDa, because it is heavily glycosylated, so it can't be used to do cross-linking.

*Gough:* Have you transfected the mouse LIF receptor into B9 cells?

*Gearing:* Not yet.

*Gough:* You spoke about both membrane-bound and soluble forms of the LIF receptor in the mouse. Do you have evidence for a soluble form of the LIF receptor in the human?

*Gearing:* We have some evidence, but it is preliminary and tentative.

*Lotem:* Do the B9 cells into which you transfect the receptor express gp130?

*Gearing:* We see a high affinity IL-6 receptor site, so I presume they contain gp130.

*Lotem:* Is it possible that the converter is gp130?

*Gearing:* From the sequence of the cloned LIF receptor this was not predicted. The IL-6 receptor complex is a heterodimer, and the gp130 component is non-ligand-binding and signal-transducing. In contrast, the LIF receptor is a ligand-binding subunit similar to gp130 and presumably signal-transducing. The G-CSF receptor on its own shows high affinity binding; Dr Nagata's group suggested that a homodimer of the G-CSF receptor is the functional unit. Thus, the G-CSF receptor has the elements required for signal transduction in the absence of gp130 and we might predict the same for the LIF receptor on the basis of the sequence similarity. [See note added in proof.]

*Heath:* Have you got a full-length, transmembrane, mouse cDNA for the LIF receptor?

*Gearing:* Yes, and this is similar to the human receptor. We are still working on it, because the mRNA is 10–12 kb.

*Sehgal:* Do you have any idea what the function of the conserved fibronectin-like domains is?

*Gearing:* I have no knowledge of their role in the LIF receptor, but deletion of these domains in the G-CSF receptor has no effect on signalling (Fukunaga et al 1991). If the active receptor is a homodimer, perhaps the units can still form that homodimer. In the case of gp130, maybe you would expect to see an effect of deletion.

*Bazan:* I would suggest that the surplus immunoglobulin and fibronectin domains in the G-CSF and LIF receptors are *not* important for cytokine binding, but are an indication of the vestigial function of these receptors as cell surface adhesion molecules.

In biology, function inevitably follows form. A complete response to the many biological questions surrounding the interactions of IL-6 and LIF with their cognate receptors requires a description of the generic molecular recognition between receptors and their cytokine ligands. The IL-6–LIF system is but a small subset of a larger family of interacting receptors and cytokines (for review, see Bazan 1990a). The receptors display an underlying simplicity of design in their extracellular binding segment. A common 200 amino acid domain is composed of repeating duplicated modules rich in β-strands that are predicted to resemble immunoglobulin folds (Bazan 1990b). The cytokine-binding pocket is then formed from amino acid loops that converge between linked modules, a region distinct from the receptor subunit dimerization site (Bass et al 1991, Cunningham et al 1991).

That we can partly decouple cytokine binding from receptor pairing is an important suggestion. It is increasingly clear that high affinity receptor signalling complexes rely on the productive association of at least two receptor subunits. Take, for example, the promiscuous interaction of GM-CSF, IL-3 and IL-5 α receptors with a common secondary receptor (for review, see Nicola & Metcalf 1991). Two recognition codes must be involved: (a) different α receptors (with low pairwise sequence identity) probably share a small number of common amino acids (clustered in the folded structures) that allow packing against the shared β receptor; (b) different ligands (with no recognizable sequence identity) specifically interact with their cognate α receptors, yet conserve a distinct tertiary epitope that is read by the shared β receptor. A 'low resolution' understanding of the first code invokes immunoglobulin-like sheet-to-sheet pairings between receptor subunits—specific mapping of residues involved awaits a high resolution crystal structure. The second code is illuminated by extensive X-ray crystallographic, spectroscopic and modelling work that suggests a common helical fold for all cytokine ligands that bind to the superfamily receptors (Bazan 1990a). The archetypal growth hormone fold (also evident in β-interferon; Senda et al 1990), an α-helix bundle with distinctive loop topology, provides a convenient scaffold for distinct tertiary epitopes on opposing faces of the cylindrical bundle (Bazan 1991). Glycosylation sites typically map to loops that protrude from the top and bottom of the bundle core, away from the receptor-binding surfaces. A preliminary crystal structure of GM-CSF (Diederichs et al 1991) may help to resolve the distinct and shared binding motifs of the promiscuous GM-CSF, IL-3 and IL-5 triad.

How is this analogy relevant to IL-6 and LIF? These two cytokines have a common spectrum of biological actions yet have distinct α receptors. There are additional functional parallels with G-CSF, Oncostatin M and ciliary neurotrophic factor (CNTF), as Dr Patterson, Professor Gauldie and Dr Gearing have described. There is a real possibility that this new grouping of homologous cytokines forms a promiscuous set with interacting receptors (Bazan 1991).

*Sehgal:* How do you imagine a multichain interaction in this model?

*Bazan:* The model proposes that the surface of interaction between the ligand and the receptor is actually not that great. For example, Jim Wells's group at

Genentech has demonstrated that you can selectively modulate the binding affinity of growth hormone by changing a small number of amino acids in the α receptor-binding helix (Cunningham & Wells 1991). Interactions between the α receptor and ligand are distinct but similar in nature; pairing between subunits is as previously discussed.

*Sehgal:* How do you envisage G-CSF binding to a high affinity homodimer? David Gearing suggested that the high affinity G-CSF receptor might be a homodimer. Do you think there is one ligand binding to two chains, or two ligands to two chains?

*Bazan:* So far we have investigated only the α receptor binding surface in detail. It is clear that helical cytokines have two distinct binding surfaces tailored to bind to α or β receptors. In some cases (and here growth hormone comes to mind; Cunningham et al 1991), α = β, and there will be a homodimeric receptor complex binding a single cytokine ligand.

*Metcalf:* That does not agree well with the model of G-CSF, which is asymmetric.

*Bazan:* In detail, that's true. The helix bundle fold is cylindrically symmetrical if we consider only the core of four helices. Cytokine binding to a homodimeric receptor will involve symmetry-breaking binding interactions.

*Nicola:* Two reports provide evidence that the elements involved in binding of GM-CSF to the α chain and the elements involved in affinity conversion can be completely separated (Shanafelt et al 1992, Lopez et al 1992). Both groups agree that Glu-21 of GM-CSF is absolutely critical for formation of a high affinity complex, but that this residue has nothing to do with interaction with the α chain.

*Bazan:* You can imagine that it will be possible to engineer chimeric cytokine ligands in which one face of the helix bundle binds one receptor of choice and the other side recognizes a different receptor. In the case of GM-CSF, Glu-21 forms part of the natural β receptor face and does not affect binding to the α receptor.

*Sehgal:* I can understand how this would work with heterodimeric receptors, but I have difficulty understanding how it would work with a homodimeric receptor complex.

*Bazan:* Such a complex involves two identical receptors contacting a ligand molecule in which structural symmetry is conserved *in spite* of a lack of tell-tale symmetry in sequence. Once again, growth hormone proves that this can occur.

*Nicola:* Something has to happen after the initial binding event. Professor Kishimoto's results show clearly that first there is binding to the 80 kDa subunit, and only after that can gp130 'see' the complex. That also seems to be true for GM-CSF. After the initial interaction, the ligand–receptor complex somehow changes, and new structures, new epitopes, can be generated which another G-CSF receptor, or β subunit, might then be able to recognize. For GM-CSF, we can map to a small region the area required for formation of a high affinity complex, but that particular region of GM-CSF doesn't recognize the β subunit until after it has bound to an α subunit.

*Gauldie:* Formation of an antigen–antibody complex changes the conformation of the antibody and possibly the antigen, such that new recognition states are made available. It's not radical to assume that once the ligand has bound to its receptor it can associate with a third molecule through a novel site.

*Ciliberto:* Professor Hirano, have you raised antibodies to the IL-6 receptor which react with the receptor alone but not with the receptor when IL-6 is bound?

*Hirano:* We have not raised such an antibody yet.

### References

Bass SH, Mulkerrin MG, Wells JA 1991 A systematic mutational analysis of hormone-binding determinants in the human growth hormone receptor. Proc Natl Acad Sci USA 88:4498–4502

Bazan JF 1990a Haemopoietic receptors and helical cytokines. Immunol Today 11:350–354

Bazan JF 1990b Structural design and molecular evolution of a cytokine receptor superfamily. Proc Natl Acad Sci USA 87:6934–6938

Bazan JF 1991 Neuropoietic cytokines in the hematopoietic fold. Neuron 7:197–208

Cunningham BC, Wells JA 1991 Rational design of receptor-specific variants of human growth hormone. Proc Natl Acad Sci USA 88:3407–3411

Cunningham BC, Ultsch M, DeVos AM, Mulkerrin MG, Wells JA 1991 Dimerization of the human growth hormone receptor by a single hormone molecule. Science (Wash DC) 254:821–825

Diederichs K, Jacques S, Boone T, Karplus PA 1991 Low-resolution structure of recombinant human granulocyte-macrophage colony stimulating factor. J Mol Biol 221:55-60

Fukunaga R, Ishizakaikeda E, Pan CX, Seto Y, Nagata S 1991 Functional domains of the granulocyte colony-stimulating factor receptor. EMBO (Eur Mol Biol Organ) J 10:2855–2865

Lopez AF, Shannon MF, Hercus T et al 1992 Residue 21 of human granulocyte-macrophage colony-stimulating factor is critical for biological activity and for high but not low affinity binding. EMBO (Eur Mol Biol Organ) J 11:909–916

Nicola NA, Metcalf D 1991 Subunit promiscuity among hemopoietic growth factor receptors. Cell 67:1–4

Senda T, Matsuda S, Kurihara Hm Nakamura KT, Mitsui Y 1990 3- dimensional structure of recombinant murine interferon-beta. Proc Jpn Acad Ser B Phys Biol Sci 66:77–80

Shanafelt AB, Miyajima A, Kitamura T, Kastelein RA 1992 The amino-terminal helix of GM-CSF and IL-5 governs high affinity binding to their receptors. EMBO (Eur Mol Biol Organ) J, in press

*Note added in proof by D. P. Gearing:* We have now demonstrated that Oncostatin M can bind to the high affinity, but not the low affinity, LIF receptor and, surprisingly, that the affinity-converting subunit of the LIF receptor is identical to that of the IL-6 receptor, gp130 (Gearing 1992 New Biol 4:61; Gearing et al 1992 Science (Wash DC) 255:1434–1437).

# General discussion IV

### Cytokines and binding proteins in the circulation

*Gauldie:* I would like to return to the issue of whether or not there are different forms of cytokines in the circulation. I have difficulty with the concept that Dr Sehgal put forward (p 225), of a molecule that's carried around in the circulation but is apparently active only in an assay system. If such a complexed molecule is circulating, at the levels he suggested, I would expect to see the same kind of response *in vivo* as he saw in the hepatocyte assay—that is, individuals should show constant raised levels of acute-phase proteins. A small amount of IL-6 generated in the body is enough to trigger the hepatic response. I cannot argue with the data, but I am struggling with the interpretation. If that material is present, does it have any biological significance, or is it on its way out of the body? Several molecules scavenge the kind of molecules Dr Sehgal showed to be present in the complexes.

*Sehgal:* I should point out that the material was obtained from the bloodstream, and may not reflect what is actually present in tissue fluids. We need to be careful to consider the compartments sampled.

*Gearing:* One might expect IL-6 injected into the bloodstream to form a high molecular weight complex. Is there any evidence for that?

*Sehgal:* We have mixed methionine-labelled natural or *E. coli*-derived IL-6 with samples of serum. This does not alter the gel filtration profile of the labelled protein. It appears that the entities that we find are endogenously produced, and cannot be generated by adding exogenous IL-6.

*Metcalf:* The soluble IL-6 receptor, when complexed with IL-6, is capable of delivering a biological stimulus; what is the efficiency of this complex compared with IL-6?

*Hirano:* In the presence of soluble IL-6 receptor, the concentration of IL-6 required for activity is lower than that required in its absence.

*Heinrich:* The addition of soluble IL-6 receptor amplifies the action of IL-6, particularly in cells with down-regulated IL-6 receptors.

*Nicola:* This might be not just a quantitative, but also a qualitative difference. Soluble receptor complexed with IL-6 has the potential to interact with a cell which expresses only gp130, and not an IL-6-binding chain. In that way, unique responses which would never be achieved with IL-6 alone might occur.

*Heinrich:* Professor Kishimoto and Professor Hirano have shown that gp130 is not the limiting protein. With the soluble IL-6 receptor–IL-6 complex acute-phase protein synthesis can be further stimulated in HepG2 cells in which the 80 kDa receptors are down-regulated.

*Hirano:* The expression of the 80 kDa ligand-binding receptor molecule is usually the limiting factor.

*Heinrich:* If you treat HepG2 cells with IL-6, the 80 kDa receptor protein is internalized and the cells become unresponsive to IL-6. If you then add the soluble receptor–IL-6 complex a further response is observed.

*Metcalf:* Is the binding of the soluble receptor–IL-6 complex to gp130 of high affinity?

*Hirano:* I don't know.

*Metcalf:* If the binding is not of high affinity, it may not elicit an efficient signal.

*Heinrich:* It's probably of high affinity but the binding of IL-6 to the soluble receptor is of lower affinity.

*Lotem:* When the IL-6 receptor is internalized together with its ligand, is gp130 is also internalized? If it is, there would be no response to the soluble receptor–IL-6 complex.

*Nicola:* There would be a response if gp130 is in excess.

*Gearing:* We have heard that there is an excess of a soluble IL-6-binding protein in serum. If a large amount of an IL-6-binding protein is circulating, and can interact with a gp130 molecule in a cell lacking the ligand-binding receptor chain, you might expect to see activities of IL-6 *in vivo* that you cannot reproduce with pure IL-6 *in vitro*. Are there any reports of such a phenomenon?

*Hirano:* I don't know of any clear evidence indicating the *in vivo* effect of the complex of IL-6 and the soluble receptor. As you said, it is possible that the complex acts on cells which do not express the receptor but do express gp130.

*Heath:* The concept of presentation is important. You could regard many of the binding proteins that have been isolated not as signalling entities but as presentation molecules. The primary function of the LIF receptor and the IL-6 receptor is to deliver a signal to another (signalling) receptor, and to control the distribution of the ligand.

*Gauldie:* Is there any evidence that the amount of soluble IL-6 receptor fluctuates in inflammatory states? Is there a steady-state level?

*Hirano:* The only patients in whom we have observed an increase in soluble IL-6 receptor in the serum are those with AIDS.

*Heinrich:* We have tested sera from 12 patients with systemic lupus erythematosus and we detected the soluble IL-6 receptor in four of them (A. Mackiewicz, H. Schooltink, P. C. Heinrich & S. Rose-John, unpublished work). What is the evidence for there being large amounts of soluble receptor in the serum of normal individuals?

*Hirano:* We measured the level of soluble receptor using an ELISA.

*Baumann:* Professor Heinrich, how much soluble receptor did you find in the four positive lupus patients?

*Heinrich:* This is very recent work and we do not know the exact levels. The soluble receptor needs to be purified to homogeneity before we can answer such questions. At the moment, we incubate the soluble IL-6 receptor that is present in patients' sera or in medium from cells which have been transfected with the soluble IL-6 receptor cDNA with an excess of $^{125}$I-labelled recombinant human IL-6. The soluble receptor–$^{125}$I-rhIL-6 complex is then immunoprecipitated with a specific antibody to the human IL-6 receptor and protein A Sepharose. The radioactivity bound to the antibody–protein A Sepharose is used as a measure of the amount of soluble IL-6 receptor.

*Metcalf:* The protein must therefore have an IL-6 recognition site, and must also have an epitope recognized by the antibody. This is probably enough to conclude provisionally that it could be the same molecule, but it is not proof.

*Baumann:* Professor Heinrich, you mentioned that in eight of the 12 lupus patients no soluble receptor was detected. Professor Kishimoto said that high levels have been found in normal individuals. That suggests that your lupus patients have depleted soluble IL-6 receptors!

*Hirano:* Novick et al (1989) reported the presence of the soluble IL-6 receptor in normal urine, and in that case they purified and sequenced the protein.

*Sehgal:* They also showed that the protein increased the biological activity of recombinant IL-6 in a variety of assays by a factor of 5–10.

*Metcalf:* That doesn't resolve the question of whether an IL-6-binding molecule of the same size in the circulation is actually related to the IL-6 receptor, because it's conceivable that the protein in urine is a degradation product of the transmembrane receptor.

*Baumann:* A soluble LIF receptor would be an inhibitor of circulating LIF, and might explain the low to non-detectable amounts of LIF in plasma.

*Gearing:* You would predict that, but we have no evidence either way.

*Heath:* It could be an inhibitor or a presentation molecule.

*Gearing:* If gp130 is the signal-transducing component, and the LIF receptor is both the signal-transducing component and the binding component, interaction of the binding component with a second subunit might give a non-functional complex.

*Heath:* Dr Nicola argued earlier that a homodimeric model is unlikely to be correct. If the function of the low affinity LIF receptor is to present LIF to the converter/signalling molecule, the soluble form of the receptor should also work, and would produce a signal in association with the converter.

*Gearing:* In the IL-6 receptor system and in the GM-CSF/IL-3/IL-5 receptor system the second subunit in each case is the non-binding subunit, and is also the one which transduces the signal.

*Heath:* That is why it is critical to know whether the cloned LIF receptor is capable of signalling.

*Gough:* Dr Gearing, what is known about expression of the soluble form of the LIF receptor *in vitro* or *in vivo*? In what sort of cells is it expressed, and at what level?

*Gearing:* We have tried looking at this by Northern analysis. The soluble receptor seems to be produced by hepatocytes. From the blot, you would predict that the soluble receptor is far in excess of the membrane-bound form. However, my ability to get high molecular weight RNAs, in the 10–12 kb range, for a Northern blot has been sporadic, so high molecular mass mRNA encoding the membrane-bound LIF receptor might be under-represented in my blots.

*Ciliberto:* This problem could be resolved by S1 mapping.

*Gearing:* Yes.

*Gough:* Is the soluble form of the LIF receptor expressed by liver tissue *in vivo*?

*Gearing:* We haven't looked at that.

*Gough:* Do you know whether hepatocytes in culture that contain mRNA for the soluble form of the receptor actually produce the protein?

*Gearing:* We haven't measured it.

*Gough:* Have you looked at any tissues *in vivo*?

*Gearing:* We have just begun investigating tissues *in vivo*.

*Lotem:* GM-CSF receptors can be down-modulated by IL-3. Is there a similar down-regulation of LIF receptors by IL-6, or of IL-6 receptors by G-CSF, or G-CSF with LIF?

*Nicola:* I don't know if we've tested all possible combinations. The only thing we have found that will down-regulate LIF receptors other than LIF is bacterial lipopolysaccharide (LPS).

*Metcalf:* Have you tested IL-6?

*Nicola:* We tested IL-6, IL-1, G-CSF, GM-CSF and multi-CSF (IL-3).

*Lotem:* With LPS is there a real down-modulation, or just masking of the receptor?

*Nicola:* It occurs after you add LPS, which does not competitively inhibit binding of LIF at 4 °C. At 37 °C LIF-binding sites are lost.

*Martin:* Is that because of the LIF that is produced in response to LPS?

*Nicola:* The down-regulation is always rapid, occurring within 2 min. If massive amounts of LIF can be produced in the first few minutes, that's a possible explanation.

*Gearing:* Was the experiment done in M1 cells?

*Nicola:* It was done in peritoneal macrophages. This doesn't happen in M1 cells, but I am not sure that M1 cells have LPS receptors.

*Lotem:* That depends on the clone. The T22 clone that I mentioned (p 82) responds to LIF but doesn't respond to LPS, whereas clone 11 shows a good response to LPS.

*Dexter:* Dr Nicola, you showed that osteoclasts don't have receptors for LIF, and you also showed that macrophages and monocytes are heterogeneous in this respect. The results suggest that osteoclasts are derived from GM-CFC progenitor cells and that it is possible to take single colonies that develop from these cells and derive what appear, from their bone-resorbing properties, to be osteoclasts. Is it possible that the osteoclasts develop from a sub-population of LIF receptor-negative macrophage precursor cells, or do you think that the LIF receptor is present in the precursor cells and is then down-modulated?

*Nicola:* In the *op/op* mouse (which lacks CSF-1 and shows osteopetrosis; Wiktor-Jedrzejczak et al 1990) there is a clear relationship between M-CSF and osteoclast formation. I don't know the answer, but there are clearly two possibilities. One is that the LIF receptor-negative monocytes give rise to osteoclasts. The other possibility is that once a cell enters the osteoclast pathway all LIF receptor expression is lost.

*Dexter:* Have you tried taking colonies derived from GM-CFC to see if there is an intrinsic heterogeneity within the colony, or whether one colony expresses the LIF receptor while others do not?

*Nicola:* That's a good question; we haven't done that.

*Gauldie:* In preliminary studies in isolated hepatocytes Margit Geisterfer has found that there seems to be cross-talk between the cytokines. IL-1 and IL-6 can up-regulate the LIF receptor.

*Metcalf:* Were those experiments on dividing populations? If experiments are extended for 24–48 hours problems of selective proliferation and selective survival arise.

*Gauldie:* We have never seen proliferation with primary cells in the hepatocyte assay system. By the time we do the assays using hepatoma cells, the cells are not proliferating; they will proliferate at lower densities, but I don't think we are looking at a proliferative effect.

*Sehgal:* Professor Heinrich, could you comment on what is currently known about the up-regulation of the IL-6 receptor by glucocorticoids in different cell types?

*Heinrich:* In HepG2 cells we find a 2–3-fold increase in levels of mRNA for the 80 kDa IL-6 receptor as well as for the functional IL-6 receptor protein (Rose-John et al 1990). Similar observations have been made with CESS human epithelial cell (Syners et al 1990). Is the LIF receptor up-regulated by glucocorticoids?

*Gauldie:* Apparently, glucocorticoids, not directly, but in combination with IL-1 and IL-6, have an enhancing effect on the expression of the LIF receptor. Corticosteroid also up-regulates mRNA for the 80 kDa IL-6 receptor in the liver *in vivo*.

*Heinrich:* If we incubate HepG2 cells with 12-*O*-tetradecanoylphorbol-13-acetate (TPA) and dexamethasone for 20 hours we get the most dramatic increase (5–8-fold) in mRNA levels for the IL-6 receptor. Because protein kinase C is down-regulated after such a long period of incubation with TPA it is possible that down-regulation of protein kinase C is important for IL-6 receptor induction.

## The role of leukaemia inhibitory factor and interleukin 6 in tumorigenesis

*Dexter:* I would like to raise a slightly different issue which we have not yet discussed. We are used to thinking about many receptors and growth factors acting as oncogenes. I would like to know if constitutive activation of gp130 has been associated with any tumours. Is constitutive activation of the LIF receptor or the IL-6 receptor associated with tumorigenesis? We have heard that production of IL-6 is perhaps associated with plasmacytomagenesis (although I think we still have to be cautious about those results), but is aberrant production of LIF or IL-6 known to be associated with any other diseases—particularly cancer?

*Stewart:* ES cells injected into adult mice will form typical teratocarcinomas, some of which are transplantable (Evans & Kaufman 1981). We are trying to determine whether the differentiated derivatives of these cells in the tumour make LIF *in vivo*. This could then bind to ES stem cells to maintain them as an undifferentiated, highly proliferating population.

*Heath:* We have looked extensively at a series of EC cell lines, which genuinely are tumorigenic, and ES cell lines, which may or may not be tumorigenic (Rathjen et al 1991). In all cases, we find that the cell lines produce low levels of LIF. The relationship between levels of LIF expression and the malignancy of embryonal carcinomas doesn't hold. However, it's interesting that a number of ES cell lines develop abnormalities of chromosome 11 in culture.

*Stewart:* You haven't established whether the tumour itself produces LIF.

*Heath:* No; one would have to argue that expression is switched on *in vivo*.

*Stewart:* That's what I would argue. There are clearly high levels of LIF production in the differentiated derivatives of ES cells grown *in vitro*.

*Heath:* You are suggesting that an environment which would maintain the stem cell population is created, as in the idea of the stem cell niche.

*Lotem:* If you induce expression of huge amounts of IL-3 *in vivo*, all you observe is something which looks like a myeloproliferative disease, but not a true malignancy. Experiments by Perkins et al (1990) have shown that after introduction of a homeobox gene, expression of the IL-3 gene results in a severe leukaemia. The expression of the growth factor by itself is not enough.

*Dexter:* I agree with that, but I have seen no results, so far, which indicate that constitutive aberrant expression of the LIF or the IL-6 receptor leads to malignant transformation. Also, it is worth bearing in mind that most oncogenes have been characterized as a consequence of retrovirus transducing the normal cellular oncogene. However, to my knowledge, no transforming retrovirus that contains LIF or IL-6 (or their receptors) has been reported. It would probably be worthwhile to put these genes into retroviruses and examine their effects in various cell types.

*Heath:* There is a report of a cytokine-like molecule which was picked up with MPSV (myeloproliferative sarcoma virus) (Souyri et al 1990).

*Gough:* That's not a LIF-related molecule or a LIF receptor-related molecule.

## References

Evans MJ, Kaufman MH 1981 Establishment in culture of pluripotential cells from mouse embryos. Nature (Lond) 292:154–156

Novick D, Englemann H, Wallach D, Rubinstein M 1989 Soluble cytokine receptors are present in normal human urine. J Exp Med 170:1409–1414

Perkins A, Kongsuwan K, Visvader J, Adams JM, Cory S 1990 Homoeobox gene expression plus autocrine growth factor production elicits myeloid leukemia. Proc Natl Acad Sci USA 87:8398–8402

Rathjen PD, Nichols J, Edwards D, Heath JK, Smith AG 1991 Developmentally programmed induction of differentiation inhibiting activity and the control of stem cell populations. Genes & Dev 4:2308–2318

Rose-John S, Schooltink H, Lenz D et al 1990 Studies on the structure and regulation of the human hepatic interleukin-6-receptor. Eur J Biochem 190:79–83

Souyri M, Vignon I, Penciolelli JF, Heard JM, Tambourin P, Wendling F 1990 A putative truncated cytokine receptor gene transduced by the myeloproliferative leukemia virus immortalizes hematopoietic progenitors. Cell 63:1137-1147

Syners L, De Wit L, Content J 1990 Glucocorticoid up-regulation of high-affinity interleukin-6 receptors on human epithelial cells. Proc Natl Acad Sci USA 87:2838–2842

Wiktor-Jedrzejczak W, Bartocci A, Ferrante AW Jr 1990 Total absence of colony-stimulating factor-1 in the macrophage deficient osteopetrotic (*op/op*) mouse. Proc Natl Acad Sci USA 87:4828–4832

# Summing-up

Donald Metcalf

*The Walter and Eliza Hall Institute of Medical Research, Post Office, Royal Melbourne Hospital, Parkville, Melbourne, Victoria 3050, Australia*

What has been clarified and what made more obscure during this symposium? A disturbing array of functional actions of IL-6 and LIF *in vitro* has been demonstrated. For many of these, there is reasonable evidence for a direct action. It's also clear that if these molecules are injected *in vivo*, certain responses can be elicited. Does this mean that IL-6 and LIF really regulate such responses *in vivo*?

Suppression experiments are really required to answer this question. What are the consequences of deleting the LIF or IL-6 genes, or of suppressing the actions of LIF or IL-6 using an antibody? I don't think we yet have that information. Some useful information has been obtained from the study of animals with excess levels of IL-6 or LIF but suppression experiments are required to prove that these two molecules have a significant role in normal health or in certain disease states.

We have discussed results indicating that increased levels of IL-6 and LIF can be found in local sites of infection and inflammation. However, many regulators are probably present in increased amounts in these situations, and the question we need to answer is, what does IL-6 or LIF contribute to the particular abnormal state? At present this question cannot be answered.

How similar are the functions of the two regulators? David Gearing summarized the evidence that they are probably more overlapping in their spectra of actions than thought previously. If it's true that IL-6 has actions on lipogenesis and on some neuronal cells, then some of the apparent differences between the two regulators no longer hold.

Do they use common signalling systems? The answer to this question is yes and no. The receptors differ but the receptor for LIF has some similarity to the β subunit of the IL-6 receptor complex. At some stage, some of the signals elicited must be common because there are similar effects on responding cells. The situation with signalling becomes complex when you start to investigate nuclear transcription factors, their binding sites, and their specificity. At the moment, there seems to be considerable non-specificity with these binding sites; the specificity comes from the context in which the nuclear transcription factor is presented. What a transcription factor is bound to, and what other transcription factors are present seem to determine the particular response exhibited by the cell. We do not know yet how IL-6 and LIF elicit common responses in liver cells.

Did we really resolve the problem posed by polyfunctionality? At the start of the meeting, I said that it seems incredible that the body would choose to use one molecule to execute such a diverse set of functions. One explanation may be that the cytokines are designed to be produced, and to act, locally. The discussion on whether IL-6 or LIF is present in the circulation is relevant to this issue, but the situation is still confused for both molecules. If they are present in the circulation, are they functionally silent until activated? For IL-6, local production and action cannot be the answer to the problem if there is plenty of IL-6 in the circulation.

Why does the body use a limited number of molecules to achieve such a diverse range of responses? Perhaps the question we should be asking is, what is so special about the receptors for these molecules? Why does the body like to use a signalling system involving a gp130-like transmembrane molecule? Is this type of molecule peculiarly efficient at eliciting the required signalling cascades? It may now be useful to examine this type of membrane receptor more closely.

I also said at the outset that we needed to consider potential clinical applications. Dr Romero's work showed that there are two parts to this question. One is the significance of the association of IL-6 or LIF with various disease states. IL-6 is likely to be relevant in plasma cell tumour development, and possibly in some autoimmune states. Although excess levels of LIF or IL-6 can be found in a variety of disease states, there is little evidence in humans for any particular disease being ascribable to abnormal LIF or IL-6 levels.

Will IL-6 or LIF ever be used as therapeutic agents? We perhaps should be developing specific antagonists in case either becomes implicated unambiguously in any particular disease state. In the meantime, both IL-6 and LIF will probably be tested clinically as agents to stimulate platelet formation. Those tests will reveal, if nothing else, the answers to some of the questions we have posed about detectable effects of these agents on neuronal function and lipid metabolism. In principle, there is no particular reason why a polyfunctional molecule cannot be used therapeutically if a favourable balance exists between side effects and the response desired. The potential uses for both agents could be quite wide, ranging from *in vitro* fertilization through to a variety of metabolic disease states. One of the more intriguing things discussed was the apparent antagonism between IL-6 and LIF during initial implantation; I look forward to hearing more about that.

We have done our best to discuss our present knowledge of IL-6 and LIF. I hope we didn't generate extra confusion. Both regulators continue to be under active investigation, and the availability of cloned cDNAs for the membrane receptors of both will accelerate progress.

I still believe that these two agents will turn out to be prototypes and that discussion of the problems posed by polyfunctionality will be a recurrent theme for many regulators yet to be discovered.

# Index of contributors

*Indexes compiled by Liza Weinkove*

269

# Subject index